Fatos Xhafa and Ajith Abraham (Eds.)

Metaheuristics for Scheduling in Industrial and Manufacturing Applications

Studies in Computational Intelligence, Volume 128

Editor-in-chief
Prof. Janusz Kacprzyk
Systems Research Institute
Polish Academy of Sciences
ul. Newelska 6
01-447 Warsaw
Poland
E-mail: kacprzyk@ibspan.waw.pl

Further volumes of this series can be found on our homepage: springer.com

Vol. 108. Vito Trianni
Evolutionary Swarm Robotics, 2008
ISBN 978-3-540-77611-6

Vol. 109. Panagiotis Chountas, Ilias Petrounias and Janusz Kacprzyk (Eds.)
Intelligent Techniques and Tools for Novel System Architectures, 2008
ISBN 978-3-540-77621-5

Vol. 110. Makoto Yokoo, Takayuki Ito, Minjie Zhang, Juhnyoung Lee and Tokuro Matsuo (Eds.)
Electronic Commerce, 2008
ISBN 978-3-540-77808-0

Vol. 111. David Elmakias (Ed.)
New Computational Methods in Power System Reliability, 2008
ISBN 978-3-540-77810-3

Vol. 112. Edgar N. Sanchez, Alma Y. Alanís and Alexander G. Loukianov
Discrete-Time High Order Neural Control: Trained with Kalman Filtering, 2008
ISBN 978-3-540-78288-9

Vol. 113. Gemma Bel-Enguix, M. Dolores Jiménez-López and Carlos Martín-Vide (Eds.)
New Developments in Formal Languages and Applications, 2008
ISBN 978-3-540-78290-2

Vol. 114. Christian Blum, Maria José Blesa Aguilera, Andrea Roli and Michael Sampels (Eds.)
Hybrid Metaheuristics, 2008
ISBN 978-3-540-78294-0

Vol. 115. John Fulcher and Lakhmi C. Jain (Eds.)
Computational Intelligence: A Compendium, 2008
ISBN 978-3-540-78292-6

Vol. 116. Ying Liu, Aixin Sun, Han Tong Loh, Wen Feng Lu and Ee-Peng Lim (Eds.)
Advances of Computational Intelligence in Industrial Systems, 2008
ISBN 978-3-540-78296-4

Vol. 117. Da Ruan, Frank Hardeman and Klaas van der Meer (Eds.)
Intelligent Decision and Policy Making Support Systems, 2008
ISBN 978-3-540-78306-0

Vol. 118. Tsau Young Lin, Ying Xie, Anita Wasilewska and Churn-Jung Liau (Eds.)
Data Mining: Foundations and Practice, 2008
ISBN 978-3-540-78487-6

Vol. 119. Slawomir Wiak, Andrzej Krawczyk and Ivo Dolezel (Eds.)
Intelligent Computer Techniques in Applied Electromagnetics, 2008
ISBN 978-3-540-78489-0

Vol. 120. George A. Tsihrintzis and Lakhmi C. Jain (Eds.)
Multimedia Interactive Services in Intelligent Environments, 2008
ISBN 978-3-540-78491-3

Vol. 121. Nadia Nedjah, Leandro dos Santos Coelho and Luiza de Macedo Mourelle (Eds.)
Quantum Inspired Intelligent Systems, 2008
ISBN 978-3-540-78531-6

Vol. 122. Tomasz G. Smolinski, Mariofanna G. Milanova and Aboul-Ella Hassanien (Eds.)
Applications of Computational Intelligence in Biology, 2008
ISBN 978-3-540-78533-0

Vol. 123. Shuichi Iwata, Yukio Ohsawa, Shusaku Tsumoto, Ning Zhong, Yong Shi and Lorenzo Magnani (Eds.)
Communications and Discoveries from Multidisciplinary Data, 2008
ISBN 978-3-540-78732-7

Vol. 124. Ricardo Zavala Yoe
Modelling and Control of Dynamical Systems: Numerical Implementation in a Behavioral Framework, 2008
ISBN 978-3-540-78734-1

Vol. 125. Larry Bull, Bernadó-Mansilla Ester and John Holmes (Eds.)
Learning Classifier Systems in Data Mining, 2008
ISBN 978-3-540-78978-9

Vol. 126. Oleg Okun and Giorgio Valentini (Eds.)
Supervised and Unsupervised Ensemble Methods and their Applications, 2008
ISBN 978-3-540-78980-2

Vol. 127. Régie Gras, Einoshin Suzuki, Fabrice Guillet and Filippo Spagnolo (Eds.)
Statistical Implicative Analysis, 2008
ISBN 978-3-540-78982-6

Vol. 128. Fatos Xhafa and Ajith Abraham (Eds.)
Metaheuristics for Scheduling in Industrial and Manufacturing Applications, 2008
ISBN 978-3-540-78984-0

Fatos Xhafa
Ajith Abraham
(Eds.)

Metaheuristics for Scheduling in Industrial and Manufacturing Applications

With 92 Figures and 86 Tables

Springer

Dr. Fatos Xhafa
Department of Languages and Informatics Systems
Polytechnic University of Catalonia
Campus Nord, Ed. Omega
C/Jordi Girona 1-3 08034 Barcelona
Spain
fatos@lsi.upc.edu

Dr. Ajith Abraham
Norwegian Center of Excellence
Center of Excellence for Quantifiable Quality of Service
Norwegian University of Science and Technology
O.S. Bragstads plass 2E
NO-7491 Trondheim
Norway
www.softcomputing.net
ajith.abraham@ieee.org

ISBN 978-3-540-78984-0 e-ISBN 978-3-540-78985-7
DOI: 10.1007/978-3-540-78985-7

Studies in Computational Intelligence ISSN 1860-949X

Library of Congress Control Number: 2008924363

Cover design: Deblik, Berlin, Germany

Printed on acid-free paper

9 8 7 6 5 4 3 2 1

springer.com

To our lovely families!

Preface

During the past decades scheduling has been among the most studied optimization problems and it is still an active area of research! Scheduling appears in many areas of science, engineering and industry and takes different forms depending on the restrictions and optimization criteria of the operating environments [8]. For instance, in optimization and computer science, scheduling has been defined as "the allocation of tasks to resources over time in order to achieve optimality in one or more objective criteria in an efficient way" and in production as "production schedule, i.e., the planning of the production or the sequence of operations according to which jobs pass through machines and is optimal with respect to certain optimization criteria."

Although there is a standardized form of stating any scheduling problem, namely "efficient allocation of n jobs on m machines –which can process no more than one activity at a time– with the objective to optimize some objective function of the job completion times", scheduling is in fact a family of problems. Indeed, several parameters intervene in the problem definition: (a) job characteristics (preemptive or not, precedence constraints, release dates, etc.); (b) resource environment (single *vs.* parallel machines, unrelated machines, identical or uniform machines, etc.); (c) optimization criteria (minimize total tardiness, the number of late jobs, makespan, flowtime, etc.; maximize resource utilization, etc.); and, (d) scheduling environment (static *vs.* dynamic, in the former the number of jobs to be considered and their ready times are available while in the later the number of jobs and their characteristics change over time).

Thus, different scheduling problems are identified in the literature (an early compendium of sequencing and scheduling problems is found in [1]; see [8] for an annotated bibliography). Among these scheduling problems we can mention Job Shop, Flow Shop, Sequencing problems, Identical Parallel Machine Scheduling, Timetabling and Multiprocessor scheduling.

On the other hand, scheduling problems have found their usefulness in a vast area of applications such as lot sizing [5], manufacturing [2] and production scheduling [10], to name a few.

Despite of large amount or research on scheduling, many researchers and practitioners from the academia and industry are devoting a considerable amount of efforts on the problem. This can be explained not only by the fact that most of the problems in the family of scheduling are computationally hard [6] and therefore there is still room for improvement in the resolution methods but also because manufacturing and production are continuously changing and introducing more and more demanding restrictions (e.g. Reconfigurable Manufacturing Systems have appeared as the next step in manufacturing, aiming the production of any quantity of highly customized products or novel scheduling problems arising from the field of environmentally conscious manufacturing); thus, "new" types of scheduling are arising. Moreover, the computational resources used for solving scheduling problems have significantly increased during the last years allowing to achieve more efficient solutions of large size scheduling problems.

This volume presents meta-heuristics approaches for scheduling problems arising in industrial and manufacturing applications. Nowadays, meta-heuristics have become a *de facto* approach to tackle in practice with the complexity of scheduling problems. Early work applied evolutionary computing methods to scheduling problems (see [3, 4, 9] and [29] for a review). The present volume is novel in many respects. First, the proposed approaches comprise a variety of meta-heuristics (Genetic Algorithms, Memetic Algorithms, Ant Colony Optimization, Particle Swarm Optimization, Tabu Search, Scatter Search, Variable Neighborhood Search). Second, in most cases, hybridization is approached as the most effective way to achieve state-of-the art results. Third, and most importantly, the scheduling problems arising in real life applications and real world data instances are solved using these meta-heuristics; these applications comprise reconfigurable manufacturing systems, lot sizing and scheduling in industry, railway scheduling and process, supply chain scheduling and scheduling problem arising in a real-world multi-commodity Oil-derivatives Pipeline. Finally, scheduling problems and meta-heuristics are presented in a comprehensive way making this volume and interesting contribution to the research on scheduling in industrial and manufacturing applications.

Chapters were selected after a careful review process by at least three reviewers on the basis of the originality, soundness and their contribution to both meta-heuristics and scheduling in industrial and manufacturing applications. Relevance of the proposed approaches was an important criterion for chapter selection, resulting in a volume where the reader will find comprehensive up-to-date surveys, novel meta-heuristic approaches and real life applications. The thirteen chapters of the volume are organized as follows.

In Chapter 1, *Zobolas et al.* present a survey on the main shop scheduling problems (flow shop, job shop, open shop, group shop and mixed shop) as well as their computational complexity. Thereafter, the most important exact, heuristic and meta-heuristic methods are presented and classified.

Iori and Martello in Chapter 2 address the identical parallel machine scheduling problem and some generalizations arising from real world situations. The authors survey the current state of the art for the most performing meta-heuristic algorithms, with special emphasis on Scatter Search.

In Chapter 3, *Czogalla and Fink* study the effectiveness of Particle Swarm Optimization (PSO) and Variable Neighborhood Descent for the Continuous Flow-Shop Scheduling Problem. The authors examine the use of different crossover operators in PSO and hybridization with variable neighborhood descent. The results of their study stress the importance of local search in increasing the performance of PSO procedures.

Lee et al. in the fourth chapter present an ACO approach for Scheduling Jobs on a Single Machine with a Common Due Date. The authors consider the version of Dynamic ACO in which the parameter of heuristic information is dynamically adjusted. Moreover, hybridization with other heuristics is implemented and evaluated.

In the fifth chapter *McGovern and Gupta* present the use of the Hunter-Killer (H-K) general purpose heuristic for solving new complex scheduling problems arising from the field of environmentally conscious manufacturing, the goal of which is to determine a product's part removal schedule. A scheduling application of the H-K heuristic is demonstrated using an electronic product case study.

Aydin and Sevkli in the sixth chapter report sequential and parallel Variable Neighborhood Search algorithms for Job Shop Scheduling problems. Starting from the observation that VNS could sometimes take long time to reach good solutions, especially when tackling large size instances of Job Shop Scheduling, the authors propose the parallelization of VNS in order to speed up computations. A number of variable neighborhood search algorithms are examined for the problem and then four different parallelization policies are presented and their performance evaluated.

In the seventh chapter *Myszkowski* uses Graph Coloring problem for modelling several scheduling problems. The author presents a new evolutionary approach to the Graph Coloring Problem, which is then applied for solving timetabling, scheduling, multiprocessor scheduling task and other assignment problems.

Ferreira et al. in the eighth chapter presents a comparison study for heuristics and meta-heuristics for lot sizing and scheduling in the soft drinks industry. The study concentrates on two-level production planning problem where, on each level, a lot sizing and scheduling problem with parallel machines, capacity constraints and sequence-dependent setup costs and times must be solved. The comparison study is accomplished for two different approaches, namely, an evolutionary technique comprising both a GA and its MA version, and a decomposition and relaxation approach.

Galan in the ninth chapter proposes hybrid heuristic approaches for scheduling problems arising in reconfigurable manufacturing systems. Besides a specific heuristic for the problem, both Tabu Search and Ant Colony Optimization have been implemented and experimentally evaluated.

Tormos et al. in the tenth chapter introduce a Genetic Algorithm for Railway Scheduling Problems. Their work is motivated by the need to solve very complex real-world problems such as timetabling in a large railway system. The authors have used real instances obtained from the Spanish Manager of Railway Infrastructure to experimentally evaluate the performance of the proposed GA.

In the eleventh chapter, *Banerjee et al.*, address a natural stigmergic computational technique, namely Bee Colony Optimization, for process scheduling arising in a milk production center, in which process scheduling, supply chain network etc. are crucial. The scheduling problem is considered in its multi-objective form and the concept of Pareto Dominance has been introduced in the form of Pareto Bee Colony Optimization. A performance simulation and comparison has been accomplished for the proposed algorithms.

García Sánchez et al. in Chapter twelve present an approach that combines simulation and Tabu Search for solving scheduling problem arising in a real-world multi-commodity Oil-derivatives Pipeline. In the proposed approach, the simulation enables an accurate assessment of particular schedules, while the Tabu Search guides the search in order to reach satisfactory schedules. The authors have applied the proposed methodology to a particular subsystem of the pipeline Spanish network.

In the last Chapter, *Abraham et al.* propose several swarm intelligence based meta-heuristics for scheduling work-flow applications in distributed data-intensive computing environments. A Variable Neighborhood Particle Swarm Optimization (VNPSO) algorithm is proposed and the performance is compared with a multi-start particle swarm optimization algorithm and multi-start genetic algorithm. Experiment results illustrate the algorithm performance and its feasibility and effectiveness for scheduling work-flow applications.

We are very much grateful to the authors of this volume and to the reviewers for their great efforts by reviewing and providing interesting feedback to authors of the chapter. The editors would like to thank Dr. Thomas Ditzinger (Springer Engineering Inhouse Editor, Studies in Computational Intelligence Series), Professor Janusz Kacprzyk (Editor-in-Chief, Springer Studies in Computational Intelligence Series) and Ms. Heather King (Editorial Assistant, Springer Verlag, Heidelberg) for the editorial assistance and excellent cooperative collaboration to produce this important scientific work. We hope that the reader will share our joy and will find it useful!

Fatos Xhafa acknowledges partial support by Projects ASCE TIN2005-09198-C02-02, FP6-2004-ISO-FETPI (AEOLUS) and MEC TIN2005-25859-E and MEFOALDISI (Métodos formales y algoritmos para el diseño de sistemas) TIN2007-66523. *Ajith Abraham* acknowledges the support by the Centre for Quantifiable Quality of Service in Communication Systems, Norwegian Centre of Excellence, appointed by The Research Council of Norway, and funded by the Research Council, Norwegian University of Science and Technology and UNINETT.

Fatos Xhafa
Departament de Llenguatges i Sistemes Informàtics
Universitat Politècnica de Catalunya
Barcelona (Spain)

Ajith Abraham
Centre for Quantifiable Quality of Service in Communication Systems
Norwegian University of Science and Technology
Trondheim (Norway)

February 2008 Barcelona (Spain), Trondheim (Norway)

References

1. Ausiello, G., Crescenzi, P., Gambosi, G., Kann, V., Marchetti-Spaccamela, A., Protasi, M. Complexity and Approximation book cover Combinatorial optimization problems and their approximability properties, Springer Verlag, 1999.
2. Baudin, M. Manufacturing Systems Analysis with Applications to Production Scheduling, Yourdon Press, 1990.
3. Cheng, R., Gen, M. and Tsujimura, Y., A tutorial survey of job-shop scheduling problems using genetic algorithms -I: representation, Comput. Ind. Eng., 30(4), 983–997, 1996.
4. Dell'Amico, M., Trubian, M., Applying tabu search to the job-shop scheduling problem, Ann. Oper. Res., 41(1-4), 231–252, 1993.
5. Drexl, A., Kimmsa, A. Lot sizing and scheduling Survey and extensions. European Journal of Operational Research, 99(2), 221-235, 1997.
6. Garey, M.R., Johnson, D.S. Computers and Intractability: A Guide to the Theory of NP-Completeness, Freeman, New York, 1979.
7. Hart, E., Ross, P. and Corne, D., Evolutionary Scheduling: A Review, Genetic Programming and Evolvable Machines, 6(2), 191–220, 2005
8. Hoogeveen, J.A., Lenstra, J.K., van de Velde, S. L. Sequencing and Scheduling: An Annotated Bibliography. Memorandum COSOR 97-02, Eindhoven University of Technology, 1997.
9. Linn R.; Zhang W. Hybrid flow shop scheduling: a survey. Computers and Industrial Engineering, 37(1), 57-61, 1999.
10. Rodammer, F.A. White, K.P., Jr. A recent survey of production scheduling. IEEE Transactions on Systems, Man and Cybernetics, 18(6), 841-851, 1988.

Contents

List of Contributors

Ajith Abraham
Norwegian Center of Excellence
Center of Excellence for Quantifiable Quality of Service
Norwegian University of Science and Technology,
O.S. Bragstads plass 2E,
NO-7491 Trondheim
Norway.
ajith.abraham@ieee.org

M. Abril DSIC, Universidad Politecnica de Valencia,
Spain. mabril@dsic.upv.es

Luis Miguel Arreche Universidad Politécnica de
Madrid. José Gutiérrez Abascal 2, 28006. Madrid, Spain.
arreche@etsii.upm.es

Mehmet E. Aydin University of Bedfordshire, Dept. of Computing and Information Systems, Luton, UK.
mehmet.aydin@beds.ac.uk

Soumya Banerjee Dept. of Computer Science and Engineering, Birla Institute of Technology, Mesra, India.
soumyabanerjee@bitmesra.ac.in

F. Barber DSIC, Universidad Politecnica de Valencia,
Spain. fbarber@dsic.upv.es

Jens Czogalla Helmut-Schmidt-University/UniBw
Hamburg, Holstenhofweg 85, 22043 Hamburg, Germany.
czogalla@hsu-hamburg.de

G.S.Dangayach Department of Mechanical Eng., Malviya National Institute of Technology, Jaipur, India.

D. Ferreira Departamento de Engenharia de Produção,
Universidade Federal de São Carlos, C.P. 676, 13565-905, São
Carlos, SP, Brazil.
degafernandes@hotmail.com

Andreas Fink Helmut-Schmidt-University/UniBw
Hamburg, Holstenhofweg 85, 22043 Hamburg, Germany.
czogalla@hsu-hamburg.de

P.M. França Departamento de Matemática,
Estatística e Computação, Universidade Estadual Paulista, C.P.

1234, 19060-400, Presidente Prudente, SP, Brazil.
paulo.morelato@gmail.com

Ricardo Galán Agua y Estructuras, S.A. (Ayesa)
Marie Curie 2, 41092 Seville, Spain.
rgalan@ayesa.es

Álvaro García-Sánchez. Universidad
Politécnica de Madrid. José Gutiérrez Abascal 2, 28006.
Madrid, Spain.
alvaro.garcia@upm.es

Surendra M. Gupta Northeastern University, Boston,
MA 02115, USA. gupta@neu.edu

L. Ingolotti DSIC, Universidad Politecnica de
Valencia, Spain.
lingolotti@dsic.upv.es

G. Ioannou Athens University of Economics and
Business, Department of Management Science & Technology
Management Science Laboratory
Evelpidon 47A & Leukados 33, 11369, Athens, Greece ioannou@aueb.gr

Manuel Iori DISMI, Università di Modena e Reggio
Emilia, Italy.
manuel.iori@unimore.it

Zne-Jung Lee Deptartment of Information Management,
Huafan University, No. 1, Huafan Rd. Shihding Township, Taipei County, 22301, Taiwan.
johnlee@hfu.edu.tw

Hongbo Liu
School of Computer Science and Engineering, Dalian Maritime University, 116026 Dalian, China.
lhb@dlut.edu.cn

Silvano Martello DEIS, Università di Bologna,
Italy. silvano.martello@unibo.it

Seamus M. McGovern U.S. DOT
National Transportation
Systems Center, Cambridge, MA 02142, USA.
mcgoverns@volpe.dot.gov

P.K. Mohanty University of New Brunswick, New
Brunswick, Canada.

S.K. Mukherjee Birla Institute of Technology, Mesra,
India.

Shih-Wei Lin Deptartment of Information Management,
Huafan University, No. 1, Huafan Rd. Shihding Township, Taipei County, 22301, Taiwan.

A. Lova DEIOAC, Universidad Politecnica de Valencia,
Spain. ptormos@eio.upv.es

R. Morabito Departamento de Matemática,
Estatística e Computação, Universidade Estadual Paulista, C.P.
1234, 19060-400, Presidente Prudente, SP, Brazil.
morabito@power.ufscar.br

Pawel B. Myszkowski Institute of Applied
Informatics, Wroclaw University of Technology, Poland.
pawel.myszkowski@pwr.wroc.pl

Miguel Ortega-Mier Universidad Politécnica de
Madrid. José Gutiérrez Abascal 2, 28006. Madrid (Spain)
miguel.ortega.mier@upm.es

M.A. Salido DSIC, Universidad Politecnica de
Valencia, Spain.
msalido@dsic.upv.es

Mehmet Sevkli Fatih University, Dept. of Industrial
Engineering, Buyukcekmece, Istanbul, Turkey.
msevkli@fatih.edu.tr

C.D. Tarantilis Athens University of Economics and
Business, Department of Management Science & Technology
Management Science Laboratory
Evelpidon 47A & Leukados 33, 11369, Athens, Greece tarantil@aueb.gr

Pilar Tormos DEIOAC, Universidad Politecnica de
Valencia, Spain.
ptormos@eio.upv.es

Kuo-Ching Ying Department of industrial engineering
and management information,
Huafan University, No. 1, Huafan Rd.
Shihding Township, Taipei County, 22301, Taiwan.

Mingyan Zhao
School of Software, Dalian University of Technology, 116620 Dalian, China. mingy_zhao@163.com

G.I. Zobolas Athens University of Economics and
Business, Department of Management Science & Technology
Management Science Laboratory
Evelpidon 47A & Leukados 33, 11369, Athens, Greece
gzobolas@aueb.gr

1

Exact, Heuristic and Meta-heuristic Algorithms for Solving Shop Scheduling Problems

G.I. Zobolas[1], C.D. Tarantilis[2], and G. Ioannou[3]

[1] Athens University of Economics and Business, Department of Management Science & Technology Management Science Laboratory Evelpidon 47A & Leukados 33, 11369, Athens, Greece gzobolas@aueb.gr
[2] Athens University of Economics and Business, Department of Management Science & Technology Management Science Laboratory Evelpidon 47A & Leukados 33, 11369, Athens, Greece tarantil@aueb.gr
[3] Athens University of Economics and Business, Department of Management Science & Technology Management Science Laboratory Evelpidon 47A & Leukados 33, 11369, Athens, Greece ioannou@aueb.gr

Summary. This chapter sets out to present a very important class of production scheduling problems and the main methods employed to solve them. More specifically, after a brief description of single and parallel machines scheduling problems, which constitute the basis of production scheduling research, the main shop scheduling problems are presented (flow shop, job shop, open shop, group shop and mixed shop) followed by an analysis of their computational complexity. Thereafter, the most important exact, heuristic and meta-heuristic methods are presented and classified. Finally, a thorough review for each shop scheduling problem is conducted where the most important methods proposed in the literature, specifically for each problem, are presented.

Key words: Production Scheduling, Single and Parallel Machines Scheduling, Shop Scheduling, Flow Shop, Job Shop, Open Shop, Group Shop, Mixed Shop, Meta-heuristics.

1.1 Introduction

The production process of manufacturing enterprises has always been a key factor for overall business success. Production scheduling problems are faced by thousands of companies worldwide that are engaged in the production of tangible goods. Thus, it is not without reason that efficiently and effectively solving production scheduling problems has attracted the interest of many practitioners and researchers from both fields of production management and

G.I. Zobolas et al.: *Exact, Heuristic and Meta-heuristic Algorithms for Solving Shop Scheduling Problems*, Studies in Computational Intelligence (SCI) **128**, 1–40 (2008)
www.springerlink.com

combinatorial optimization. This interest is further amplified by the similarity of production scheduling problems with problems arising in other scientific areas (e.g. packet scheduling in telecommunication networks, PCB design, routing, timetabling etc) [97] and, therefore, the applicability of the developed methods in these areas as well. It should be also mentioned that due to the complex nature of scheduling problems, many new computational methods for their solution have emerged which can also be applied to a wide range of combinatorial optimization problems.

Due to the virtually unlimited number of different production environments, many variations of production scheduling problems exist. However, academic research and solution methodology development have focused mainly on a limited number of classical problems which, most of the times, cannot be directly applied to complex manufacturing structures. Thus, the development of flexible solution methodologies, which can be modified and applied to several different cases, is of critical importance for production management practice.

Production scheduling problems have detained thousands of researchers worldwide during the 20th century. Initially, research focused on simplified and generic problems with limited applicability in real cases [78]. Most such problems dealt with optimizing a single objective in one-machine environments and featured many simplified assumptions. Since then, a wide variety of scheduling problems (and their variants) has been identified and an even wider range of solution methodologies has been proposed. At the very beginning, research focused on exact methods, i.e. methods that guaranteed the optimum solution of a given problem. Due to the lack of computational resources and the need to solve large scale scheduling problems, it was soon realized that exact methods were impractical and thus research focused on problem specific heuristics. On the other hand, the need for more robust solution methodologies led to the development and application of the first meta-heuristic methods (the term meta-heuristic was proposed by Glover in 1986 [57]) whose performance is continuously improving. Contemporary methodologies usually combine several heuristic and meta-heuristic algorithms (hybrid algorithms) in an effort to overcome the inherent limitations of single meta-heuristic components.

This chapter focuses on the most important shop production scheduling problems and the solution methodologies employed to solve them. The remainder is organized as follows: Section 1.2 is devoted to the presentation of the most classical production scheduling problems focusing on shop scheduling. Section 1.3 describes the main categories of solution methodologies emphasizing the latest meta-heuristic algorithms. Section 1.4 addresses the Flow Shop Scheduling Problem (FSSP) and relative research on solution methodologies. Similarly, Section 1.5 considers the Job Shop Scheduling Problem (JSSP) and finally, Section 1.6 covers the Open Shop Scheduling Problem (OSSP).

1.2 Production Scheduling Problems and their Classification

Although production scheduling problems have overlapping characteristics, they can be classified based on several facets [30]. Among them, the most widely used are the job arrival process, the inventory policy, various shop and job attributes and shop configuration.

Based on the job arrival process, production scheduling problems can be either ***static*** if all jobs arrive at the same time or ***dynamic*** if jobs arrive intermittently. Based on the inventory policy, a problem might be ***open*** if all products are made-to-order or ***closed*** if all products are made-to-stock. Hybrids of open and closed systems are very frequently observed in real cases. Production scheduling problems can also be classified as ***deterministic*** if job processing times and machine availability is known a priori or, conversely, ***probabilistic***. Among other job attributes based on which such problems can be classified is whether the production environment is single or multi stage. In a single stage environment, each job goes through one machine, whereas in multi stage environments, each job consists of several operations that might be processed in different machines. Finally, the number of jobs and machines and the flow pattern of jobs among machines also constitute a major classification facet.

It should be mentioned that during the last decades, academic research has focused mainly on static, deterministic, multi stage shop scheduling problems. Therefore, this chapter also focuses on these problems and their solution methodologies.

1.2.1 Single Machine Scheduling Problem

The simplest production environment is the one machine or single machine environment where all operations go through the same resource. The single machine scheduling problem was the first to be addressed academically and its characteristics and findings have been applied to more complex problems. They are also very useful for studying more complex serial structures where one machine is the bottleneck of the whole process and thus, generating a good schedule for the bottleneck machine is essential for the overall schedule performance. Generally, researchers focus on a specific performance measure and they try to develop methods to schedule operations in order to optimize the specific performance criterion. Finding the optimum solution with respect to the specific performance measure is not always feasible (see Section 1.2.4 about computational complexity). Analytically presenting all findings on single machine scheduling is out of the scope of this chapter and therefore research findings are summarized in Table 1.1 [144].

The Flowtime of a job is calculated as its completion time minus the release time, in other words it is the total time the job remains in the system. In case all jobs are not equally important, weights can be introduced so as the

Table 1.1: Single machine scheduling.

Performance Measure	Optimum solution
Min Flowtime	Shortest processing time dispatching rule
Min weighted flowtime	Weighted shortest processing time dispatching rule
Min Total Lateness	Shortest processing time dispatching rule
Min Max Tardiness	Earliest due date dispatching rule
Min number of tardy jobs	Hodgson's algorithm
Min Tardiness	Heuristically optimized
Min weighted number of tardy jobs	Heuristically optimized

scheduler can pay special attention to 'important' jobs (weighted Flowtime criterion). The Lateness of a job is calculated as its completion time minus its due date. If this value is positive then the Lateness is also called Tardiness while if this value is negative, Lateness is also denoted as Earliness. If there are specific due dates for all jobs, a very important performance measure is the 'number of tardy jobs' optimized by a special procedure proposed by Hodgson [144]. Also, see Table 1.2 for a list of the most widely used priority dispatching rules.

A very important variant of the single machine problem is the single machine scheduling problem with sequence dependent setups. This problem is equivalent to the notorious travelling salesman problem (TSP) which is NP-Hard. Small instances of these problems can be solved efficiently but in general, both problems are considered computationally intractable and medium-large instances can only be approximated with heuristic or meta-heuristic methodologies.

1.2.2 Parallel Machine Scheduling Problem

Most of the times, real life scheduling problems consider multiple machines. Multiple machines may occur in parallel or in series or both. In parallel machine scheduling we consider that each job can be processed on any of the machines and processing times are independent to the machine (identical machines). The scheduling decisions involved are: a) which machine processes each job, b) in what order [144]. Similarly to the single machine case, some problems can be solved optimally while others are approximated. The minimum flowtime problem can be solved with the shortest processing time list dispatching rule (job with the shortest processing time assigned to the least loaded machine). On the other hand, large instances of makespan minimization problem cannot be solved with exact algorithms in practical computational times. Further details on parallel machine scheduling are also beyond of the scope of this chapter.

1.2.3 Major Shop Scheduling Problems

As the single and parallel machine problems rarely represent actual production environments, academic research soon focused on more complex problems with multiple jobs and resources to be considered. Based on the flow pattern of jobs towards the resources, one can distinguish among several types of shops. The most studied problems in the literature are the flow shop scheduling problem (FSSP), the job shop scheduling problem (JSSP) and the open shop scheduling problem (OSSP), which are analyzed in the following subsections. Apart from these main three shop scheduling problem cases, the mixed shop (MSSP) and group shop (GSSP) scheduling problems have recently been proposed but have attracted much less academic interest and their solution methodologies are usually variants of the meta-heuristic methods proposed for the three main problems.

Flow Shop Scheduling Problem

The general flow shop scheduling problem (FSSP) consists of a set of N jobs $(1, 2, \ldots, n)$ to be processed on a set of M machines $(1, 2, \ldots, m)$. In the FSSP, all jobs are processed sequentially on multiple machines in the same order. Additionally, each job can be processed on one and only machine at a time and each machine can process only one job at a time respectively. Additionally, all operations are assumed non-preemptable and setup times are included in the processing times and are independent to sequencing decisions. The scheduling problem lies in finding a sequence of jobs that optimizes a specific performance criterion (usually makespan, number of tardy jobs, total tardiness or total flowtime). According to Conway's [33] notation the FSSP with makespan criterion can be shown by $n/m/F/C_{max}$ and according to Graham et al [65], it can be shown by $F//C_{max}$.

Many variants of the general FSSP exist, like the zero-buffer flow shops or flow shop with blocking, the no-wait flow shops and the hybrid flow shops [74]. In the zero-buffer flow shop, a job i having been processed on machine j cannot advance to machine $j+1$ if this machine is still processing the predecessor of job i. In this case, job i must remain at machine j, also delaying job i's successor to be processed on machine j. Generally, in the flow shop with blocking, there is no intermediate buffer and, therefore, a job cannot proceed to the next machine until this machine is free. In a no-wait flow shop, once a job is started on the first machine it has to be continuously processed through completion at the last machine without interruptions. If this is not possible, the start of a job on a given machine must be delayed so that the completion of the operation coincides the starting of the operation on the next machine. Finally, in the hybrid flow shop, there are K serial workstations and there are one or more identical parallel machines at each workstation.

Although the variants of the FSSP have extensive industrial applications, academic research has focused mainly on a reduced version of the general

FSSP, the permutation flow shop scheduling problem (PFSP) with the added assumption that jobs must be processed in the same sequence by each of the M machines. Due to the extensive academic research on the PSFP, the problem will be analyzed in the rest of this chapter.

Job Shop Scheduling Problem

The general job shop scheduling problem (JSSP) consists of a set of N jobs $(1, 2, \ldots, n)$ to be processed on a finite set of M machines $(1, 2, \ldots, m)$. In the general JSSP, each job must be processed on every machine and consists of a series of m_i operations which have to be scheduled in a predetermined order, different for each job. These precedence constraints differentiate the JSSP from FSSP. Similarly to the FSSP, each job can be processed on one and only machine at a time and each machine can process only one job at a time respectively. Additionally, all operations are assumed non-preemptable and setup times are included in the processing times and are independent to sequencing decisions. The scheduling problem lies in finding a sequence of jobs for each machine that optimizes a specific performance criterion (usually the total makespan) while, at the same time, ensuring the observance of all problem constraints. According to Conway's [33] notation the JSSP with makespan criterion can be shown by $n/m/J/C_{max}$ and according to Graham et al [65], it can be shown by $J//C_{max}$.

At first, a solution to a job shop scheduling problem was represented with the use of Gantt charts. Although the Gantt chart is an excellent monitoring tool that displays operation processing throughout the time horizon, its limitations concerning the problem representation itself led to the development of other representation methods. Among them, the one that prevailed is the disjunctive graph representation proposed by Roy and Sussmann [138]. However, it should be mentioned that Gantt charts are still widely used in user interfaces to represent a solution. Fig. 1.1 displays a disjunctive graph for a 4x3 job-shop problem.

In the node-weighted disjunctive graph of Fig. 1.1, a vertex corresponds to each operation. The set of nodes (N) represents operations to be processed on the set of machines M. The fictitious initial node S is called the source and the final fictitious node F the sink respectively. Each operation's processing time $t_{O_{kl}}$ is represented by the positive weight of each node j (thus $t_S = t_F = 0$). O_{kl} denotes that the specific operation belongs to job k and is processed on machine l. Let A be the set of directed conjunctive arcs (shown by complete lines) representing each job's precedence constraints, such that $(O_{kl}, O_{km}) \in A$ indicates that operation O_{kl} is an immediate predecessor of operation O_{km} within the subset job's k operations. Capacity constraints are represented by the set E of uni-directed orientable edges (shown by dotted lines - one colour for each machine) where each member of E is linked with a pair of disjunctive arcs sharing a common machine. Thus, two operations O_{kl} and O_{il} to be processed by the same machine l cannot be executed simultaneously.

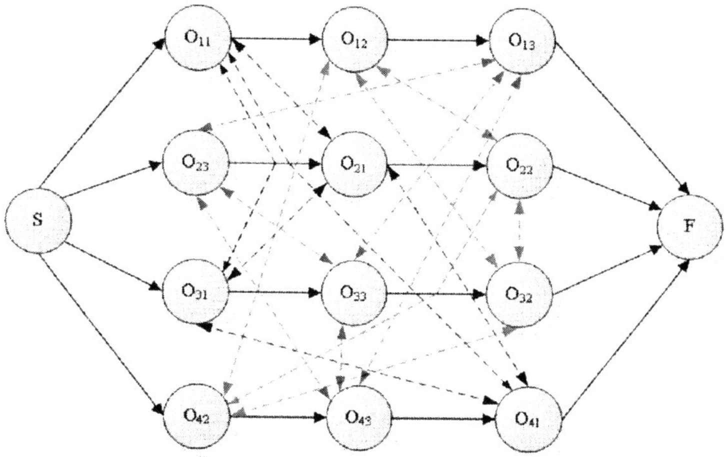

Fig. 1.1: The disjunctive graph representation for a 4x3 job shop problem.

Open Shop Scheduling Problem

A special type of the JSSP is the open shop scheduling problem (OSSP). In the OSSP there is no predefined sequence of operations among jobs and, therefore, the OSSP has a considerably larger solution space than a JSSP with similar dimensions (nxm). Probably, the best example of open shop is a car repair shop where the operation/repair sequence is not strictly defined. Just like the FSSP and JSSP, the OSSP consists of a set of N jobs $(1, 2, \ldots, n)$ to be processed on a finite set of M machines $(1, 2, \ldots, m)$. In the general OSSP, each job must be processed on every machine and consists of a series of m_i operations which have to be scheduled in any order. Similarly to the general FSSP and JSSP, each job can be processed on one and only machine at a time and each machine can process only one job at a time respectively. Additionally, all operations are assumed non-preemptable and setup times are included in the processing times and are independent to sequencing decisions. The scheduling problem lies in finding a sequence of jobs for each machine that optimizes a specific performance criterion (usually the total makespan) while, at the same time, ensuring the observance of all problem constraints. According to Conway's [33] notation, the JSSP with makespan criterion can be shown by $n/m/O/C_{max}$ and according to Graham et al [65], it can be shown by $O//C_{max}$.

Mixed Shop and Group Shop Scheduling Problems

In 1985, the mixed shop scheduling problem (MSSP) was introduced by Masuda et al [108] and some years later, in 1997, the group shop scheduling problem (GSSP) was proposed in the context of a mathematical competition [80]. In the MSSP, the machine routes of jobs can either be fixed or unrestricted [103]. The problem can also be regarded as a mix of the three aforementioned main shop scheduling problems. Similarly, the GSSP shares many characteristics of the three main shop scheduling problems. More specifically, let O be a set of operations that is partitioned into subsets $J = \{J_1, .., J_n\}$, and subsets $M = \{M_1, .., M_m\}$, where n is the number of jobs, m is the number of machines. Let J_i be the set of operations which belong to job i and M_k the set of operations which have to be processed on machine k. Each job's i operations belong to g groups $G = \{G_1, ..., G_g\}$. Within each group, operations are not restricted while, on the other hand, operations that belong to different groups must satisfy some precedence relationship between the groups imposed by the problem. In the special case where each operation constitutes a group, the GSSP is equivalent to the JSSP. Correspondingly, if for all jobs, all operations of jobs i belong to the same group, the GSSP is equivalent to the OSSP.

Considering that the mixed shop and group shop scheduling are special cases of the three main shop scheduling problems and that the meta-heuristic solution methodologies used to solve them share many characteristics and in essence are simple variations of the ones proposed for the main shop scheduling problems, the MSSP and GSSP are not further analyzed in this chapter.

1.2.4 Computational Complexity of Shop Scheduling Problems

All shop scheduling problems belong to the NP class [95]. Although some special cases of these problems might be solved by specific algorithms in polynomial time, in the general case, they are computationally intractable when the number of machines is greater than three, which means that they can only be solved with deterministic algorithms with exponential behaviour. More specifically, the time required to solve shop scheduling problems increases exponentially with the size of the input [85]. In the following subsections, the computational complexity of the three major shop scheduling problems is analyzed.

Complexity of the FSSP

Small instances of flow shops can be solved optimally. For example, the minimum makespan model for two machines can be solved with Johnson's algorithm [86] which also solves special cases with three machines. More specifically, Johnson's algorithm may be applied to a flow shop scheduling problem when the middle machine (machine 2) is dominated by the other two (a machine l dominates a machine k if for each job i the processing time of job i on machine l is greater than or equal to the processing time of job i on machine k). However, many researchers [63,96] have proved that the n-job m-machine flow shop sequencing problems belong to the class of NP-Hard problems, and therefore, computational time for obtaining an optimal solution increases exponentially with increasing problem size. As a result, academic research has focused on the development of heuristic and meta-heuristic methods.

For an n-job, m-machine general flow shop scheduling instance, the maximum number of possible solutions is $(n!)^m$. Thus, even for a relatively small instance, the number of possible solutions is considerably large (e.g., for a 10x10 instance the maximum number of possible solutions is $(10!)^{10} = 3.96 \times 10^{65}$. Concerning the PFSP, a reduced version of the general flow shop, the maximum number of possible solutions is considerably smaller, i.e. $n!$. However, Garey et al [55] proved that the $F_3/prmu/C_{max}$ is also strongly NP-Hard (F_3 indicates the three-machine case and $prmu$ denotes the permutation flow shop variant).

Complexity of the JSSP

The reason for the computational intractability of the JSSP is the fact that many different and conflicting factors must be taken into account. Such factors are due date requirements, cost restrictions, production levels, machine capacity, alternative production processes, order characteristics, resource characteristics and availability etc. However, it is the combinatorial nature of the job shop problem which determines its computational complexity [30]. Most job shop scheduling problems belong to the NP class [33]. Lenstra and Rinnooy Kan [95] proved that the general JSSP is NP-Hard for shops when the number of machines is greater than three. The only efficiently solvable cases of the JSSP are [17]:

- The number of jobs is equal to two [22].
- The two-machine JSSP where each job consists of at most two operations [83].
- The two-machine JSSP with unit processing times [72].
- The two-machine JSSP with a fixed number of jobs (repetitious processing of jobs on the machines). This case was solved by Brucker [23].

For an n-job, m-machine job shop scheduling instance, the complexity is equal to the FSSP case ($(n!)^m$ possible solutions).

Complexity of the OSSP

Various polynomial algorithms have been proposed for some cases of OSSP. For the two-machine problem, a polynomial time algorithm has been proposed by Gonzalez and Sahni [62] while Adiri and Aizikowitz [4] presented a linear time algorithm for the three-machine OSSP, provided that one machine dominates one of the other two. Although these problems with special structures are polynomially solvable, Gonzalez and Sahni [62] proved that in the general case and if the number of machines is greater than three, the OSSP is

NP-Complete. For an n-job, m-machine open shop scheduling instance, the maximum number of possible solutions is $(nm)!$. Thus, even for a relatively small instance, the number of possible solutions is considerably large (e.g., even for a small 10x10 instance the maximum number of possible solutions is $100! = 9.33 \times 10^{157}$). As a result, research on OSSP solution methodology focused on heuristic and meta-heuristic methodologies.

1.2.5 Optimization of Production Scheduling Problems

Shop scheduling problems are typical representatives of the Combinatorial Optimization class of problems where solutions are encoded with discrete variables. Generally, combinatorial optimization problems are optimization problems where the set of feasible solutions is or can be reduced to a discrete one, and the goal is to find the best possible solution. According to Blum and Roli [19], a combinatorial optimization problem $P = (S, f)$ can be defined as:

- A set of variables $X = \{x_1, \ldots, x_n\}$.
- Variable domains $D_1, \ldots, D_n$.
- Constraints among variables.
- An objective function f to be minimized/maximized where $f : D1 * \ldots * Dn \rightarrow \Re^+$.

In a combinatorial optimization problem, the set of all possible and feasible assignments is:

$$S = \{s = \{(x_1, u_1), \ldots, (x_n, u_n)\}\} \mid u_i \in D_i, s. \tag{1.1}$$

S is called a search or solution space and consists of every possible solution to the specific problem. In order to find an optimum solution, the solution space has to be explored effectively and efficiently. This optimum solution minimizes or maximizes (depending on the problem) the objective function while satisfying all constraints of the specific problem. The optimum solution is also called global optimum.

Apart from the solution space, the coding space represents the set of possible assignments of a problem's variables after being encoded so that a specific method can be used. For example, Cheng et al [32] have conducted an extensive review of possible encoding methods when using evolutionary algorithms in JSSP. Among the representations mentioned, the operation-based one encodes each possible solution with a vector that corresponds to the operation sequence. In this scheme, the coding space consists of all possible operation permutations. On the other hand, the solution space of JSSP consists of detailed operation schedules for each machine and may also include non feasible schedules, i.e. schedules that do not satisfy all constraints of the problem. Other possible representations for such problems are the job-based, the preference-list based, the priority rule-based, the completion time-based, the machine-based and the random keys. In the job-based, the solution to a given problem is given by a simple job permutation. More specifically, following this representation, the operations that belong to the first job of the sequence are scheduled in the earliest possible position and so on, so forth. The preference-list based representation is a priority rule permutation and at each position, the operation that complies with the given priority rule is scheduled. The completion time-based representation is based on a vector of completion times for all operations while the machine-based representation generates m vectors with job permutations where m is the number of machines. Finally, the random keys representation is mainly used with continuous optimization methods where an operation permutation is extracted from a series of real values with a heuristic procedure (usually, the position with the lowest real value is scheduled first and so on, so forth).

Ideally, a representation of a problem should link a single encoded solution (coding space) to a single feasible solution (solution space) and additionally, all possible and feasible solutions should be able to be encoded (full coverage of the solution space). However, in some cases, various encoding schemes not only direct to infeasible solutions but to illegal ones as well, i.e. solutions outside the solution space. Therefore, each problem's representation should be carefully chosen and always in conjunction with the solution methodology to be developed.

1.3 Solution Methodologies for Shop Scheduling Problems

Combinatorial optimization problems can be solved/optimized using two different approaches that are presented in the rest of this Section. The first approach is by using complete or exact algorithms. The second approach is applied to more complex or larger problems, leads to near-optimum solutions and is characterized by the use of approximate algorithms. The second approach can be further divided to heuristic and meta-heuristic methods. The scope of this chapter is to present the latest meta-heuristic trends on solving shop scheduling problems and thus, the analysis focuses on meta-heuristic methods.

1.3.1 Exact Algorithms

Exact or complete algorithms are guaranteed to find for every finite size instance of a combinatorial optimization problem an optimal solution in bounded time. However, for the typical combinatorial optimization problems, like the shop scheduling problems which are usually NP-Hard, no algorithms exist to solve these problems in polynomial time. Therefore exact algorithms need exponential computation time in most cases which leads to impractical computational burden for real large scale applications. The family of exact methods is considerably large but the most common exact/complete methods for scheduling problems are branch and bound algorithms, mixed integer programming and decomposition methods.

1.3.2 Heuristic Algorithms

As seen, the use of complete/exact methods to solve complex combinatorial optimization problems often leads to impractical computational times. This phenomenon has led the vast majority of researchers on such problems to approximation methods. In approximation methods, the guarantee of finding optimal solutions is sacrificed in order to get near-optimum solutions in reasonable and practical computational times. The basic form of approximation algorithms is called 'heuristics', a name derived from the Greek verb '$\epsilon\upsilon\varrho\iota\sigma\kappa\varepsilon\iota\nu$' which means 'to find'. The usual classification of heuristic methods is constructive and local search methods [19].

Constructive

Starting from scratch, constructive algorithms generate solutions by gradually adding parts of the solution to the initially empty partial solution. In typical shop scheduling problems for example, these parts are usually operations. Constructive heuristics are generally the fastest approximate algorithms although some special implementations may induce high computational load. Their advantage in computational time requirements is counterbalanced by generally inferior quality solutions when compared to local search techniques. Among the most widely used constructive heuristics for shop scheduling problems are the various 'Dispatching Rules'. Table 1.2 summarizes the most common dispatching rules for shop scheduling problems [17].

Dispatching (or Priority) Rules are the most common heuristics for shop scheduling problems due to their easy implementation and low requirements in computational power. Although they perform very well in certain cases, no rule exists that can be applied to all shop problems and perform satisfactorily. Even worse, there is no way to estimate the performance of a dispatching rule for a specific instance a priori. It should be mentioned that some priority rules generate the optimum solution in certain simple problems (e.g. the minimization of flowtime in single-machine scheduling where the SPT priority rule generates the global optimum solution).

Local Search

According to [19], a neighborhood of a solution s may be defined as a function $N : S \rightarrow 2^S$ where each s is assigned a set of neighboring solutions $N(s) \subseteq S$. $N(s)$ represents all neighboring solutions of s. For every defined neighborhood of solutions, the solution or solutions of highest quality (i.e. the best objective

Table 1.2: Dispatching rules.

Rule	Description
SOT	An operation with the shortest processing time on the machine considered
LOT	An operation with longest processing time on the machine considered
LRPT	An operation with longest remaining job processing times
SRPT	An operation with shortest remaining job processing times
LORPT	An operation with highest sum of tail and operation processing time
Random	The operation for the considered machine is randomly chosen
FCFS	The first operation in the queue of jobs waiting for the same machine
SPT	A job with smallest total processing time
LPT	A job with longest total processing time
LOS	An operation with longest subsequent operation processing time
SNRO	An operation with smallest number of subsequent operations
LNRO	An operation with largest number of subsequent operations

function value) are called locally optimum solutions within the defined neighborhood. In the case where there is only one solution with the best objective function value, this local optimum is called strict locally optimum solution.

Local search algorithms start from an initial solution (most of the times generated by a constructive heuristic or randomly) and iteratively try to replace part or the whole solution with a better one in an appropriately defined set of neighboring solutions. In order to replace parts of an initial solution, local search methods perform a series of moves leading to the formation of new solutions in the same neighborhood. The most common moves are the $2-Opt$, the $1-1$ *exchange* and the $1-0$ *exchange* moves. The $2-Opt$ move reverses a set of tasks of random length in a machine while the $1-1$ *Exchange* move swaps two tasks from the same machine. Finally, the $1-0$ *Exchange* move transfers a task from its position in one machine to another position in the same machine [161] Of course, the number of possible moves and the corresponding neighborhoods are virtually unlimited.

The main drawback of basic local search methods is that they get easily trapped in local optima as they are myopic in nature. More specifically, local search with appropriate moves can be very effective in exploring a neighborhood of an initial solution but no mechanism exists that can lead to other distant neighborhoods of the solution space where the global optimum may exist. To remedy this weakness, new modern local search methods (explorative local search) have been developed with embedded meta-strategies to guide the search process. Such methods are presented in Section 1.3.3.

1.3.3 Meta-heuristic Algorithms

During the last decades, a new family of approximate algorithms has emerged and has dominated the combinatorial optimization problem solution research.

This new type of algorithm basically combines heuristic methods in higher level frameworks. The aim of the new methodology is to efficiently and effectively explore the search space driven by logical moves and knowledge of the effect of a move [19] facilitating the escape from locally optimum solutions. These methods are nowadays called meta-heuristics, a term that was first introduced by Glover [57]. Meta-heuristic methods have an advantage over simpler heuristics in terms of solution robustness; however they are usually more difficult to implement and tune as they need special information about the problem to be solved to obtain good results. Due to the computational complexity of combinatorial optimization problems, the moderate results acquired by heuristic methods and the time limitations for application of exact algorithms, the application of meta-heuristic methods to solve such problems is a well established field of research.

Definition

There are many definitions of meta-heuristic algorithms and methods. Perhaps, the most thorough definition was given by Stützle in 1999 and is cited in [19] (p. 270):

> *"Meta-heuristics are typically high-level strategies which guide an underlying, more problem specific heuristic, to increase their performance. The main goal is to avoid disadvantages of iterative improvement and, in particular, multiple descent by allowing the local search to escape from local optima. This is achieved by either allowing worsening moves or generating new starting solutions for the local search in a more intelligent way than just providing random initial solutions. Many of the methods can be interpreted as introducing bias such that high quality solutions are produced quickly. This bias can be of various forms and can be cast as descent bias (based on the objective function), memory bias (based on previously made decisions) or experience bias (based on prior performance). Many of the meta-heuristic approaches rely on probabilistic decisions made during the search. But, the main difference to pure random search is that in meta-heuristic algorithms randomness is not used blindly but in an intelligent, biased form."*

Based on the above definition, it could be said that meta-heuristics are an intelligent way to explore the solution space. Exact algorithms are too slow for even small to medium search spaces and simple heuristics blindly or myopically program the next move based on rules that cannot be characterized robust and logically defined. Based on recent findings, the best performing methods for optimizing shop scheduling problem are those encompassing hybrid systems such as local search techniques embedded within metastrategies that overcome local optimality by accepting non improving moves and, thus, inferior solutions. Although allowing non-improving moves seems contradicting at first, such strategies help meta-heuristic algorithms to escape local optimality as the new inferior solutions might be in the neighborhood of the global optimum and therefore, a simple local search may thereafter lead to the global optimum solution.

Concerning the search direction strategy, the diversification and intensification strategies can be defined [19]. When a meta-heuristic method uses the diversification strategy, the main aim is the effective exploration of all possible neighborhoods of the solutions space. On the other hand, the intensification strategy focuses on the use of the gathered search knowledge and the exploration of a narrower solution subspace.

Classification of Meta-heuristic Methods

Several meta-heuristic algorithm classification schemes exist based on various properties of these methods. Among them, the most meaningful and the most widely used is based on the quantity of solutions these algorithms deal with. More specifically, meta-heuristic algorithms can be divided in population-based and single point search. Population based meta-heuristic methods combine a number of solutions in an effort to generate new solutions that share good merits of the old ones and are expected to have better fitness. Such methods are iterative procedures that gradually replace solutions with better found ones. On the other hand, single point search methods improve upon a specific solution by exploring its neighborhood with a set of moves.

Another important and widely used classification scheme is based on the memory used during the search process. Concerning memory usage, it has nowadays become implicit in modern meta-heuristic methods. Memory usage constitutes a main characteristic of effective meta-heuristic methods and is the indication of the intelligence employed during the search process. Memory usage may be further divided in short-term and long-term. Short-term memory keeps track of recent moves and helps avoiding cycling around specific solutions. On the other hand, in the long-term memory, information is gradually stored and employed at specifically defined stages of the algorithm, depending on the implementation. Memory-less algorithms are nowadays rarely employed for complex combinatorial optimization problems.

Meta-heuristic Algorithms used for Shop Scheduling Problems

Almost all meta-heuristic algorithms proposed in the literature have been employed to solve shop scheduling problems. For a thorough description of these methods, the reader may refer to Blum and Roli's work [19]. Epigrammatically, the most popular meta-heuristic methods for solving shop scheduling problems are:

- **Evolutionary Computation** algorithms that fall mainly into three main categories: Genetic Algorithms [77], Evolutionary Strategies [133] and Evolutionary Programming [49]. One of the newest algorithms of this category

is the Differential Evolution Algorithm (DEA) introduced by Storn and Price [146].
- **Particle Swarm Optimization** (PSO) [42].
- **Ant Colony Optimization** (ACO) introduced by Dorigo [40].
- **Scatter Search and Path Relinking** proposed by Glover et al [60].
- **Neural Networks** (NN) that constitute an advanced artificial intelligence technology that mimics the brain's learning and decision making process [51, 52].
- **Basic Local Search** where a neighborhood of a solution is explored with a set of moves and the local optimum is returned.
- **Explorative Local Search** mainly represented by the *Greedy Randomized Adaptive Search Procedure* (GRASP) proposed by Feo and Resende [45], *Variable Neighborhood Search* (VNS) proposed by Hansen and Mladenović [70] and *Iterated Local Search* (ILS) proposed by Stützle [149].
- **Simulated Annealing** (SA) proposed by Kirkpatrick et al [87].
- **Tabu Search** (TS) proposed by Glover [59].
- **Threshold Accepting** proposed by Dueck and Scheuer [41].

Hybrid Methods

The development and application of hybrid meta-heuristic algorithms is increasingly attracting academic interest. Hybrid meta-heuristic algorithms combine different concepts or components from various meta-heuristics [19] and towards this end, they attempt to merge the strengths and eliminate the weaknesses of different meta-heuristic concepts. Therefore, the effectiveness of the solution space search may be further enhanced and new opportunities emerge which may lead to even more powerful and flexible search methods. Talbi [158] has proposed a taxonomy for hybrid meta-heuristics. Generally, we can distinguish three main forms of hybridization [19]. The first form is called *component exchange* among meta-heuristic methods and its most typical representative is the hybridization of population-based methods with local search methods. The second form is called *cooperative search* and typically involves the exchange of information between two or more different meta-heuristic algorithms. The information exchange level can vary from implementation to implementation as well as during the same implementation. The third form is called *integrating meta-heuristic algorithms and systematic methods* and has produced very promising results in real-world cases. Among the successful implementations of this form is the combination of meta-heuristic algorithms and constraint programming [128].

Most hybrid methods for shop scheduling methods belong to the first form of hybridization where various components and solution characteristics are shared among two or more heuristic and meta-heuristic methodologies. The power of hybrid meta-heuristic methods is indicated by the fact that, as we will see in the following sections of this chapter, current state-of-the-art methods for shop scheduling problems are all hybrid meta-heuristic algorithms.

Adaptive Memory Programming - The Unified View

Judging from the brief review of meta-heuristic algorithms in the previous section, it is obvious that the family of meta-heuristic methods is large and keeps growing as new forms and hybrids are still presented. Taillard et al [157] however mentioned that meta-heuristic methods at their present form share many common characteristics and could be unified under the term adaptive memory programming (AMP), firstly proposed by Glover [58]. More specifically, Taillard et al [157] mentioned that most meta-heuristic components follow a generalized procedure: at first, they memorize parts, whole solutions or characteristics of solutions during the search process. In a second stage, this stored information is used to generate new solutions and at the third stage, this new solution's neighborhood is explored by means of a local search method.

Thus, whether a meta-heuristic is a simple population-based genetic algorithm or a sophisticated hybrid incorporating various meta-heuristic components, the essence is that it usually follows the same three-step procedure. Therefore, under the unified view of adaptive memory programming, the various meta-heuristic methodologies might not be very different from each other after all. On the other hand, some methods (like simulated annealing and threshold accepting) cannot be generalized under the AMP scheme as their basic versions are virtually memory-less. However, their core methodology can be used in the improvement stage of the AMP framework (step three of AMP).

1.4 The Flow Shop Scheduling Problem

The flow shop scheduling problem is, as mentioned, NP-Hard. Due to the complexity of the problem, exact algorithms developed for the general FSSP failed to achieve high quality solutions for problems of increased size in reasonable time and thus academic research focused on heuristic and meta-heuristic methods. Regarding solution methodologies, Johnson [86] proposed an $O(nlogn)$ complexity algorithm which optimally solves the $F_2//C_{max}$ problem. Under the special circumstance where the middle machine is dominated by the other two, Johnson's algorithm may solve to optimality the $F_3//C_{max}$ problem. Among exact methods proposed for the general FSSP, we can distinguish among dynamic programming [75], branch and bound [37, 81, 90, 104] elimination rules [6] and, of course, complete enumeration which is the most time consuming exact method. Considering that these methods may be applied in small instances of FSSP, it is beyond the scope of this chapter to further analyze them.

1.4.1 Heuristics for the FSSP

A large number of heuristics has been proposed for the FSSP. The most important constructive ones are the Palmer [126], CDS [26], RA [38], Gupta [68],

NEH [118] and HFC [88]. Palmer [126] and Gupta [68] construction heuristics are based on the utilization of a criterion to generate the sequence of jobs. This criterion is a slope index that is calculated for each job based on the processing times and their pattern in terms of machine sequence. The CDS [26] algorithm on the other hand, generates $m-1$ artificial two-machine schedules (where m is the number of machines) and solves them with Johnson's rule. The best $m-1$ solution is then chosen as the best sequence for the m-machine problem. According to the RA heuristic [38], a virtual two machine problem is defined (just like the CDS heuristic) but instead of directly applying Johnson's algorithm over the processing times, weighting schemes for each machine are calculated before.

The NEH heuristic [118] is much more complex than the aforementioned heuristics. Initially, jobs are arranged in a descending order of their total processing time. Then, based on this mentioned order, an increasingly larger partial sequence is generated at each step by introducing one job from the unscheduled order into the partial sequence (until all jobs are scheduled). At each step, a new job is scheduled in all possible positions ($k+1$ positions where k is the size of last step's partial sequence) and after choosing the best place for this job, regarding the obtained makespan, this new partial sequence is fixed for the remaining procedure. The HFC heuristic [88] can only be applied to FSSP and not the reduced version of it, the PFSP. It is a two-stage method where Johnson's rule is used extensively in the first stage, while in the second stage, improvement of the initial schedule is performed by allowing job passing between machines (non permutation schedules).

In addition to the aforementioned constructive heuristics, a few improvement heuristics have been proposed by Dannenbring [38], Ho and Chang [76] and Suliman [150]. Contrary to constructive heuristics, improvement heuristics start from an already built schedule and try to improve it by some given procedure. Dannenbring [38] proposed two simple improvement heuristics based on the constructive heuristic RA proposed by the same author for initial solution generation: Rapid Access with Close Order Search (RACS) and Rapid Access with Extensive Search (RAES). RACS works by swapping every adjacent pair of jobs in a sequence. The best schedule among the $n-1$ generated is then given as a result. In RAES heuristic, RACS is repeatedly applied while improvements are found. Both RACS and RAES heuristics start from a schedule generated with the RA constructive heuristic.

Ho and Chang [76] developed a method that aims at the minimization of the elapsed times between the end of the processing of a job in a machine and the beginning of the processing of the same job in the following machine in the sequence. The authors refer to this time as 'gap' which is calculated for every possible pair of jobs and machines. Starting from the CDS heuristic solution, a series of calculations is conducted and then, the heuristic swaps jobs based on the corresponding gap value. Suliman [150] developed a two-phase improvement heuristic for the FSSP. In the first phase, similarly to Ho and Chang's improvement heuristic, a schedule is generated with the CDS heuristic method. In the second phase, the schedule generated is improved with a job pair exchange mechanism. In an effort to reduce the computational load of an exhaustive job pair exchange, the number of possible moves is reduced by introducing a directionality constraint. More specifically, if a better schedule is acquired by moving a specific job forward, then this job is not allowed to move backward in the job sequence string.

Although FSSP heuristics perform very fast in most cases, they generally fail to produce high quality solutions. Even the NEH heuristic, which is considered the most powerful construction heuristic [154], fails to reach solutions even 5-7% worse than the optimum in some difficult instances due to Taillard [155]. Therefore, it is not without reason that most of the academic effort has been put towards the development of meta-heuristic and most recently, powerful hybrid meta-heuristic methods.

1.4.2 Meta-heuristics for the FSSP

A very large number of meta-heuristic algorithms, hybrid or not, have been proposed for the FSSP and PSFP. The application list for the FSSP includes most known meta-heuristic algorithms including evolutionary algorithms, particle swarm optimization, ant colony optimization, scatter search, neural networks, simulated annealing, tabu search and many forms of explorative local search. Most of these methods use simpler heuristic algorithms to generate an initial population of solutions.

Concerning the implementation of ***Genetic Algorithms*** (GA) in FSSP and PFSP, some early noteworthy research was performed by Reeves [134]. In his implementation, the offsprings generated at each step of the algorithm do not replace their parents but solutions with a fitness value below average. Among the innovations presented in this paper, Reeves used an equivalent to the One Point Order Crossover operator, called C1. Moreover the algorithm uses an adaptive mutation rate where the position of one job is simply changed by the shift mutation operator. A common feature of many meta-heuristic methods for the FSSP that is also found in this implementation is the use of the NEH heuristic to seed the initial population with a good sequence among randomly generated ones. Finally, another innovative feature of Reeves' implementation is that during selection, parent one is selected using a fitness rank distribution while parent two is chosen using a uniform distribution. During the same year, Chen et al [31] presented their own GA for the PFSP. Their implementation was basically a simple genetic algorithm with some added features and enhancements. More specifically, in this implementation the initial population is generated with the CDS and RA heuristics and also from simple job exchanges of some individuals. The crossover operator used is the Partially Mapped Crossover (PMX) and it is noteworthy that no mutation is applied.

Murata et al [115] presented their own version of GA for the PSFP where the two-point crossover operator and a shift mutation along with an elitist strategy to obtain good solutions are used. Based on initial results, the algorithm failed to achieve results competitive to other meta-heuristic methodologies and therefore, the authors developed two hybrid versions; genetic simulated annealing and genetic local search. As their name implies, in these algorithms an improvement is conducted before the selection and crossover phases by means of a simulated annealing and local search algorithms respectively. These two hybrid algorithms performed better than the non-hybrid genetic algorithm and the simple implementations of tabu search, simulated annealing and local search. Reeves and Yamada [135] presented another hybrid genetic algorithm. The innovation in this algorithm was the use of a Multi-Step Crossover Fusion (MSXF) operator which combined crossover with a local search procedure. This operator uses one parent as a reference to conduct a biased local search to the other. The calculation of new upper bounds for some Taillard's [155] benchmark instances is indicative of the algorithm's high performance.

Another genetic algorithm was presented by Ponnambalam et al [130]. Their algorithm features the Generalised Position Crossover (GPX), shift mutation and a randomized initial solution. A hybrid implementation based on genetic algorithms and simulated annealing was presented by Wang and Zheng [167]. The powerful NEH heuristic is also used for population initialization and multi-crossover operators are used for solution recombination. However, the mutation operator is replaced by a simulated annealing component. Finally, one of the most recent genetic algorithms for the PFSP and a hybrid version of it were proposed by Ruiz et al [140]. Ruiz et al used a modified NEH heuristic to generate a diversified initial population, proposed four new crossover operators, used shift mutation and implemented specialized restart techniques for population renewal. Finally, they hybridized their genetic algorithm with a local search scheme.

Similarly to the Genetic Algorithm, the ***Differential Evolution Algorithm*** (DEA) has also been proposed, in a hybrid form, to solve the PFSP. More specifically, Tasgetiren et al [162] have proposed a DEA coupled with local search to optimize the PFSP. The authors borrowed the random key representation by Bean [14] to convert the real variables to discrete ones and used a repairing technique so a basic local search could be applied to the resulting solutions. Another attempt to apply the DEA to PSFP for minimizing makespan, total flowtime and tardiness has been conducted by Onwubolu and Davendra [124]. The authors compared the proposed method with the classic genetic algorithm and concluded that DEA demonstrates competitive performance and very easy implementation for the specific problem. Considering that the DEA in its canonical form operates with real variables, the authors proposed a transformation scheme to convert real-coded solutions to discrete ones and vice versa.

Particle Swarm Optimization (PSO) has only recent and thus limited number of applications. The two major papers in this field have been presented by Lian et al [98] and Tasgetiren et al [163]. Lian et al [98] proposed a conversion technique to apply the PSO algorithm in discrete problems like the FSSP and compared the PSO with a traditional GA. The authors concluded that the PSO performed better than the traditional GA. Similarly, Tasgetiren et al [163] proposed a hybrid PSO algorithm for the PFSP with both makespan and total flowtime minimization criteria. More specifically, they hybridized the basic PSO algorithm with an explorative local search technique, the Variable Neighborhood Search (VNS). They also proposed a scheme to allow the PSO to operate in the discrete variable environment of PFSP. This heuristic scheme is named Shortest Position Value (SPV) and it is borrowed from the random key representation of Bean [14]. Computational tests conducted by the author revealed the strength of the proposed method for both performance criteria and in both Taillard and Watson benchmark instances.

The ***Ant Colony Optimization*** (ACO) algorithm has also been used for solving the FSSP/PFSP. One of the first attempts to apply ACO to PFSP was done by Stützle [147] and some years later by Ying and Liao [177]. Ying and Liao's computational experiments on Taillard benchmark instances revealed that the ACO approach is a very effective meta-heuristic for the PFSP. Among other presented implementations, T'kindt et al [152] proposed a version of ACO to solve the two-machine flow shop with two criteria. More recently, Rajendran and Ziegler [132] proposed two versions of ACO to solve the PFSP with the minimum makespan and minimum total flowtime objectives. The first algorithm presented is an extension of Stützle's [147] ideas of the ant colony algorithm (MMAS algorithm) by including the summation rule suggested by Merkle and Middendorf [112] and a local search technique.

Nowicki and Smutnicki [121] presented a ***Scatter Search and Path Relinking*** version (algorithm MSSA) for the PFSP and have conducted extensive research on the properties on the problem's solutions space. The strength of their implementation is attested by some new upper bounds found in Taillard's benchmark instances for minimizing the total makespan. More specifically, the authors have used their TSAB algorithm and extended it using a modified scatter search algorithm (MSSA). Finally, one of the most important characteristics of their algorithm is that its good properties remain scalable with increasing instance size.

There are quite a few works about ***Neural Networks*** (NN) applied to FSSP, although most of them are hybrid implementations with other meta-heuristics. One of the first works in this field was conducted by Lee and Shaw [94]. The authors developed a hybrid neural network-genetic algorithm for the FSSP showing good performance. A few years later, Solimanpur et al [145] developed a hybrid tabu search-neural network approach called EXFS. In their implementation, the NEH heuristic is initially used to construct a solution. Thereafter, tabu search is used to improve the solution and special neurons are employed to penalize some moves. A more recent implementation was developed by El-bouri et al [44] for the PFSP. More specifically, they used solved instances of PFSP instances to train some neurons. Afterwards, the neurons were employed to guide a local search procedure to the most promising job assignments.

Concerning ***Basic Local Search*** and ***Explorative Local Search*** methods, there are very few implementations where such methods are used alone (i.e. not as a part of a hybrid algorithm). It should be mentioned though that, when hybridized, such methods greatly improved the performance of other methods, especially of evolutionary methods. One of the few implementations was the iterated local search (ILS) procedure proposed by Stützle [148] which was later recoded by Ruiz and Maroto [139]. The latter proved that ILS performed very satisfactorily in the PFSP.

Simulated Annealing (SA) was one of the first meta-heuristic methods proposed [87] and therefore it is not without reason that a very large number of SA algorithms have been proposed for the FSSP/PFSP. One of the first works in this area has been conducted by Osman and Potts [125]. The authors proposed a simple SA algorithm using a shift neighborhood and a random neighborhood search. Ogbu and Smith [123] proposed an SA algorithm for the PFSP which involved an initialization with the Palmer [126] and Dannenbring's [38] RA heuristics. In a later work by the same authors [122], their initial approach was compared to that of Osman and Potts [125] where Ogbu and Smith's [122] algorithm was found slightly more efficient. Gangadharan and Rajendran [54] also applied the SA method to solve the FSSP. Two SA algorithms comparable in performance with Osman and Potts's [125] approach were developed by Ishibuchi et al [82].

Zegordi et al [179] developed a hybrid simulated annealing algorithm called SAMDJ with the incorporation of problem domain knowledge in the basic SA scheme. More specifically, they used a specially formulated table with several rules concerning the biased movement of jobs towards specific positions or directions in the sequence. It was empirically proven that this table facilitated the annealing process and lessened the SA control parameters. Performance wise, SAMDJ proved to be slightly inferior in terms of solution quality compared to Osman and Potts's [125] SA algorithm but considerably faster. Murata et al [115], as already mentioned in the genetic algorithm section, developed a hybrid genetic-simulated annealing algorithm for the PFSP which, according to their experiments, outperformed the single meta-heuristic components when used individually. The same procedure was followed by Moccellin and dos Santos [114] who presented a hybrid tabu search - simulated annealing heuristic. The hybrid algorithm was then compared to simple tabu search and simple simulated annealing algorithms, showing the superiority of the hybrid approach. Wodecki and Božejko [170] proposed an SA algorithm that was run in a parallel computing environment. The algorithm was tested against the classic NEH heuristic and superior results were recorded. Finally, Hasija and Rajendran [71] proposed an algorithm based on simulated annealing for the PFSP with total tardiness of jobs criterion and two new solution perturbation schemes. The authors concluded that their algorithm performed better than existing tabu search and simulated annealing techniques.

Tabu Search (TS) has also been extensively used to solve the FSSP/PFSP in single or hybrid forms. One of the first attempts towards this direction was conducted by Widmer and Hertz [169] who presented the 'SPIRIT' method, a two-stage heuristic. In the first stage, an initial solution is calculated with an insertion method in direct relation to the Open Travelling Salesman Problem (OTSP). In the second stage, a standard TS meta-heuristic with exchange neighborhood is used to improve the initial solution. Taillard [154] also presented a similar procedure to that of Widmer and Hertz [169]. In Taillard's implementation, an improved version of the classic NEH heuristic is used to obtain good initial schedules which are then improved by a tabu search procedure. After having tested various types of neighborhoods, the one-job change of position proved to be superior for the specific implementation.

Another meta-heuristic algorithm based on Widmer and Hertz's [169] SPIRIT heuristic is the TS method of Moccellin et al [113]. Their implementation resembles Widmer and Hertz's procedure with the main difference being the calculation of the initial solution. One of the most important works in this field was conducted by Nowicki and Smutnicki [120]. The authors proposed a TS meta-heuristic where only reduced parts of the possible neighborhoods are evaluated along with a fast method for obtaining the makespan after performing moves. Although this method is only suitable for the minimization of makespan criterion (and not the also widely used total flowtime of jobs) it quickly became the benchmark for other methods due to its high performance in terms of solution quality and computational speed. This fact is also demonstrated by the new best upper bounds found in Taillard's benchmark instances. The innovation in Nowicki and Smutnicki's work is that instead of moving single jobs during the search procedure, whole blocks of jobs are moved.

A few years later, Ben-Daya and Al-Fawzan [15] implemented a TS algorithm with extra intensification and diversification schemes that enabled better moves in the TS process. The algorithm proposed provides similar results to the TS algorithm of Taillard [154], indicating a successful implementation. Finally, as already mentioned in the SA section, a successful hybrid TS-SA method was proposed by Moccellin and dos Santos [114]. More specifically, a classic TS was hybridized with an SA procedure and the hybrid algorithm outperformed the single meta-heuristic components. One of the most recent implementations for the PFSP with tabu search was proposed by Solimanpur et al [145]. The authors developed a hybrid tabu search - neural network approach and they implemented block properties like Taillard [154]. Finally, Grabowski and Wodecki [64] proposed a fast TS approach.

Threshold Accepting has also been used to solve PFSP by Gupta et al [69]. More specifically, the authors developed and compared various local search schemes for the two-machine PFSP with two criteria: the main criterion being the optimization of makespan and a secondary criterion being either the total flow time, total weighted flow time, or total weighted tardiness. Among the methods tested are simulated annealing, genetic algorithms, threshold accepting, tabu search, and multi-level search algorithms. The authors conclude that the multi-level search heuristics were most appropriate for the flow time related secondary objective, while simulated annealing proved to be superior best for the due date related objective. The authors also concluded that the genetic algorithm performed poorly if not coupled with some other local search methods.

1.5 The Job Shop Scheduling Problem

The history of the job shop scheduling problem, starting more than 40 years ago, is closely linked to the history of the most known benchmark instance introduced by Fisher and Thompson [46]. This particular 10-job, 10-machine instance (also known as MT10 or FT10) detained researchers for over 25 years and signaled a continuous competition among researchers for the most powerful solution procedure. The JSSP solution history is characterized by circles concerning researchers' preferences. More specifically, at first, researchers proposed some exact methods followed by an era of heuristic methodologies development. Thereafter, researchers focused again on exact methods and JSSP complexity until the new era of sophisticated heuristic and meta-heuristic algorithms began.

Branch and bound algorithms are the most efficient exact methods for JSSP and they explore specific knowledge about the critical path of the problem. The main principle behind them is the enumeration of all possible feasible solutions of the problem so that properties or attributes not shared by any optimal solution are detected as early as possible. A branch of the enumeration tree defines a subset of the feasible solutions where each element of the subset satisfies this branch. For each subset, the objective value of its best solution is estimated by a lower bound. In case this lower bound is greater than the best known upper bound the subset can be dropped from further consideration. The best known upper bound can be a heuristic solution of the original problem [17]. Although branch and bound algorithms guarantee the optimal solving of the problem, they tend to be very slow in combinatorial optimization problems and rather impractical to use. Their main drawback is the lack of strong lower bounds in order to cut branches of the tree as early as possible and reduce the computational time required. Researchers are focused on finding ways to improve lower bounds.

One of the earliest works in the field of exact methods was performed by Brooks and White [21] and Greenberg [66]. These methods were based on Manne's [106] integer programming formulation. Other papers in this field are presented by Balas [7], Florian et al [48] and Fisher [47]. A very efficient method was proposed by McMahon and Florian [111] which turned out to be the best performing exact method for several years. The authors combined the bounds for the single-machine scheduling problem with the objective function and operation release dates in order to minimize maximum lateness with the enumeration of active schedules. They were based on the fact that among active schedules, the optimal ones existed. In more recent methods, the neighborhood structure used in some recent local search implementations was employed as a branching structure [12]. The optimum solution of the infamous FT10 instance was first found by Lageweg in 1984 as reported by Lawler et al [91] although no proof of optimality was given. In this work, multi-machine lower bounds were used. Optimality of the found solution to FT10 (with an optimum makespan of 930) was first proven by Carlier and Pinson [28]. The latter were based on one-machine bounds and several simple inference rules on operation subsets. Current research status concerning branch and bound schemes is characterized by the very well performing methods of Applegate and Cook [5] and Martin and Shmoys [107]. Moreover, the use of advanced inference rules characterize the very well performing branch and bound methods of Baptiste et al [10, 11], Carlier and Pinson [29] and Brucker et al [25]. Finally, it should be mentioned that an emerging field of research concerning branch and bound techniques is their coupling with meta-heuristic strategies to form advanced hybrid algorithms.

1.5.1 Heuristics for the JSSP

Early works for the JSSP can be found as early as 1956 when Jackson [83] generalized Johnson's [86] rule for the FSSP and applied it to the two-machine JSSP. Since then, a large number of heuristic methodologies have been proposed. Perhaps the most frequently applied heuristics for JSSP are the various dispatching rules, presented in Section 1.3.2. One of the most important implementations of priority dispatching rules is Giffler and Thompson's algorithm [56] which can be generally considered as a common basis of all priority rule based heuristics. Their algorithm is a constructive heuristic where at each step, an unscheduled operation is chosen randomly from a conflict set of operations that 'compete' for the same machine. This machine corresponds to the machine that the operation with the lowest processing time among all unscheduled operations is to be processed at.

Apart from the priority dispatching rules, one of the most powerful heuristics for the JSSP is the shifting bottleneck heuristic originally proposed by Adams et al [3] and later improved by Balas et al [8]. The main concept of this algorithm is the solution, for each machine, of a one-machine scheduling problem to optimality, under the assumption that a lot of arc directions in the optimal one-machine schedules coincide with the optimal job shop schedule. The algorithm optimizes the schedule of the bottleneck machine first. The bottleneck machine is defined as the machine whose one machine optimum schedule has the longest makespan. The algorithm continues until all machines are scheduled. The power of this method is that although one-machine schedules are also NP-Hard, algorithms exist to efficiently solve them [27]. However, the assumption that the lot of arc directions in the optimal one machine schedules coincide with the optimal job shop schedule is not true and thus this method does not lead to the global optimum solution. However, the shifting bottleneck algorithm has proven to be a very effective heuristic with many successful implementations in JSSP. This heuristic is also widely used nowadays for efficient hybrid algorithm implementations.

1.5.2 Meta-heuristics for the JSSP

Similarly to the FSSP, a very large number of hybrid and non hybrid meta-heuristic algorithms have been proposed for the JSSP. The application list for the FSSP includes most known meta-heuristic algorithms including evolutionary algorithms, particle swarm optimization, ant colony optimization, scatter search, neural networks, simulated annealing, tabu search and many forms of explorative local search. Most of these methods use simpler heuristic algorithms to generate an initial population of solutions.

Concerning the implementation of ***Genetic Algorithms*** (GA) in JSSP, many researchers have concluded that GAs are not suitable for fine-tuning of solutions close to optimal ones. In other words, genetic algorithms fail to intensify the search to the most promising regions of a neighborhood. This is why successful implementations of genetic algorithms usually incorporate a local search procedure for search intensification. A large number of early genetic algorithm applications may be traced in the literature. Among, the most well-known methods are those of Davis [39], Nakano and Yamada [117], Yamada and Nakano [172], Tamaki and Nishikawa [159], Mattfeld et al [110], Croce et al [34], Bierwirth et al [16] and Cheng et al [32]. Among more recent works in this field, Zhou et al [184] proposed a hybrid genetic algorithm with a neighborhood search procedure which also used a series of priority dispatching rules in the genetic evolution process. A very successful implementation of hybrid genetic algorithms can be found in the work of Wang and Zheng [166]. The authors also mention the drawback of genetic algorithm operators to disrupt the search in areas close to local or global optima and thus, they hybridized their GA with a simulated annealing algorithm which enhanced the intensification operation of their hybrid method. Park et al [127] also proposed a genetic algorithm and used Giffler and Thompson's algorithm [56] for the initial chromosomes generation. Murovec and Šuhel [116] presented a hybrid genetic-tabu search implementation with exceptional results in all benchmarks tested. This is also indicated by the three new upper bounds found for the ABZ9, YN1 and YN2 benchmark instances. Gonçalves et al [61] have presented an efficient hybrid GA implementation. The authors utilized the random keys representation [14] and hybridized their genetic algorithm with a local search scheme. Finally, Zobolas et al [185] developed a three-component hybrid algorithm where a Genetic Algorithm, a Differential Evolution and a Variable Neighborhood Search procedure were employed.

The ***Particle Swarm Optimization*** (PSO) method applications in JSSP are relatively very recent, indicating the great interest of the academic community in this optimization method. Although PSO can be generally used with continuous variables and, therefore, is unsuitable for discrete variables, most researchers have presented special transformation methods or variations of the original PSO so that this method can be also applied to JSSP. Lian et al [99] presented a similar PSO algorithm with the added feature of converting a continuous domain to a discrete domain. Xia and Wu [171] hybridized the PSO algorithm with a classic simulated algorithm and formed their HPSO algorithm. Zhao and Zhang [181] also presented a hybrid particle swarm - simulated annealing algorithm and compared the hybrid version with the single algorithm implementations. Their hybrid method was found to be more effective and more efficient than the single meta-heuristic components. Finally, Sha and Hsu [143] also used a transformation routine but unlike most researchers, utilized the preference list-based representation and hybridized their PSO implementation with a tabu search procedure to improve the solutions acquired.

The use of ***Ant Colony Optimization*** (ACO) algorithm in solving the JSSP is rather limited. Just like the PSO algorithm, the implementations found in literature are very recent. Huang and Liao [79] proposed a hybrid ant colony - tabu search algorithm. The authors employed a novel decomposition method inspired by the shifting bottleneck procedure and a mechanism of occasional re-optimizations of partial schedules instead of using the more conventional feasible solutions construction approach. Heinonen and Pettersson [73] presented a hybrid ant colony - local search algorithm and focused on the infamous FT10 benchmark instance so that the ACO performance can be evaluated.

The application of ***Neural Networks*** (NN) to the JSSP can be characterized quite extensive. However, the last few years there seems to be a decreasing interest towards neural network implementations for JSSP. Among early works conducted in this field are the works of Foo and Takefuji [51–53], Zhou et al [182, 183] and Dagli et al [35]. However, mediocre results have been obtained and only the work of Sabuncuoglu and Gurgun [141] produced good results in benchmark tests. Jain and Meeran [84] also presented a powerful neural network scheme for the JSSP. There are quite a few implementations of hybrid neural networks with genetic algorithms including the algorithms of Dagli and Sittisathanchai [36], Lee and Dagli [93] and Yu and Liang [178]. Among more recent neural network hybrid methods is the work of Luh et al [105]. The authors combined recurrent neural network optimization ideas with Lagrangian relaxation for constraint handling to solve the JSSP. According to their tests, they managed to outperform most neural network implementations presented up to that date. Yang and Wang [176] have presented an adaptive neural network and heuristics hybrid approach for JSSP. More specifically they implemented two heuristics that operate in the neural network framework and are used to accelerate the solving process of the neural network and guarantee its convergence and to obtain non-delay schedules from the feasible solutions gained by the neural network. Finally, one of the most recent works in the domain of neural networks for JSSP is the work of

Fonseca and Navaresse [50] concerning job shop simulation. The proposed scheme managed to satisfactorily estimate the manufacturing lead times for orders simultaneously processed in a four-machine job shop. The authors compared their findings with data generated from three well-known simulation packages (Arena, SIMAN and ProModel) and found out that the manufacturing lead times produced by their scheme turned out to be equally valid.

Concerning ***Basic Local Search*** and ***Explorative Local Search*** methods, similarly to the FSSP/PFSP, there are very few implementations where such methods are used alone (i.e. not as a part of a hybrid algorithm). It should be mentioned though that, when hybridized, such methods greatly improve the performance of other methods, especially of evolutionary methods. One of the few implementations was the Greedy Randomized Adaptive Search Procedure (GRASP) of Resende [136] and the Guided Local Search - Shifting Bottleneck hybrid of Balas and Vazacopoulos [9]. Both algorithms and especially the one of Balas and Vazacopoulos, have demonstrated very good performance in the JSSP.

Simulated Annealing, as mentioned in the FSSP section, is one of the earliest proposed meta-heuristic [87]. The most known implementations of simulated annealing have been proposed by Matsuo et al [109], Van Laarhoven et al. [164,165], Aarts et al [1,2], Yamada et al. [175], Sadeh and Nakakuki [142] and Yamada and Nakano [173, 174]. However, as Jain and Meeran [85] observed, simulated annealing is unable to quickly achieve good solutions to JSSP problems, perhaps because it is is a generic and memory-less technique. As a result, academic research focused on hybrid versions of simulated annealing some of which are already mentioned in the previous subsections of meta-heuristics for the JSSP (e.g. the work of Wang and Zheng [166]). Zhang et al [180] recently proposed a hybrid simulated annealing - tabu search approach whose main principle is using simulated annealing to find the elite solutions inside big valleys so that tabu search can re-intensify search from the promising solutions. The effectiveness of this method is demonstrated by the 17 new upper bounds found in benchmark test instances. Finally, a very recent work on a hybrid simulated annealing algorithm has been conducted by El-Bouri et al [43]. More specifically, the authors developed a hybrid algorithm that apart from simulated annealing, it consists of a tabu search and adaptive memory programming framework. The proposed framework uses two distinct memories modules: the first memory temporarily prevents further changes in the configuration of a solution in an effort to maintain the presence of good solution elements while the second memory aims at tracking good solutions found during an iteration, so that the best ones can be used as the starting point in a subsequent iteration. The authors compared their algorithm with the traditional simulated annealing procedure and realized substantial performance improvements.

Tabu Search (TS) is one of the most powerful meta-heuristics for the JSSP. This is affirmed by the fact that most state-of-the-art algorithms for the JSSP include some sort of tabu search functionality. The main strength of tabu search is the use of memory that speeds up the solution space search process. Early implementations of tabu search in JSSP have been conducted by Taillard [153,156], Barnes and Chambers [13] and Sun et al [151]. One of the most powerful implementations was proposed by Nowicki and Smutnicki [119]. The computational effectiveness of their TSAB method is demonstrated by the fact that the notorious FT10 instance was solved in just 30 seconds using a PC of that era. A later work conducted by Watson et al [168] is based on Nowicki and Smutnicki's TSAB algorithm and aims at developing an understanding of why tabu search is so effective on JSSP. Later hybrid implementations include Pezzella and Merelli's [129] hybrid tabu search-shifting bottleneck procedure where the shifting bottleneck heuristic is initially used to generate schedules and then to refine solutions in subsequent iterations. Finally, as mentioned in the simulated annealing for JSSP section, a very recent hybrid tabu search - simulated annealing method was proposed by Zhang et al [180] which managed to find new upper bounds for many benchmark instances.

Threshold Accepting algorithms have also been used for JSSP optimization but their application is rather limited. One of the first implementations was proposed by Aarts et al [1] in their computational study of local search algorithms for JSSP. However, poor results in comparison to other local search methods were recorded. A variant of the classic threshold accepting algorithm was proposed by Tarantilis and Kiranoudis [161]. The variant used is named 'List-Based Threshold Accepting method' (LBTA) and the authors demonstrated competitive results. A very important characteristic of this algorithm is that tuning of only one parameter is needed (namely the list size), thus making this implementation very easy to tune for a wide variety of problems. Based on the LBTA method of Tarantilis and Kiranoudis [161], Lee et al [92] proposed a new version and adopted the probabilistic steepest neighbor selection strategy to the original LBTA method to compromise between searching time and neighborhood quality, a strategy aiming at speeding up the search process.

1.6 The Open Shop Scheduling Problem

The open shop scheduling problem has attracted considerably less attention from academic researchers than flow shop and job shop scheduling problems, a fact outlined by the noticeably smaller number of papers in this field. Concerning exact methods for the OSSP, a polynomial time algorithm has been proposed by Gonzalez and Sahni [63] for the two-machine problem. Brucker et al [24] developed a branch and bound algorithm for the general m-machine problem based on the disjunctive graph representation of the problem. The method was very efficient and some benchmark instances were solved to optimality for the first time. Adiri and Aizikowitz [4] presented a linear time algorithm for the three-machine OSSP, provided that one machine dominates one of the other two. Algorithms for arbitrary m-machine problems with one or two dominating machines are analyzed in Tanaev et al [160]. More recently, Kyparissis and Koulamas [89] presented a polynomial time algorithm for the two-machine OSSP with the objective of minimizing the total completion time subject to minimum makespan. They also extended their algorithm to the three-machine OSSP.

1.6.1 Heuristics for the OSSP

Röck and Schmidt [137] introduced a machine aggregation algorithm for the general m-machine OSSP based on the fact that the two-machine cases is polynomially solvable. Some more promising methods focus on 'dense' schedules where no machine is idle unless there is no operation available to be processed on that machine. Bräsel et al [20] developed some efficient constructive heuristic algorithms based on a structure analysis of the feasible combinations of job and machine orders. Guéret and Prins [67] presented list scheduling heuristics with two priorities for each operation, and matching heuristics which are followed by a local search improvement procedure. Finally, Liaw [100] introduced an iterative improvement approach based on a decomposition technique.

1.6.2 Meta-heuristics for the OSSP

Although there are quite a few exact and heuristic methods for the OSSP, the great difference between the OSSP and the JSSP/FSSP is observed in the number of meta-heuristic methodologies proposed. Liaw [101] proposed a robust tabu search algorithm for the OSSP. In this work, Liaw also developed a dispatching rule for initial solution generation. One year later, a hybrid genetic algorithm - tabu search algorithm (HGA) for the OSSP was proposed by the same author [102]. This work was based on the previous work of Liaw on tabu search for the OSSP and the author managed to improve the results of the first paper to the point that some benchmark instances were solved to optimality for the first time. Liaw also compared the results acquired by HGA to those obtained with list scheduling heuristic, insertion heuristic (IH), simulated annealing and pure TS algorithms and HGA outperformed them all. The same year, Prins [131] also proposed a genetic algorithm for the OSSP with slightly worse results than Liaw's implementation in Taillard's benchmark instances, except for some 20x20 problems. A more recent meta-heuristic implementation for the OSSP was proposed by Blum [18]. More specifically, Blum proposed a hybrid ant colony optimization with beam search algorithm which outperformed both genetic algorithm implementations of Liaw [102] and Prins [131], rendering the beam-ACO method a state-of-the-art method for OSSP.

Conclusions

In this chapter, the main shop scheduling problems were presented along with a review of exact, heuristic and meta-heuristic methodologies used to solve them. Exact methods fail to solve large instances of these problems in reasonable computational time while heuristic methods lack the robustness required, although they induce much less computational effort. Thus, it is not without reason that late research has mainly focused on meta-heuristic methods for these hard combinatorial optimization problems.

Concerning current and future research on this field, there are mainly two trends: the hybridization of single meta-heuristic components to form more powerful search methods and the use of advanced computational equipment (i.e. parallel processing). These two trends can be also unified. However, most importantly, future research should also concentrate on the applicability of any proposed method in real life scheduling problems. Considering the expansion of Enterprise Resource Planning (ERP) systems and their applicability in most production environments, the incorporation of advanced shop scheduling problem solving methods in ERP systems is of great importance for both academic research and industrial practice.

References

1. Aarts, E.H.L, Van Laarhooven, P.J.M., Lenstra, J.K., Ulder, N.L.J. (1994) A computational study of local search algorithms for job-shop scheduling. ORSA J Comput 6:118–125
2. Aarts, E.H.L., Van Laarhooven, P.J.M., Ulder, N.L.J. (1991) Local search based algorithms for job-shop scheduling. Working Paper, Department of Mathematics and Computer Science, Eindhoven University of Technology, Eindhoven, The Netherlands
3. Adams, J., Balas, E., Zawack, D. (1988) The shifting bottleneck procedure for the job-shop scheduling. Manage Sci 34:391–401
4. Adiri, I., Aizikowitz, N. (1989) Open shop scheduling problems with dominated machines. Nav Res Logist Q 36:273–281
5. Applegate, D., Cook, W. (1991) A computational study of the job-shop scheduling problem. ORSA J Comput 3:149–156
6. Baker, K.R. (1975) A comparative survey of flowshop algorithms. Oper Res 23:62–73
7. Balas, E. (1969) Machine sequencing via disjunctive graphs: An implicit enumeration algorithm. Oper Res17:941–957
8. Balas, E., Lenstra, J.K., Vazacopoulos, A. (1995) The one machine problem with delayed precedence constraints and its use in job shop scheduling. Manage Sci 41:94–109
9. Balas, E., Vazacopoulos, A. (1998) Guided local search with shifting bottleneck for job-shop scheduling. Manage Sci 44:262–275
10. Baptise, P., Le Pape, C., Nuijten, W. (1995a). Constrained based optimization and approximation for job-shop scheduling. In Proc. AAAI-SIGMAN Workshop on Intelligent Manufacturing Systems, 14th Int. Joint Conference on Artificial Intelligence, IJCAI, Montreal, Canada
11. Baptiste, P., Le Pape, C., Nuijten, W. (1995b) Incorporating efficient operations research algorithms in constraint-based scheduling. In Proc. 1st Joint Workshop on Artificial Intelligence and Operations Research, Timberline Lodge, OR

12. Barker, J.R., McMahon, G.B. (1985) Scheduling the general job-shop. Manage Sci 31:594–598
13. Barnes, J.W., Chambers, J.B. (1995) Solving the job shop scheduling problem using tabu search. IIE Trans 27:257–263
14. Bean, J.C. (1994) Genetic algorithm and random keys for sequencing and optimization, ORSA J Comput 6:154–160
15. Ben-Daya, M., Al-Fawzan, M. (1998) A tabu search approach for the flow shop scheduling problem. Eur J Oper Res 109:88–95
16. Bierwirth, C., Mattfeld, D.C., Kopfer, H. (1996) On permutation representations for scheduling problems. In: Voigt H M et al (Eds.), PPSN'IV Parallel Problem Solving from Nature, Springer, Berlin 310–318
17. Blazewicz, J., Domschke, W., Pesch, E. (1996) The job-shop scheduling problem: Conventional and new solution techniques. Eur J Oper Res 93:1–33
18. Blum, C. (2005) Beam-ACO-hybridizing ant colony optimization with beam search: an application to open shop scheduling. Comput Oper Res 32: 1565–1591
19. Blum, C., Roli, A. (2003) Meta-heuristics in Combinatorial Optimization: Overview and Conceptual Comparison. ACM Comput Surv 35:268–308
20. Bräsel, H., Tautenhan, T., Werner, F. (1993) Constructive Heuristic Algorithms for the Open Shop Problem. Computing 51:95–110
21. Brooks, G.H., White, C.R. (1965) An algorithm for finding optimal or near-optimal solutions to the production scheduling problem. J Ind Eng 16:34–40
22. Brucker, P. (1988) A polynomial algorithm for the two machine job-shop scheduling problem with a fixed number of jobs. OR Spektrum 16:5–7
23. Brucker, P. (1994) An efficient algorithm for the job-shop problem with two jobs. Computing 40:353–359
24. Brucker, P., Hurink, J., Jurish, B., Wostmann, B. (1997) A branch and bound algorithm for the open-shop problem, Discrete Appl Math 76: 43–59
25. Brucker, P., Jurisch, B., Sievers, B. (1994) A branch and bound algorithm for the job-shop scheduling problem. Discrete Appl Math 49:107–127
26. Campbell, H.G., Dudek, R.A., Smith, M.L. (1970) A heuristic algorithm for the n-job, m-machine problem. Manage Sci 16:B630–B637.
27. Carlier, J. (1982) The one-machine sequencing problem. Eur J Oper Res 11: 42–47
28. Carlier, J., Pinson, E. (1989) An algorithm for solving the job-shop problem. Manage Sci 35:164–176
29. Carlier, J., Pinson, E. (1994) Adjustments of heads and tails for the job-shop problem. Eur J Oper Res 78:146–161
30. Charalambous, O. (1991) Knowledge based job-shop scheduling. PhD thesis, University of Manchester Institute of Science and Technology, Manchester
31. Chen, C.L., Vempati, V.S., Aljaber, N. (1995) An application of genetic algorithms for flow shop problems. Eur J Oper Res 80:389–396
32. Cheng, R., Gen, M., Tsujimura, Y. (1996) A tutorial survey of job-shop scheduling problems using genetic algorithms, part I: representation. Comput Ind Eng 30:983–997
33. Conway, R.W., Maxwell, W.L., Miller, L.W. (1967) Theory of scheduling, Addison-Wesley
34. Croce, F., Tadei, R., Volta, G. (1995) A genetic algorithm for the job shop problem. Comp Oper Res 22: 15–24

35. Dagli, C.H., Lammers, S., Vellanki, M. (1991) Intelligent scheduling in manufacturing using neural networks. J Neural Networ Comput Tech Design Applic 2:4–10
36. Dagli, S.H., Sittisathanchai, S. (1995) Genetic neuro-scheduler: A new approach for job shop scheduling. Int J Prod Econ 41:135–145
37. Daniels, R.L., Mazzola, J.B. (1994) Flow shop scheduling with resource flexibility. Oper Res 42:1174–1182
38. Dannenbring, D.G. (1977) An evaluation of flow shop sequencing heuristics. Manage Sci 23:1174–1182
39. Davis, L. (1985) Job-shop scheduling with genetic algorithm. In: Grefenstette J J (Ed.), Proceedings of the First International Conference on Genetic Algorithms and their Applications. Lawrence Erlbaum, Pittsburg, PA, USA, 136–140
40. Dorigo, M. (1992) Optimization, Learning and Natural Algorithms (in Italian). PhD thesis, Dipartimento di Elettronica, Politecnico di Milano, Milan, Italy
41. Dueck, G., Scheuer, T. (1990) Threshold accepting. A general purpose optimization algorithm appearing superior to simulated annealing, J Comput Phys 90:161–175
42. Eberhart, R., Kennedy, J. (1995) A new optimizer using particle swarm theory, in: Proceedings of the Sixth International Symposium on Micro Machine and Human Science, Nagoya, Japan, 39–43
43. El-Bouri, A., Azizi, N., Zolfaghari, S. (2007) A comparative study of a new heuristic based on adaptive memory programming and simulated annealing: The case of job shop scheduling. Eur J Oper Res 177:1894–1910
44. El-Bouri, A., Balakrishnan, Popplewell, N. (2005) A neural network to enhance local search in the permutation flowshop. Comput Ind Eng 49:182–196
45. Feo, T.A., Resende, M.G.C. (1995) Greedy randomized adaptive search procedures. J Global Optim 6:109–133
46. Fisher, H., Thompson, G.L. (1963) Probabilistic learning combinations of local job-shop scheduling rules. In: Muth J F, Thompson G L (Eds.). Industrial Scheduling, Prentice- Hall, Englewood Cliffs, NJ
47. Fisher, M.L. (1973) Optimal solution of scheduling problems using Lagrange multipliers: Part I. Oper Res 21:1114–1127
48. Florian, M., Trepant, P., McMahon, G. (1971) An implicit enumeration algorithm for the machine sequencing problem. Manage Sci 17:B782–B792
49. Fogel, L.J. (1962) Toward inductive inference automata. In Proceedings of the International Federation for Information Processing Congress. Munich, 395–399
50. Fonseca, D.J., Navaresse, D. (2002) Artificial neural networks for job shop simulation. Adv Eng Inform 16:241–246
51. Foo, S.Y., Takefuji, Y. (1988a) Stochastic neural networks for solving job-shop scheduling: Part 1. Problem representation. In: Kosko B (Ed.), IEEE International Conference on Neural Networks, San Diego, CA, USA 275–282
52. Foo, S.Y., Takefuji, Y. (1988b) Stochastic neural networks for solving job-shop scheduling: Part 2. Architecture and simulations. In: Kosko B (Ed.), IEEE International Conference on Neural Networks, San Diego, CA, USA 283–290
53. Foo, S.Y., Takefuji, Y. (1988c) Integer linear programming neural networks for job-shop scheduling. In: Kosko B (Ed.), IEEE International Conference on Neural Networks, San Diego, CA, USA 341–348
54. Gangadharan, R., Rajendran, C. (1994) A simulated annealing heuristic for scheduling in a flowshop with bicriteria. Comput Ind Eng 27:473–476

55. Garey, M.R.D., Johnson, D.S., Sethi, R. (1976) The complexity of flowshop and job shop scheduling. Math Oper Res 1:117–129
56. Giffler, B., Thompson, G.L. (1960) Algorithms for solving production scheduling problems. Oper Res 8:487–503
57. Glover, F. (1986) Future paths for integer programming and links to artificial intelligence, Comput Oper Res 13:533–549
58. Glover, F. (1997) Tabu search and adaptive memory programming - advances, applications and challenges. In: Barr, Helgason Kennington (Eds.), Advances in Meta-heuristics, Optimization and Stochastic Modelling Technologies, Kluwer Academic Publishers, Boston, MA 1–75
59. Glover, F., Greenberg, H.J. (1989) New approaches for heuristic search: A bilateral linkage with artificial intelligence. Eur J Oper Res 39:119–130
60. Glover, F., Laguna, M., Marti, R. (2000) Fundamentals of scatter search and path relinking. Control 29:653–684
61. Gonçalves, J.F., Mendes, J.J.d.M., Resende, M.G.C. (2005) A hybrid genetic algorithm for the job shop scheduling problem. Eur J Oper Res 167:77–95
62. Gonzalez, T., Sahni, S. (1976) Open shop scheduling to minimize finish time, J Assoc Comput Mach 23:665–679
63. Gonzalez, T., Sahni, S. (1978) Flowshop and Jobshop Schedules: Complexity and approximation, Oper Res 20:36–52
64. Grabowski, J., Wodecki, M. (2004) A very fast tabu search algorithm for the permutation flow shop problem with makespan criterion. Comput Oper Res 31:1891–1909
65. Graham, R.L., Lawler, E.L., Lenstra, J.K., Rinnooy Kan, A.H.G. (1979) Optimization and approximation in deterministic sequencing and scheduling: a survey. Ann Discrete Math 5:287–326
66. Greenberg, H.H. (1968) A branch and bound solution to the general scheduling problem. Oper Res 16:353–361
67. Guéret, C., Prins, C. (1998) Classical and new heuristics for the open-shop problem: A computational evaluation. Eur J Oper Res 107:306–314
68. Gupta, J.N.D. (1971) A functional heuristic algorithm for the flow-shop scheduling problem. Oper Res 22:39–47
69. Gupta, J., Hennig, K., Werner, F. (2002) Local search heuristics for two-stage flow shop problems with secondary criterion. Comput Oper Res 29:123–149
70. Hansen, P., Mladenović, N. (2001) Variable neighborhood search: Principles and Applications. Eur J Oper Res 130:449–467
71. Hasija, S., Rajendran, C. (2004) Scheduling in flowshops to minimize total tardiness of jobs. Int J Prod Res 42:2289–2301
72. Hefetz, N., Adiri, I. (1982) An efficient optimal algorithm for the two-machine, unit-time, jobshop, schedule-length, problem. Math Oper Res 7:354–360
73. Heinonen, J., Pettersson, F. (2007) Hybrid ant colony optimization and visibility studies applied to a job-shop scheduling problem. Appl Math Comput 187:989–998
74. Hejazi, S.R., Saghafian, S. (2005) Flowshop-scheduling problems with makespan criterion: a review. Int J Prod Res 43:2895–2929
75. Held, M., Karp, R.M. (1962) A Dynamic Programming Approach to Sequencing Problems. J Soc Ind and Appl Math 10:196–210
76. Ho, J.C., Chang, Y-L. (1991) A new heuristic for the n-job, m-machine flowshop problem. Eur J Oper Res 52:194–202

77. Holland, J.H. (1975) Adaption in natural and artificial systems. The University of Michigan Press, Ann Harbor, MI
78. Hopp, W.J., Spearman, M.L. (1996) Factory Physics: Foundations of Manufacturing Management, Irwin
79. Huang, K-L., Liao, C-J. (2007) Ant colony optimization combined with taboo search for the job shop scheduling problem. Comput Oper Res, in press
80. Hurkens C, http://www.win.tue.nl/whizzkids/1997, accessed April 2007
81. Ignall, E., Schrage, L. (1965) Application of the branch and bound technique to some flow-shop scheduling problems. Oper Res 11:400–412
82. Ishibuchi, H., Misaki, S., Tanaka, H. (1995) Modified simulated annealing algorithms for the flow shop sequencing problem. Eur J Oper Res 81:388–398
83. Jackson, J.R. (1956) An extension of Johnson's result on job lot scheduling. Int J Prod Res 36:1249–1272
84. Jain, A.S., Meeran, S. (1998) Job-shop scheduling using neural networks. Int J Prod Res 36:1249–1272
85. Jain, A.S., Meeran, S. (1999) Deterministic job-shop scheduling: Past, present and future. Eur J Oper Res 113:390–434
86. Johnson, S.M. (1954) Optimal two- and three-stage production schedules with set-up times included. Nav Res Logist Q 1:61–68
87. Kirkpatrick, S., Gelatt, C.D., Vecchi, M.P. (1983) Optimization by simulated annealing. Science 4598:671–680
88. Koulamas C (1998) A new constructive heuristic for the flowshop scheduling problem. Eur J Oper Res 105:66–71
89. Kyparisis, G.J., Koulamas, C. (2000) Open shop scheduling with makespan and total completion time criteria. Comput Oper Res 27:15–27
90. Lageweg, B.J., Lenstra, J.K., Rinnooy Kan, A.H.G. (1978) A general bounding scheme for the permutation flow-shop problem. Oper Res 26:53–67
91. Lawler, E.L., Lenstra, J.K., Rinnooy Kan, A.H.G., Shmoys, D.B. (1993) Sequencing and scheduling: algorithms and complexity. In: Graves S C, Rinnooy Kan A H G, Zipkin P H (Eds.). Handbooks in Operations Research and Management Science, 4, Logistics of Production and Inventory, Elsevier, Amsterdam
92. Lee, D.S., Vassiliadis, V.S., Park, J.M. (2004) A novel threshold accepting meta-heuristic for the job-shop scheduling problem. Comput Oper Res 31: 2199–2213
93. Lee, H.C., Dagli, C.H. (1997) A parallel genetic-neuro scheduler for job-shop scheduling problems. Int J Prod Econ 51:115–122
94. Lee, I., Shaw, M.J. (2000) A neural-net approach to real-time flow-shop sequencing. Comput Ind Eng 38:125–147
95. Lenstra, J.K., Rinnooy Kan, A.H.G. (1979) Computational complexity of discrete optimization problems. Ann Discrete Math 4:121–140
96. Lenstra, J.K., Rinnooy Kan, A.H.G., Brucker, P. (1977) Complexity of machine scheduling problems. Ann Discrete Math 7:343–362
97. Leung, J.Y.-T. (2004) Handbook of scheduling, Chapman & Hall/CRC, Boca Raton 78
98. Lian, Z., Gu, X., Jiao, B. (2006) A similar particle swarm optimization algorithm for permutation flowshop scheduling to minimize makespan. Appl Math and Computation 175:773–785
99. Lian, Z., Jiao, B., Gu, X. (2006) A similar particle swarm optimization algorithm for job-shop scheduling to minimize makespan. Appl Math Comput 183:1008–1017

100. Liaw, C.-F. (1998) An iterative improvement approach for the nonpreemptive open shop scheduling problem. Eur J Oper Res 111:509–517
101. Liaw, C.-F. (1999) A tabu search algorithm for the open shop scheduling problem. Comput Oper Res, 26:109–126
102. Liaw, C.-F. (2000) A hybrid genetic algorithm for the open shop scheduling problem. Eur J Oper Res 124:28–42
103. Liu, S.Q., Ong, H.L., Ng, K.M. (2005) A fast tabu search algorithm for the group shop scheduling problem. Adv Eng Softw 36:533–539
104. Lomnicki, Z.A. (1965) A branch-and-bound algorithm for the exact solution of the three-machine scheduling problem. Oper Res Q 16:89–100
105. Luh, P.B., Zhao, X., Wang, Y., Thakur, L.S. (2000) Lagrangian Relaxation Neural Networks for Job Shop Scheduling. IEEE T Robotic Autom 16:78–88
106. Manne, A.S. (1960) On the job shop scheduling problem. Oper Res 8:219–223
107. Martin, P., Shmoys, D. (1995) A new approach to computing optimal schedules for the job shop scheduling problem. Extended Abstract, Comell University, Ithaca
108. Masuda, T., Ishii, H., Nishida, T. (1985) The mixed shop scheduling problem. Discrete Appl Math, 11:175–86
109. Matsuo, H., Suh, C.J., Sullivan, R.S. (1988) A controlled search simulated annealing method for the general job-shop scheduling problem. Working Paper, 03-04-88, Graduate School of Business, University of Texas at Austin, Austin, Texas, USA
110. Mattfeld, D.C., Kopfer, H., Bierwirth, C. (1994) Control of parallel population dynamics by social-like behaviour of GA-individuals. In: Proceedings of the Third International Conference on Parallel Problem Solving from Nature (PPSN3). Springer, Berlin 15–25
111. McMahon, G.B., Florian, M. (1975) On scheduling with ready times and due dates to minimize maximum lateness. Oper Res 23:475–482
112. Merkle, D., Middendorf, M. (2000) An ant algorithm with a new pheromone evaluation rule for total tardiness problems. In Proceedings of the EvoWorkshops, Springer, Berlin, 287–296
113. Moccellin, J.a.V. (1995) A new heuristic method for the permutation flow shop scheduling problem. J Oper Res Soc 46:883–886
114. Moccellin, J.a.V., dos Santos, M.O. (2000) An adaptive hybrid meta-heuristic for permutation flowshop scheduling. Control Cybern 29:761–771
115. Murata, T., Ishibuchi, H., Tanaka, H. (1996) Genetic algorithms for flowshop scheduling problems. Comput Ind Eng 30:1061–1071
116. Murovec, B., Šuhel, P. (2004) A repairing technique for the local search of the job-shop problem. Eur J Oper Res 153:220–238
117. Nakano, R., Yamada, T. (1991) Conventional genetic algorithm for job-shop problems. In: Kenneth M K, Booker L B (Eds.), Proceedings of the Fourth International Conference on Genetic Algorithms and their Applications, San Diego, California, USA, 474–479
118. Nawaz, M., Enscore, Jr E., Ham, I. (1983) A heuristic algorithm for the m-machine, n-job flow-shop sequencing problem. Omega-Int Journal Manage S 11:91–95
119. Nowicki, E., Smutnicki, C. (1996) A fast taboo search algorithm for the job-shop problem. Manage Sci 42:797–813
120. Nowicki, E., Smutnicki, C. (1996) A fast tabu search algorithm for the permutation flow-shop problem. Eur J Oper Res 91:160–175

121. Nowicki, E., Smutnicki, C. (2006) Some aspects of scatter search in the flow-shop problem. Eur J Oper Res 169:654–666
122. Ogbu, F., Smith, D. (1990b) Simulated annealing for the permutation flowshop problem. OMEGA-Int J Manage S 19:64–67
123. Ogbu, F., Smith, D. (1990a) The application of the simulated annealing algorithms to the solution of the n=m=Cmax flowshop problem. Comput Oper Res 17:243–253
124. Onwubolu, G.C., Davendra, D. (2006) Scheduling flow-shops using differential evolution algorithm, Eur J Oper Res 171:674–692
125. Osman, I., Potts, C. (1989) Simulated annealing for permutation flow-shop scheduling. OMEGA-Int J Manage S 17:551–557
126. Palmer, D.S. (1965) Sequencing jobs through a multistage process in the minimum total time: a quick method of obtaining a near optimum. Oper Res Q 16:101–107
127. Park, B.J., Choi, H.R., Kim, H.S. (2003) A hybrid genetic algorithm for the job shop scheduling problems. Comput Ind Eng 45:597–613
128. Pesant, G., Gendreau, M. (1996) A view of local search in Constraint Programming. In Principles and Practice of Constraint Programming- CP'96. Lect Notes in Computer Science 1118:353–366, Springer-Verlag
129. Pezzella, F., Merelli, E. (2000) A tabu search method guided by shifting bottleneck for the job shop scheduling problem. Eur J Oper Res 120:297–310
130. Ponnambalam, S.G., Aravindan, P., Chandrasekaran, S. (2001) Constructive and improvement flow shop scheduling heuristics: An extensive evaluation. Prod Plan Control 12:335–344
131. Prins, C. (2000) Competitive genetic algorithms for the open-shop scheduling problem. Math Method Oper Res 52:389–411
132. Rajendran, C., Ziegler, H. (2004) Ant-colony algorithms for permutation flowshop-scheduling to minimize makespan/total flowtime of jobs. Eur J Oper Res 155:426–438
133. Rechenberg, I. (1973) Evolutionstrategie: Optimierung technischer Systeme nach Prinzipien der biologischen Evolution. Frommann-Holzboog
134. Reeves, C.R. (1995) A genetic algorithm for flowshop sequencing. Comput Oper Res 22:5–13
135. Reeves, C., Yamada, T. (1998) Genetic algorithms, path relinking, and the flowshop sequencing problem. Evol Comput 6:45–60
136. Resende, M.G.C. (1997) A GRASP for job shop scheduling. INFORMS Spring Meeting, San Diego, CA
137. Röck, H., Schmidt, G. (1983) Machine aggregation heuristics in shop-scheduling. Math Oper Res 45:303–314
138. Roy, B., Sussmann, B. (1964) Les problèmes d'ordonnancement avec contraintes disjonctives. Note D.S. no. 9 bis, SEMA, Paris
139. Ruiz, R., Maroto, C. (2005) A comprehensive review and evaluation of permutation flowshop heuristics, Eur J Oper Res 165:479–494
140. Ruiz, R., Maroto, C., Alcaraz, J. (2006) Two newrobust genetic algorithms for the flowshop scheduling problem, Omega-Int Journal Manage S 34:461–476
141. Sabuncuoglu, I., Gurgun, B. (1996) A neural network model for scheduling problems. Eur J Oper Res 93:288–299
142. Sadeh, N., Nakakuki, Y. (1996) Focused simulated annealing search an application to job-shop scheduling. Ann Oper Res 63:77–103

143. Sha, D.Y., Hsu, C-Y. (2006) A hybrid particle swarm optimization for job shop scheduling problem. Comput Ind Eng 51:791–808
144. Sipper, D., Bulfin, R.L. (1997) Production Planning, Control and Integration, McGraw Hill
145. Solimanpur, M., Vrat, P., Shankar, R. (2004) A neuro-tabu search heuristic for the flow shop scheduling problem. Comput Oper Res 31:2151–2164
146. Storn, R., Price, K. (1997) Differential Evolution - A simple and efficient heuristic for global optimization over continuous spaces. J Global Optim 11:341–359
147. Stützle, T. (1998a) An ant approach for the flow shop problem. In EUFIT'98, Aaeban, Germany, 1560–1564
148. Stützle, T. (1998b) Applying iterated local search to the permutation flow shop problem. Technical Report, AIDA-98-04, FG Intellektik, TU Darmstadt
149. Stützle, T. (1999) Iterated local search for the quadratic assignment problem. Technical report, aida-99-03, FG Intellektik, TU Darmstadt.
150. Suliman, S. (2000) A two-phase heuristic approach to the permutation flowshop scheduling problem. Int J Prod Econ 64:143–152
151. Sun, D.K., Batta, R., Lin, L. (1995) Effective job-shop scheduling through active chain manipulation. Comput Oper Res 22:159–172
152. T'kindt, V., Monmarch, N., Tercinet, F., Laugt, D. (2002) An Ant Colony optimization algorithm to solve a (2) machine bicriteria flowshop-scheduling problem. Eur J Oper Res 142:250–257
153. Taillard, E. (1989) Parallel taboo search technique for the jobshop scheduling problem. Internal Research Report ORWP89/11, Départment de Mathématiques (DMA), École Polytechnique Fédérale de Lausanne, 1015 Lausanne, Switzerland
154. Taillard, E. (1990) Some efficient heuristic methods for the flow shop sequencing problem. Eur J Oper Res 47:65–74
155. Taillard, E. (1993) Benchmarks for basic scheduling problems. Eur J Oper Res 64:278–285
156. Taillard, E. (1994) Parallel taboo search techniques for the jobshop scheduling problem. ORSA J Comput 16:108-117
157. Taillard, E.D., Gambardella, L.M., Gendreau, M., Potvin, J.-Y. (2001) Adaptive memory programming: A unified view of meta-heuristics, Eur J Oper Res 135:1–16
158. Talbi, E-G. (2002) A Taxonomy of Hybrid Meta-heuristics. J Heuristics 8: 541–564
159. Tamaki, H., Nishikawa, Y. (1992) A paralleled genetic algorithm based on a neighbourhood model and its application to the job-shop scheduling. In: M?nner R, Manderick B (Eds.), Proceedings of the Second International Workshop on Parallel Problem Solving from Nature (PPSN'2), Brussels, Belgium, 573–582
160. Tanaev, V.S., Sotskov, Y.N., Strusevich, V.A. (1994) Scheduling Theory: Multi-Stage Systems, Kluwer Academic Publishers, Printed in Dordrecht
161. Tarantilis, C.D., Kiranoudis, C.T. (2002) A list-based threshold accepting method for the job-shop scheduling problems. Int J Prod Econ 77:159–171
162. Tasgetiren, M.F., Sevkli, M., Liang, Y.C., Gencyilmaz, G. (2004) Differential Evolution Algorithm for Permutation Flowshop Sequencing Problem with Makespan Criterion. In Proceedings of the 4th International Symposium on Intelligent Manufacturing Systems (IMS 2004) Sakarya, Turkey, 442–452

163. Tasgetiren, M.F., Liang, Y.C., Sevkli, M., Gencyilmaz, G. (2007) A particle swarm optimization algorithm for makespan and total flowtime minimization in the permutation flowshop sequencing problem. Eur J Oper Res 177: 1930–1947
164. Van Laarhoven, P.J.M., Aarts, E.H.L., Lenstra, J.K. (1988) Job-shop scheduling by simulated annealing. Report OSR8809. Centrum voor Wiskunde en Informatica, Amsterdam, The Netherlands
165. Van Laarhoven, P.J.M., Aarts, E.H.L., Lenstra, J.K. (1992) Jobshop scheduling by simulated annealing. Oper Res 40: 113-125
166. Wang, L., Zheng, D.Z. (2001) An effective hybrid optimization strategy for job-shop scheduling problems. Comput Oper Res 28:585–596
167. Wang, L., Zheng, D.Z. (2003) An Effective Hybrid Heuristic for Flow Shop Scheduling. The Int J Adv Manuf Tech 21:38–44
168. Watson, J-P., Howe, A.E., Whitley, L.B. (2006) Deconstructing Nowicki and Smutnicki's i-TSAB tabu search algorithm for the job-shop scheduling problem. Comput Oper Res 33:2623–2644
169. Widmer, M., Hertz, A. (1989) A new heuristic method for the flow shop sequencing problem. Eur J Oper Res 41:186–193
170. Wodecki, M., Božejko, W. (2002) Solving the flow shop problem by parallel simulated annealing. In: Wyrzykowski R, Dongarra J, Paprzycki M, Waàsniewski J (Eds.), Parallel Processing and Applied Mathematics, 4th International Conference, PPAM 2001. In: Lect Notes Comput Sc 2328:236–244 Springer-Verlag, Berlin
171. Xia, W-j., Wu, Z-m. (2006) A hybrid particle swarm optimization approach for the job-shop scheduling problem. Int J Adv Manuf Tech 29:360–366
172. Yamada, T., Nakano, R. (1992) A genetic algorithm applicable to large-scale job-shop problems. In: M?nner R, Manderick B (Eds.), Proceedings of the Second International Workshop on Parallel Problem Solving from Nature (PPSN'2), Brussels, Belgium 281–290
173. Yamada, T., Nakano, R. (1995) Job-shop scheduling by simulated annealing combined with deterministic local search. In: Meta-heuristics International Conference (MIC'95), Hilton, Breckenridge, Colorado, USA 344–349
174. Yamada, T., Nakano, R. (1996) Job-shop scheduling by simulated annealing combined with deterministic local search. Meta-heuristics: Theory and Applications. Kluwer Academic Publishers, Hingham, MA 237–248
175. Yamada, T., Rosen, B.E., Nakano, R. (1994) A simulated annealing approach to job-shop scheduling using critical block transition operators. In: IEEE International Conference on Neural Networks (ICNN'94), Orlando, Florida, USA 4687–4692
176. Yang, S., Wangm D, (2001) A new adaptive neural network and heuristics hybrid approach for job-shop scheduling. Comput Oper Res 28:955–971
177. Ying, K.C., Liao, C.J. (2004) An ant colony system for permutation flow-shop sequencing. Comput Oper Res 31:791–801
178. Yu, H., Liang, W. (2001) Neural network and genetic algorithm-based hybrid approach to expanded job-shop scheduling. Comput Ind Eng 39:337–356
179. Zegordi, S.H., Itoh, K., Enkawa, T. (1995) Minimizing makespan for flowshop scheduling by combining simulated annealing with sequencing knowledge. Eur J Oper Res 85:515–531
180. Zhang, C., Li, P., Rao, Y., Guan, Z. (2007) A very fast TS/SA algorithm for the job shop scheduling problem. Comput Oper Res, in press

181. Zhao, F., Zhang, Q. (2006) An Improved Particle Swarm Optimization-Based Approach for Production Scheduling Problems. In Proceedings of the 2006 IEEE International Conference on Mechatronics and Automation June 25 - 28, 2006, Luoyang, China
182. Zhou, D.N., Cherkassky, V., Baldwin, T.R., Hong, D.W. (1990) Scaling neural networks for job-shop scheduling. In: International Joint Conference on Neural Networks(IJCNN'90), San Diego, California 889–894
183. Zhou, D.N., Cherkassky, V., Baldwin, T.R., Olson, D.E. (1991) A neural network approach to job-shop scheduling. IEEE T Neural Networ 2: 175–179
184. Zhou, H., Feng, H., Han, L. (2001) The hybrid heuristic genetic algorithm for job shop scheduling. Comput Ind Eng 40:191–200
185. Zobolas, G., Tarantilis, C.D., Ioannou, G. (2007) A Hybrid Evolutionary Algorithm for the Job Shop Scheduling Problem. J Oper Res Soc, doi: 10.1057/palgrave.jors.2602534

2

Scatter Search Algorithms for Identical Parallel Machine Scheduling Problems

Manuel Iori[1] and Silvano Martello[2]

[1] DISMI, Università di Modena e Reggio Emilia, Italy. manuel.iori@unimore.it
[2] DEIS, Università di Bologna, Italy. silvano.martello@unibo.it

Summary. We address the *Identical Parallel Machine Scheduling Problem*, one of the most important basic problems in scheduling theory, and some generalizations of it arising from real world situations. We survey the current state of the art for the most performing meta-heuristic algorithms for this class of problems, with special emphasis on recent results obtained through Scatter Search. We present insights in the development of this heuristic technique, and discuss the combinatorial difficulties of the problems through the analysis of extensive computational results.

Key words: Identical Parallel Machine Scheduling, Scatter Search, Local Search.

2.1 Introduction

Given a set of n *jobs* j, each having an associated processing time p_j ($j = 1, \ldots, n$), and a set of m parallel identical *machines* i ($i = 1, \ldots, m$), each of which can process at most one job at a time, the *Identical Parallel Machine Scheduling Problem* calls for the assignment of each job to exactly one machine, so as to minimize the maximum completion time of a job (*makespan*). We assume, as is usual, that $1 < m < n$, and that the processing times are positive integers. The problem is denoted as $P||C_{\max}$ in the three-field classification by Graham, Lawler, Lenstra and Rinnooy Kan [23], and is NP-hard in the strong sense. It is one of the most intensively studied problems in combinatorial optimization, since it has considerable theoretical interest and arises as a sub-problem in many real world applications.

The problem is related to another famous combinatorial optimization problem. In the *Bin Packing problem* (BPP) we are given n items, each having an associated size p_j ($j = 1, \ldots, n$), and an unlimited number of identical bins of capacity c: we want to assign each item to one bin, without exceeding its capacity, so that the number of bins used is minimized. Hence BPP can be

M. Iori and S. Martello: *Scatter Search Algorithms for Identical Parallel Machine Scheduling Problems*, Studies in Computational Intelligence (SCI) **128**, 41–59 (2008)
www.springerlink.com

seen as a "dual" of P||C_{max} in which the objective is to minimize the number of machines needed not to exceed a prefixed makespan c.

In this survey we review recent Scatter Search algorithms for P||C_{max} and two generalizations of it that arise from real world situations. In the next section we describe the three considered problems. In Section 2.3 we discuss a general framework for applying Scatter Search to identical parallel machine scheduling problems, focusing on the most effective components of this meta-heuristic technique, and detail its use for these problems. The inherent combinatorial difficulty of the problems and the performance of the Scatter Search algorithms is analyzed in Section 2.4 through the results of extensive computational experiments.

2.2 The problems

We first introduce the basic P||C_{max} problem, and then derive its two generalizations.

2.2.1 Identical Parallel Machine Scheduling Problem

The problem P||C_{max} can be formally stated as an Integer Linear Programming model by introducing a binary variable x_{ij}, taking value one if and only if job j is assigned to machine i ($j = 1, \ldots, n$; $i = 1, \ldots, m$):

$$\min z \tag{2.1}$$

$$\text{s.t.} \quad \sum_{j=1}^{n} p_j x_{ij} \leq z \quad (i = 1, \ldots, m), \tag{2.2}$$

$$\sum_{i=1}^{m} x_{ij} = 1 \quad (j = 1, \ldots, n), \tag{2.3}$$

$$x_{ij} \in \{0, 1\} \quad (i = 1, \ldots, m; j = 1, \ldots, n), \tag{2.4}$$

where z is the optimum makespan value.

Although this model can be strengthened by means of valid inequalities (as shown in Mokotoff [30]), its computational behavior is unsatisfactory from a practical point of view, due to the high number of variables that take a fractional value in the solution of its linear programming relaxation.

Exact algorithms with better computational behavior were obtained from implicit enumeration techniques. Dell'Amico and Martello [11] proposed combinatorial lower bounds, dominance criteria and a depth-first branch-and-bound algorithm based on the following enumeration scheme. The jobs are sorted so that

$$p_1 \geq p_2 \geq \cdots \geq p_n, \tag{2.5}$$

and are assigned to machines by increasing index. Let $\tilde{z}$ denote the incumbent solution value, and c_i ($i = 1, \ldots, m$) the sum of the processing times currently assigned to machine i. At level j of the branch-decision tree, at most m nodes are generated by assigning job j to those machines i such that $c_i + p_j < \tilde{z}$ (but for each subset of machines having identical c_i value, only the one with minimum index is considered). Computational experiments (see Dell'Amico and Martello [11,13]) showed a good practical behavior, definitely better than the one of the algorithm in [30].

Further improvements were obtained by Dell'Amico, Iori, Martello and Monaci [10] through an algorithm that exploits the dual relationship with the BPP. Given lower and upper bound L and U ($L < U$) on the optimal solution value, the algorithm performs a binary search which, at each iteration, checks wether there exists a solution with makespan at most $c = \lfloor (L + U)/2 \rfloor$ (and updates L or U correspondingly). The core problem is the decision version of a BPP, which is formulated as a set covering problem and solved through LP relaxation, column generation and branch-and-price.

Computational evidence shows however that all exact algorithms fail when addressing large-size instances, hence the problem has received a large attention from the heuristic point of view. Among the recent relevant contributions we cite the heuristic algorithm based on a simple exchange neighborhood by França, Gendreau, Laporte and Müller [16], the multiple exchanges neighborhoods studied by Frangioni, Necciari and Scutellà [17], the Tabu Search approach by Alvim and Ribeiro [1], the multi-start local search method by Haouari, Gharbi and Jemmali [24] and finally the Scatter Search algorithm by Dell'Amico, Iori, Martello and Monaci [10] which is discussed in Section 2.3.1.

2.2.2 Cardinality Constrained Parallel Machine Scheduling Problem

A well-studied generalization of the $P||C_{\max}$ arises when the number of jobs that can be assigned to each machine cannot exceed a given integer k. This can be modeled by adding to the previous model the constraint:

$$\sum_{j=1}^{n} x_{ij} \leq k \quad (i = 1, \ldots, m). \tag{2.6}$$

The model defined by (2.1)–(2.4) and (2.6) describes what is known in the literature as the *Cardinality Constrained Parallel Machine Scheduling Problem*, denoted as $P|\# \leq k|C_{\max}$. Problem $P||C_{\max}$ is thus the special case of $P|\# \leq k|C_{\max}$ arising when $k = n - m + 1$ (as there always exists an optimal solution in which each machine processes at least one job).

Possible applications of this problem arise when m machines (e.g., cells of a Flexible Manufacturing System, robots of an assembly line) have to perform n different types of operation. In real world contexts, each machine can have a limit k on the number of different types of operation it can perform, coming, e.g., from the capacity of the cell tool inventory or the number of robot feeders. If it is imposed that all operations of type j ($j = 1, \dots, n$) have to be performed by the same machine, and p_j is the total time they require, then $\mathrm{P}|\# \le k|\mathrm{C}_{\max}$ models the problem of performing all operations with minimum makespan. A real world $\mathrm{P}|\# \le k|\mathrm{C}_{\max}$ case was presented by Hillier and Brandeau [25], who studied a printed circuit board assembly process inspired by an application at Hewlett-Packard. The dual of $\mathrm{P}|\# \le k|\mathrm{C}_{\max}$ is a generalization of BPP in which a limit k is imposed to the number of items that can be packed into each bin.

Lower and upper bounds for $\mathrm{P}|\# \le k|\mathrm{C}_{\max}$ were provided by Babel, Kellerer and Kotov [2] and by Dell'Amico and Martello [12], who also proposed a truncated branch-and-bound algorithm. Felinskas [15] developed genetic algorithms and tested them on the real world case proposed by Hillier and Brandeau [25]. The Scatter Search algorithm by Dell'Amico, Iori and Martello [8] is reviewed in Section 2.3.2.

2.2.3 k_i-Partitioning Problem

Another interesting generalization (of both $\mathrm{P}||\mathrm{C}_{\max}$ and $\mathrm{P}|\# \le k|\mathrm{C}_{\max}$) is the *$k_i$-Partitioning Problem* (*k_i-PP*), introduced by Babel, Kellerer and Kotov [2]. In this case each machine i has a specific cardinality limit k_i ($i = 1, \dots, m$). The problem is thus modeled by (2.1)–(2.4), and

$$\sum_{j=1}^{n} x_{ij} \le k_i \quad (i = 1, \dots, m). \tag{2.7}$$

In the special case where $k_i = k$ for $i = 1, \dots, m$ the problem coincides with $\mathrm{P}|\# \le k|\mathrm{C}_{\max}$.

Possible applications of this problem arise again in Flexible Manufacturing System, when the cells are not identical. The dual of *k_i-PP* is a generalization of BPP in which the bins are numbered by consecutive integers, the first m bins have limits k_i ($i = 1, \dots, m$) on the number of items that can be packed into them, and all other bins have this limit set to one.

Lower bounds, reduction methods, constructive heuristics and a particular lower bound computation based on column generation were developed by Dell'Amico, Iori, Martello and Monaci [9]. The Scatter Search algorithm proposed by the same authors is described in Section 2.3.3.

2.3 Scatter Search

Scatter Search originates from heuristics for integer programming developed by Glover [18] in the Seventies. Several books and surveys on this methodology can be found in the literature. The reader is referred to the recent survey by Martí, Laguna and Glover [28] and to the special issue of *European Journal of*

Operational Research edited by Martí [14]. Scatter Search can be defined as a population based meta-heuristic that operates on a set of "good" solutions, the *reference set* $\mathcal{RS}$, and iteratively creates new solutions by combining subsets of $\mathcal{RS}$. These new solutions are possibly used to periodically update $\mathcal{RS}$ and the final outcome is the best solution obtained during the search process.

Most implementations of Scatter Search algorithms are based on the template formalized for the first time in Glover [19]. Candidate solutions for the reference set are evaluated on the basis of two criteria: quality and diversity. The *quality* of a solution s coincides or is strictly related to its value, denoted in the following as $z(s)$, while its *diversity* is a relative measure indicating how much its structure differs from that of the solutions that are currently in the reference set. Since the new solutions are created by combining solutions of $\mathcal{RS}$, diversity is a crucial tool for giving a Scatter Search algorithm the possibility of continuously diversifying the search to efficiently explore the solution space.

A concise formulation of the template is as follows:

1. generate a starting population $\mathcal{P}$ of good solutions;
2. create the initial *reference set* $\mathcal{RS}$ by selecting from $\mathcal{P}$ a number of solutions on the basis of their quality and their diversity;
3. **while** a stopping criterion is not met **do**
   3.1 generate a family of subsets of solutions from $\mathcal{RS}$;
   3.2 **for each** subset S of the family **do**
      3.2.1 combine the solutions in S so as to obtain a set of new solutions;
      3.2.2 improve each new solution;
      3.2.3 add to $\mathcal{RS}$ the solutions that meet a quality or diversity criterion
      end for
   end while.

Five main issues have to be dealt with, when developing a Scatter Search algorithm:

A. *Starting population* (Step 1.). The population should consist of high quality solutions that differ consistently from one another. It can be constructed in a totally random way or using heuristics. It is generally convenient to apply an improvement algorithm to each such solution. For the scheduling problems we are considering, Scatter Search appears to be quite robust with respect to the initial population, provided that enough CPU time is given to the overall algorithm.
B. *Improvement* (Steps 1. and 3.2.2). Local search and post optimization are frequently used to improve the solutions quality. Similarly to what happens in the intensification phase of other meta-heuristic approaches, the balance between computational effort and effectiveness may affect dramatically (in positive or in negative) the behavior of the overall algorithm.
C. *Reference set update* (Steps 2. and 3.2.3). The reference set is typically the union of a set Q of high quality solutions and a set D of solutions

highly different from those in Q and from one another. The reference set is usually quite small, typically containing 20 to 30 solutions. Two updating methods are common. A *dynamic* method updates $\mathcal{RS}$ as soon as a solution meets the required quality or diversity criterion, while in a *static* method updating only occurs when all generated subsets have been handled.

D. *Subset generation* (Step 3.1). The method adopted to select a subset of solutions to be combined together can have a relevant effect on the computational effort spent at each iteration of Step 3. Selecting a high number of subsets generally produces, at Step 3.2.1, a high number of new candidate solutions, hence a deep exploration of the current reference set.

E. *Solution combination* (Step 3.2.1). For each subset of solutions, one or more new candidate solutions are produced through combination. This is frequently a crucial aspect of Scatter Search. Specifically tailoring the combination method to the problem can considerably improve the convergence to high quality solutions.

The initial generation method, the improvement algorithms and the solution combination method developed for the three problems we are considering were specifically tailored to the problems, hence are discussed in the next sections. For the reference set update and for the subset generation instead, although several approaches were attempted, it turned out that the most efficient choices were common to the three cases.

Concerning the reference set update (issue C. above), the dynamic method was always used, with different values for $|Q|$ and $|D|$.

The subset generation method (issue D. above) always followed the classical approach proposed by Martí, Laguna and Glover [28], which consists in generating: (i) all the two-solution subsets; (ii) the three-solution subsets obtained by adding to each two-solution subset the solution of highest quality not already contained in it; (iii) the four-solution subsets obtained by the three-solution subsets in the same way as for (ii); (iv) the $|\mathcal{RS}| - 4$ subsets containing the best s solutions, for $s = 5, 6, \ldots, |\mathcal{RS}|$.

In the next sections we detail the specific methods adopted, for issues A., B., and E. above, in the three considered problems.

2.3.1 Scatter Search for $P||C_{max}$

In the scatter search algorithm proposed by Dell'Amico, Iori, Martello and Monaci [10], a starting population of $|\mathcal{P}| = 40$ solutions was generated through a number of approximation algorithms (see Mokotoff [29], Brucker [3], Leung [27] and Chen [5] for recent surveys on the huge literature on heuristics for $P||C_{max}$). The algorithms used were in particular the well-known *Longest Processing Time* algorithm by Graham [21,22], algorithm *Multi-Subset*, a two-phase approach by Dell'Amico and Martello [11] and other two-phase algorithms proposed by Mokotoff, Jimeno and Gutiérrez [31].

The improvement method adopted was a $k - \ell$ *swap* procedure, used to swap up to k jobs assigned to a machine with up to ℓ jobs assigned to another one, when this leads to decrease the makespan restricted to such two machines. Extensive computational experiments showed that exchanging single jobs ($k = \ell = 1$) gives modest improvements to the solutions quality, exchanging single jobs and pairs of jobs ($k = \ell = 2$) has the best balance between CPU time and efficacy, while higher values of k and/or ℓ produce limited improvements at the expenses of a considerable running time increase.

The reference set $\mathcal{RS}$ was composed by the $|Q| = 10$ solutions with minimum makespan and by the $|D| = 8$ solutions with highest diversity. The diversity of a solution was measured as follows. Given two jobs j and l ($l > j$) and two solutions $s \notin \mathcal{RS}$ and $t \in \mathcal{RS}$, let

$$\delta_{jl}(s,t) = \begin{cases} 1 & \text{if } j \text{ and } l \text{ are processed on the same machine in } s \\ & \text{and on different machines in } t \text{, or vice versa;} \\ 0 & \text{otherwise,} \end{cases} \tag{2.8}$$

and define the *diversity* of s with respect to $\mathcal{RS}$ as

$$d(s) = \min_{t \in \mathcal{RS}} \left\{ \sum_{j=1}^{\tilde{n}-1} \sum_{l=j+1}^{\tilde{n}} \delta_{jl}(s,t) \right\}, \tag{2.9}$$

where $\tilde{n} = \min(n, 4m)$ is used to limit the evaluation to the largest (hence, most critical) processing times. An accurate definition of the diversity can help in avoiding useless computations. In this case, if $\mathcal{RS}$ contains a solution $\overline{s}$ equivalent to s (i.e., that can be obtained from s by just permuting the machines) equation (2.8) gives $\delta_{ij}(s, \overline{s}) = 0$ for all i and j, so, in (2.9), $d(s)$ has its minimum (of value 0) for $t = \overline{s}$ and s is not added to $\mathcal{RS}$.

From each subset S generated at Step 3.1, new solutions are generated as follows. For each pair of jobs (j, l) $(j, l \leq \tilde{n})$, let S_{jl} be the subset of S containing those solutions in which j and l are processed on the same machine, and define the sum of the inverse relative errors of such solutions:

$$\varphi_{jl} = \sum_{s \in S_{jl}} \frac{L}{z(s) - L}, \tag{2.10}$$

where L denotes the best lower bound value available. (Observe that φ_{jl} tends to be high when j and l are processed on the same machine in many good solutions.) A pair (j, l) is then selected with probability proportional to φ_{jl}, and assigned to the machine of lowest index i for which the current completion time c_i satisfies $c_i + p_j + p_l \leq L$. If no such machine exists, j and l are assigned to the machine i with minimum c_i. The process is iterated until a complete solution s is obtained.

2.3.2 Scatter Search for P|# $\leq k$|C$_{\max}$

A scatter search algorithm for P|# $\leq k$|C$_{\max}$ was proposed by Dell'Amico, Iori and Martello [8]. The starting population was obtained by generating $|\mathcal{P}| = 80$ totally random solutions. Attempts to initialize it through a heuristic algorithm by Babel, Kellerer and Kotov [2] or through adaptations of the Multi-Subset algorithm by Dell'Amico and Martello [11] gave limited improvements.

The improvements were obtained through: (i) the $k - \ell$ swap procedure described in Section 2.3.1 with $k = \ell = 1$ (implying that only shifts of single jobs from one machine to another, and one to one job exchanges between pairs of machines were attempted); (ii) a re-optimizing procedure that iteratively fixes the jobs assigned to one machine, and runs a heuristic to schedule the remaining jobs on $m-1$ machines. Additional specially tailored heuristics were also executed for special instances satisfying $n = mk$ (known as *k-partitioning*, see Babel, Kellerer and Kotov [2]), which are particularly difficult to solve in practice.

The reference set $\mathcal{RS}$ was composed by the $|Q| = 8$ solutions with minimum makespan and by the $|D| = 7$ solutions with highest diversity. The evaluation of the diversity of a solution was implemented in a simpler way with respect to P||C$_{\max}$. Let $y_j(s)$ be the machine job j is assigned to in solution s. Then the *diversity* of s with respect to $\mathcal{RS}$ is

$$d(s) = \min_{t \in \mathcal{RS}} \left| \left\{ j \in \{1, \ldots, \tilde{n}\} : y_j(s) \neq y_j(t) \right\} \right| \tag{2.11}$$

with $\tilde{n} = 2m$. Using this definition, the reference set update is much faster than using (2.9), but the risk exists of having equivalent solutions in $\mathcal{RS}$.

The combination method was as follows. Given a subset $S \subseteq \mathcal{RS}$ generated at Step 3.1, let $S(i,j) \subseteq S$ be the set of solutions of S in which job j is assigned to machine i, and define an $m \times n$ matrix F with

$$F_{ij} = \sum_{s \in S(i,j)} \frac{z(s)}{z(s) - L} \tag{2.12}$$

A job-machine pair (i, j) has thus an high F_{ij} value if j is processed on i in many high quality solutions. The meaning of F_{ij} in (2.12) is similar to that of φ_{jl} in (2.10). Three solutions are then created through the following random process. For $j = 1, \ldots, n$, job j is assigned to to machine i with probability $F_{ij} / \sum_{h=1}^{m} F_{hj}$. If this makes machine i to have k jobs assigned, F_{il} is set to zero for $l = 1, \ldots, n$ in order to prevent the selection of i at the next iterations. The solution of minimum makespan among the three new solutions is then improved through the local search procedures mentioned above.

2.3.3 Scatter Search for k_i-*PP*

The scatter search algorithm by Dell'Amico, Iori, Martello and Monaci [9] starts by randomly generating and improving an initial population of $|\mathcal{P}| = 100$ solutions.

Concerning the improvement, the algorithms adopted for $\mathrm{P}|\# \leq k|\mathrm{C}_{\max}$ were generalized to the considered case of different k_i values.

The reference set $\mathcal{RS}$ had $|Q| = 10$ solutions with lowest makespan, and $|D| = 8$ solutions with high diversity.

The *diversity* of a solution s with respect to $\mathcal{RS}$ was computed as in (2.11). In this case, the advantage of a quick reference set update was preserved with a very limited risk (with respect to $\mathrm{P}|\# \leq k|\mathrm{C}_{\max}$) of having equivalent solutions in $\mathcal{RS}$, as in k_i-*PP* the machines are not identical, due to the different k_i values. Also the combination method was very similar to the one adopted for $\mathrm{P}|\# \leq k|\mathrm{C}_{\max}$, the only difference being in the number of random solutions generated.

2.4 Computational Results

We give here a concise exposition of the computational results obtained by the Scatter Search algorithms described in Section 2.3. In Section 2.4.1 we summarize the main results presented in [8–10] on the three problems addressed, comparing the Scatter Search algorithms with greedy heuristics, local search and exact methods. In Section 2.4.2 we concentrate on $\mathrm{P}||\mathrm{C}_{\max}$, and compare the Scatter Search algorithm with the most successful meta-heuristics in the literature. In Section 2.4.3 we evaluate the behavior of the three Scatter Search algorithms on the same set of $\mathrm{P}||\mathrm{C}_{\max}$ instances (remind that the two other problems are generalizations of $\mathrm{P}||\mathrm{C}_{\max}$). We conclude with Section 2.4.4, where we give some details on the parameters tuning.

2.4.1 Scatter Search vs simple heuristics and exact algorithms

We compare in this section the performance of the three Scatter Search algorithms introduced in Section 2.3 with that of the most well-known greedy heuristics for these three parallel machine scheduling problems, and with three exact approaches consisting in executing, after the Scatter Search, an implicit enumeration algorithm. The heuristics can be classified into three categories: List Scheduling, Threshold and Mixed.

List Scheduling Heuristics initially sort the jobs according to a prespecified criterion, and then assign them to the machines, one at a time, following a given rule. For $\mathrm{P}||\mathrm{C}_{\max}$ the most famous heuristic of this kind is the *Longest Processing Time* (LPT) by Graham [21], which sorts the jobs according to non-increasing processing time, and then iteratively assigns the next job to the machine having the minimum current completion time. It is known from the probabilistic analysis by Coffman, Lueker and Rinnooy Kan [7] that, if certain conditions on the processing times are satisfied, the solution produced by LPT is asymptotically optimal. In [12] and [9], LPT was generalized to $P|\# \leq k|C_{max}$ and k_i-*PP*, respectively, by imbedding in it the cardinality constraints.

Threshold Heuristics exploit the "duality" with BPP defined in Section 2.1. These algorithms consider a tentative value c for the optimum makespan, and solve the corresponding BPP instance, with bin capacity c, using one or more BPP heuristics. If not all items are assigned to the m bins, the remaining items are added through a greedy method. The process is possibly iterated by adjusting the threshold value on the basis of the solution obtained in the current attempt. Examples of threshold algorithms for $P||C_{max}$ are the *Multi-Subset* (*MS*) method by Dell'Amico and Martello [11], the *Multi Fit* algorithm (*MF*) by Coffman, Garey and Johnson [6], and the ϵ-dual method by Hochbaum and Shmoys [26]. Algorithm MS was adapted to $P|\# \leq k|C_{max}$ and k_i-*PP*, respectively in [12] and [9].

Mixed Heuristics combine list scheduling and threshold techniques, by switching from one to another during the construction process. Several mixed heuristics were proposed by Mokotoff, Jimeno and Gutiérrez [31] for $P||C_{max}$, and generalized to k_i-*PP* in [9].

In Tables 2.1–2.3 we compare the results given by the heuristics above and by the Scatter Search. The last column of each table shows the improvement that was obtained by executing an exact algorithm on the instances that had not been solved to optimality by the Scatter Search. For each algorithm we give the percentage of cases where the algorithm found the best solution with respect to the other algorithms (*%best*), the percentage of cases where the algorithm found the optimal solution (*%opt*) and the percentage gap between the solution found by the algorithm and the best lower bound (*%gap*).

In Table 2.1 we compare the results obtained on $P||C_{max}$ instances by the Scatter Search algorithm of Section 2.3.1 and by heuristics from the literature. The table summarizes the results obtained on two classes of classical $P||C_{max}$ benchmarks: *uniform instances*, proposed by França, Gendreau, Laporte and Müller [16], and *non-uniform instances*, proposed by Frangioni, Necciari and Scutellà [17]. For a given range $[a, b]$, in the former class each processing time

Table 2.1: Overall performance of heuristics, Scatter Search, and Scatter Search followed by branch-and-price on 780 $P||C_{max}$ instances.

	List Scheduling	Mixed Heuristics	Threshold Heuristics	Scatter Search	Scatter Search + Branch-and-Price
%best	21.9	30.9	79.1	97.7	100.0
%opt	21.9	30.9	79.1	97.7	100.0
%gap	0.54	0.36	0.05	0.01	0.00

is uniformly randomly generated in $[a, b]$, while in the latter class 98% of the processing times are uniformly randomly generated in $[0.9(b-a), b]$ and the remaining ones in $[a, 0.2(b-a)]$. The considered values were, for both classes, $a = 1$ and $b \in \{100, 1000, 10000\}$. The test instances were obtained by considering all pairs (m, n), with $m \in \{5, 10, 25\}$, $n \in \{10, 50, 100, 500, 1000\}$ and $m < n$, and generating ten instances per pair, resulting in a total of 780 instances[3]).

The experiments were performed on a Pentium IV 3 GHz. The heuristics usually needed less than one CPU second. The Scatter Search was allowed a maximum time limit of 30 seconds for $n \leq 50$, and of 120 seconds for $n > 50$. The branch-and-price algorithm never exceeded one hour CPU time. The table shows that the Scatter Search clearly outperforms all the classical heuristics with respect to the percentages of best and optimal solutions and to the percentage gap from the best lower bound. The last column shows that the improvement obtained by executing the exact branch-and-price algorithm (see [10]) after the Scatter Search is limited, but it allows to solve all instances to optimality.

In Table 2.2 we compare heuristics for $P|\# \leq k|C_{max}$ with the Scatter Search algorithm of Section 2.3.2. The experiments were performed on a large set of fifteen classes of randomly generated instances (twelve proposed by Dell'Amico and Martello [12], and three added by Dell'Amico, Iori and Martello [8]). The first nine classes were obtained by generating the p_j values according to, respectively: (i) uniform distributions in $[10, 1000]$, $[200, 1000]$ and $[500, 1000]$ (Classes 1-3); (ii) exponential distributions with average value $\mu = 25$, 50, 100 (Classes 4-6); (iii) normal distributions with average value $\mu = 100$ and standard deviation $\sigma = 33$, 66, 100 (Classes 7-9). The other six classes were generated as $k-$partitioning instances (with $n = km$): (iv) in Classes 10-12 the p_j values were uniformly distributed in $[500, 10000]$, $[1000, 10000]$ and $[1500, 10000]$; (v) in the additional Classes 13-15 they were generated as "perfect packing" instances (i.e., instances for which the optimal solution value z satisfies $z = \sum_{j=1}^{n} p_j/m$), with $z = 1000$, 5000, 10000. For each class, different instances were created, involving up to 50 machines and 400 jobs, and having cardinality limit values up to 50. In total, 9420 instances were generated.

The algorithms were run on a Pentium IV 2.4 GHz. For most of the instances the complete execution of all the algorithms (including branch-and-bound) needed few seconds in total, and it never exceeded 4 CPU minutes. The Scatter Search was halted if during the last iteration no new solution entered the reference set, or after a maximum of 10 iterations of Step 3. (see the template of Section 2.3). In this case too the Scatter Search clearly outperforms the classical heuristics in terms of percentage of best and optimal solutions found and percentage gap. Running a branch-and-bound algorithm

---

[3] http://www.inf.puc-rio.br/~alvim/adriana/tese.html

after the Scatter Search only leads to an improvement of 0.5% in %*best*, and of 0.2% in %*opt*.

In Table 2.3 we finally compare the Scatter Search algorithm of Section 2.3.3 with classical heuristics for k_i-*PP*. The algorithms were run on 81 classes of test problems obtained by combining in all possible ways nine weight classes for the generation of the p_j values and nine cardinality classes for the generation of the k_i values. The weight classes were the first nine classes adopted in [12] for $\mathrm{P}|\# \leq k|\mathrm{C}_{\max}$, and described above. The first six cardinality classes were characterized by a limited range of cardinality limits, obtained by uniformly randomly generating the k_i values so as to ensure $\lceil n/m \rceil - 1 \leq k_i \leq \lceil n/m \rceil + 3$ for $i = 1, \ldots, m$. The last three cardinality classes were characterized by more sparse k_i values, obtained through a particular generation procedure (see [9] for a more detailed description). The algorithms were tested on instances with up to 400 jobs and 50 machines, for a total of 25110 test problems.

The computational experiments were performed on a Pentium III 1.133 GHz. The Scatter Search was halted if either (i) no reference set update occurred during the last iteration, or (ii) Step 3. was executed α times (with $\alpha = 10$ for $n < 100$, $\alpha = 5$ for $100 \leq n < 400$ and $\alpha = 1$ for $n \geq 400$). The algorithms usually needed no more than 10 CPU seconds in total, although this limit could occasionally increase to 500 CPU seconds for particularly difficult instances. The considerations seen for $\mathrm{P}|\# \leq k|\mathrm{C}_{\max}$ apply to this case too. The Scatter Search consistently improves the behavior of the classical heuristics. Adding an exact approach (in this case, a column generation algorithm) only leads to small improvements in the solution quality.

2.4.2 Comparison of meta-heuristic algorithms for $\mathrm{P}||\mathrm{C}_{\max}$

In this section we compare the performance of the Scatter Search algorithm for $\mathrm{P}||\mathrm{C}_{\max}$ (see Section 2.3.1) with that of other meta-heuristic algorithms. The reason for restricting our attention to problem $\mathrm{P}||\mathrm{C}_{\max}$ is that this is the only problem, among the three considered here, for which a clear computational comparison with other meta-heuristics from the literature is possible.

According to our knowledge, the other most successful meta-heuristic algorithms for $\mathrm{P}||\mathrm{C}_{\max}$ are the Tabu Search approach by Alvim and Ribeiro [1]

Table 2.2: Overall performance of heuristics, Scatter Search and Scatter Search followed by branch-and-bound on 9420 $\mathrm{P}|\# \leq k|\mathrm{C}_{\max}$ instances.

	List Scheduling	Threshold Heuristics	Scatter Search	Scatter Search + Branch-and-Bound
%*best*	20.2	32.5	99.5	100.0
%*opt*	20.1	32.5	82.2	82.4
%*gap*	1.65	6.40	0.05	0.05

Table 2.3: Overall performance of heuristics, Scatter Search and Scatter Search followed by column generation on 25110 k_i-PP instances.

	List Scheduling	Mixed Heuristics	Threshold Heuristics	Scatter Search	Scatter Search + Column Generation
%*best*	4.1	8.3	62.6	99.1	100.0
%*opt*	4.1	8.2	61.2	88.9	89.7
%*gap*	3.17	1.12	28.5	0.01	0.01

and the multi-start local search method by Haouari, Gharbi and Jemmali [24]. The algorithm in [1] operates by iteratively defining a tentative threshold for the optimum makespan and solving the associated bin packing problem (see Section 2.1) through a specialized Tabu Search. The algorithm proposed in [24] is based on iterated solutions of a subset-sum problem for assigning jobs to one machine at a time. (Given a set S of n integers, the *subset-sum problem* is to find a subset $S' \subseteq S$ such that the sum of the values in S' is closest to, without exceeding, a given integer c.)

Other interesting results in the meta-heuristic field are the simple exchange neighborhood by França, Gendreau, Laporte and Müller [16], and the multiple exchange neighborhoods studied by Frangioni, Necciari and Scutellà [17]. We do not refer explicitly to these results, as the quality of the solutions they provide is generally worse than that given by the meta-heuristics above.

In Tables 2.4 and 2.5 we compare the three meta-heuristics on the uniform and non-uniform benchmark instances for $P||C_{\max}$ described in Section 2.4.1. The entries give: (i) the average percentage gap between the solution value found and the best lower bound (%*gap*); (ii) the average CPU time spent (*sec*); (iii) the total number of optimal solutions found (#*opt*). Each entry refers to 50 instances for $m = 5$ (10 instances each for n = 10, 50, 100, 500, 1000) and to 40 instances for $m > 5$ (10 instances each for n = 50, 100, 500, 1000). We also give the average and total values over the 130 instances of each range, and over the complete set of 390 instances.

The algorithm by Alvim and Ribeiro [1] was run on a Pentium II 400 MHz with 256 MB RAM. The algorithm by Haouari, Gharbi and Jemmali [24] was run on a Pentium IV 3.2 GHz with 1.5 GB RAM, and the %*gap* values were evaluated with respect to a better lower bound than the one used for the two other algorithms. The Scatter Search algorithm by Dell'Amico, Iori, Martello and Monaci [10] was run on a Pentium IV 3 GHz with 512 MB RAM.

The two competitors are faster than the Scatter Search, especially for uniform instances (although all CPU times are very reasonable), but the solutions provided by the Scatter Search are clearly better: the number of optimal solutions found is considerably higher and the percentage gap lower.

2.4.3 Comparison among Scatter Search algorithms

Since both $P|\# \leq k|C_{\max}$ and k_i-*PP* are generalizations of $P||C_{\max}$, the three Scatter Search algorithms we considered can all be executed on $P||C_{\max}$ instances. In order to evaluate the relative merits of the general structure of these algorithms (which is quite similar for the three cases) and of the different specializations adopted to handle the specific constraints, we performed a new series of experiments by executing the Scatter Search algorithms for $P|\# \leq k|C_{\max}$ and k_i-*PP* on the same instances used in Table 2.1 for the evaluation of the $P||C_{\max}$ Scatter Search. The results obtained are presented in Tables 2.6 and 2.7, as average values on a total of 390 instances per table. The information provided is the same as in the tables of the previous section.

The results for the $P||C_{\max}$ Scatter Search were directly taken from [10]. The results for the two other Scatter Search algorithms were obtained by running the corresponding codes on the same computer used in [10], namely a Pentium IV 3 GHz, without varying the termination criteria adopted as best choices for the particular problems addressed.

The tables highlight the relevance of the specializations induced in the algorithms by the cardinality constraints. In Table 2.6 the Scatter Search for $P||C_{\max}$ clearly outperforms the ones for $P|\# \leq k|C_{\max}$ and k_i-*PP* in terms of solution quality (19 not-optimal solutions versus, respectively, 30 and 29), although it needs a larger CPU time. All the percentage gaps are very small. In Table 2.7 the differences in the solution quality become more evident, since

Table 2.4: Meta-heuristic results on 390 $P||C_{\max}$ uniform instances.

		Alvim and Ribeiro			Haouari et al.			Scatter Search		
Range	*m*	*%gap*	*sec*	*#opt*	*%gap*	*sec*	*#opt*	*%gap*	*sec*	*#opt*
	5	0.0000	0.00	50	0.0000	0.02	50	0.0000	0.00	50
$[1, 10^2]$	10	0.0000	0.00	40	0.0000	0.03	40	0.0000	0.00	40
	25	0.0227	0.00	39	0.0453	0.04	38	0.0227	0.75	39
Average/Total		0.0070	0.00	129	0.0139	0.03	128	0.0070	0.23	129
	5	0.0000	0.01	50	0.0000	0.08	50	0.0000	0.03	50
$[1, 10^3]$	10	0.0019	0.02	38	0.0000	0.09	40	0.0000	0.04	40
	25	0.0322	0.02	37	0.0897	0.10	32	0.0311	5.86	38
Average/Total		0.0105	0.02	125	0.0276	0.09	122	0.0096	1.82	128
	5	0.0408	0.04	48	0.0406	0.42	48	0.0334	0.05	49
$[1, 10^4]$	10	0.0026	0.08	30	0.0021	0.32	30	0.0004	4.64	36
	25	0.0536	0.17	29	0.0257	0.19	28	0.0479	30.79	29
Average/Total		0.0330	0.09	107	0.0242	0.32	106	0.0277	10.92	114
Overall		0.0168	0.04	361	0.0219	0.15	356	0.0147	4.32	371

Table 2.5: Meta-heuristic results on 390 $P||C_{max}$ non-uniform instances.

		Alvim and Ribeiro			Haouari et al.			Scatter Search		
Range	*m*	*%gap*	*sec*	*#opt*	*%gap*	*sec*	*#opt*	*%gap*	*sec*	*#opt*
	5	0.0000	0.01	50	0.0000	0.06	50	0.0000	0.04	50
$[1, 10^2]$	10	0.1981	0.13	32	0.0000	0.09	40	0.0000	0.04	40
	25	0.0334	0.20	38	0.0000	0.14	40	0.0000	0.37	40
Average/Total		0.0712	0.11	120	0.0000	0.09	130	0.0000	0.14	130
	5	0.0000	0.01	50	0.0000	0.15	50	0.0000	0.04	50
$[1, 10^3]$	10	0.0000	0.12	40	0.0000	0.19	40	0.0000	0.06	40
	25	0.0335	0.41	38	0.0000	0.62	40	0.0000	0.29	40
Average/Total		0.0103	0.17	128	0.0000	0.31	130	0.0000	0.12	130
	5	0.0000	0.01	50	0.0000	0.74	50	0.0000	0.05	50
$[1, 10^4]$	10	0.0001	0.16	38	0.0002	0.56	37	0.0000	0.05	40
	25	0.0338	1.91	33	0.0007	4.11	30	0.0002	10.95	37
Average/Total		0.0104	0.64	121	0.0003	1.72	117	0.0001	3.40	127
Overall		0.0306	0.31	369	0.0001	0.71	377	0.0000	1.22	387

the Scatter Search for $P||C_{max}$ only misses 3 optimal solutions, against the 19 and 18 missed by the two other Scatter Search algorithms. The largest average CPU times were required by the Scatter Search for k_i-*PP*, while the two other algorithms needed on average about one CPU second. The uniform instances appear to be a more difficult test bed than the non-uniform ones, showing a larger average gap and a smaller total number of optimal solutions.

On the other hand we can observe that the Scatter Search template we adopted appears to be very robust. The Scatter Search algorithms for $P|\# \leq k|C_{max}$ and k_i-*PP*, when executed on $P||C_{max}$ instances, obtain very good results, not too far from the best performance obtained, as it could be expected, by the Scatter Search specifically tailored for this problem.

2.4.4 Parameters Tuning

We finally comment the parameters tuning within the Scatter Search framework, by particularly focusing on the $P||C_{max}$ algorithm of Section 2.3.1, and reviewing the tuning process with reference to the five main points outlined in Section 2.3.

Starting Population. The size of the initial pool of solutions was set to 40, after having tried the values 30, 50, 60, 70 and 100. The algorithm is very robust with respect to this parameter, i.e., all values produced results of comparable quality.

Improvement. The algorithm is very sensitive to this parameter. Limiting the $k - \ell$ swap to $k = \ell = 1$ lead to a non satisfactory performance in which

Table 2.6: Scatter Search algorithms on 390 $P||C_{max}$ uniform instances.

		Scatter Search for $P\|\|C_{max}$			Scatter Search for $P\|\# \le k\|C_{max}$			Scatter Search for k_i-PP		
Range	m	*%gap*	*sec*	*#opt*	*%gap*	*sec*	*#opt*	*%gap*	*sec*	*#opt*
	5	0.0000	0.00	50	0.0000	0.00	50	0.0000	0.00	50
$[1, 10^2]$	10	0.0000	0.00	40	0.0000	0.00	40	0.0000	0.00	40
	25	0.0227	0.75	39	0.0225	0.01	39	0.0225	0.02	39
Average/Total		0.0070	0.23	129	0.0069	0.00	129	0.0069	0.01	129
	5	0.0000	0.03	50	0.0000	0.00	50	0.0000	0.00	50
$[1, 10^3]$	10	0.0000	0.04	40	0.0021	0.01	38	0.0021	0.23	38
	25	0.0311	5.86	38	0.0320	0.05	38	0.0320	0.75	38
Average/Total		0.0096	1.82	128	0.0105	0.02	126	0.0105	0.30	126
	5	0.0334	0.05	49	0.0330	0.01	46	0.0329	0.08	47
$[1, 10^4]$	10	0.0004	4.64	36	0.0138	0.05	30	0.0091	1.02	30
	25	0.0479	30.79	29	0.0695	0.39	29	0.0658	5.95	29
Average/Total		0.0277	10.92	114	0.0383	0.14	105	0.0357	2.18	106
Overall		0.0147	4.32	371	0.0186	0.05	360	0.0177	0.83	361

only 734 optimum values out of 780 instances were found. Using $k = \ell = 2$ produced the final result (758 optima). Further enlarging the local search with $k = \ell = 3$ very slightly decreased the percentage gap without finding new optima, but required a consistently higher CPU time.

Reference Set Update. Updating the reference set as soon as a solution with high quality or high diversity is found lead to better solutions than updating it at the end of the iteration. Varying the sizes of the two subsets Q and D in the ranges $8 \le |Q| \le 12$ and $6 \le |D| \le 10$ did not produce relevant variations.

Subset Generation Method. The final choice reported in [10] was to adopt the method proposed by Glover, Laguna and Martí [20] (see Section 2.3). Considering a more limited number of subsets lead to slightly worse performances.

Solution combination. For each subset, the policy of obtaining a single new solution by combination proved to be better than that of generating more (2, 3 or 4) new solutions. The diversity function (2.9) outperformed (2.11), since solutions in which the same groups of jobs are assigned to different machines are not identified as identical by the latter function. Similar arguments hold for the combination method, as the one adopted in [10] outperformed the one in [8]. Other methods such as, e.g., using matrix φ (see (2.10)) to choose only one job at a time, lead to slightly worse results. The solution combination method turned out to be very important, since the Scatter Search algorithm is particularly sensitive to this aspect.

Table 2.7: Scatter Search algorithms on 390 $P||C_{max}$ non-uniform instances.

		Scatter Search for $P\|\|C_{max}$			Scatter Search for $P\|\# \leq k\|C_{max}$			Scatter Search for k_i-PP		
Range	*m*	*%gap*	*sec*	*#opt*	*%gap*	*sec*	*#opt*	*%gap*	*sec*	*#opt*
	5	0.0334	0.05	49	0.0330	0.01	46	0.0329	0.08	47
$[1, 10^2]$	10	0.0004	4.64	36	0.0138	0.05	30	0.0091	1.02	30
	25	0.0479	30.79	29	0.0695	0.39	29	0.0658	5.95	29
Average/Total		0.0000	0.14	130	0.0000	1.02	130	0.0000	2.07	130
	5	0.0000	0.04	50	0.0000	0.01	50	0.0000	0.18	50
$[1, 10^3]$	10	0.0000	0.04	40	0.0000	0.17	40	0.0000	1.69	40
	25	0.0000	0.37	40	0.0000	3.14	40	0.0000	4.81	40
Average/Total		0.0000	0.12	130	0.0000	0.52	130	0.0000	2.21	130
	5	0.0000	0.04	50	0.0000	0.02	50	0.0000	0.36	50
$[1, 10^4]$	10	0.0000	0.06	40	0.0000	0.26	40	0.0000	2.70	40
	25	0.0000	0.29	40	0.0000	1.42	40	0.0000	4.03	40
Average/Total		0.0001	3.40	127	0.0006	1.43	111	0.0004	3.96	112
Overall		0.0000	1.22	387	0.0002	0.99	371	0.0001	2.75	372

2.5 Conclusions

We presented a survey on heuristic results for the well-known Identical Parallel Machine Scheduling Problem and for two generalizations of practical relevance, known as the Cardinality Constrained Parallel Machine Scheduling Problem and the k_i-Partitioning Problem. The problems are particularly challenging from the heuristic point of view, since they present very low percentage gaps between the lower bounds and the upper bounds found by the classical heuristics from the literature. Hence the room for improvement is quite limited.

We described three Scatter Search approaches for these problems, highlighting their common components and their differences. We evaluated the behavior of these algorithms by summarizing the results of extensive computational experiments from the literature, and by presenting new results. The Scatter Search algorithms consistently improve on the results found by the classical heuristics. Using different exact methods leads to limited improvements. A test of the three algorithms on the same set of instances shows that the general approach is very robust.

A possible extensions could be to include Scatter Search in frameworks for parallel machine computing, such as, e.g., the one proposed by Cahon, Melab and Talbi [4].

Acknowledgements

We thank the Ministero dell'Università e della Ricerca (MIUR) Italy, for the support given to this project. Thanks are also due to two anonymous referees for helpful comments. The computational experiments have been executed at the Laboratory of Operations Research of the University of Bologna (Lab.O.R.).

References

1. A.C.F. Alvim and C.C. Ribeiro. A hybrid bin-packing heuristic to multiprocessor scheduling. In C.C. Ribeiro and S.L. Martins, editors, *Lecture Notes in Computer Science*, volume 3059, pages 1–13. Springer-Verlag, Berlin, 2004.
2. L. Babel, H. Kellerer, and V. Kotov. The k-partitioning problem. *Mathematical Methods of Operations Research*, 47:59–82, 1998.
3. P. Brucker. *Scheduling Algorithms*. Springer-Verlag, New York, 2001.
4. S. Cahon, N. Melab, and E.-G. Talbi. Paradiseo: A framework for the reusable design of parallel and distributed meta-heuristics. *Journal of Heuristics*, 10(3):357–380, 2004.
5. B. Chen. Parallel scheduling for early completion. In J.Y.T. Leung, editor, *Handbook of Scheduling: Algorithms, Models, and Performance Analysis*, chapter 9, pages 175–184. CRC Press, Boca Raton, FL, 2004.
6. E.G. Coffman, M.R. Garey, and D.S. Johnson. An application of bin-packing to multiprocessor scheduling. *SIAM Journal on Computing*, 7:1–17, 1978.
7. E.G. Coffman, G.S. Lueker, and A.H.G. Rinnooy Kan. Asymptotic methods in the probabilistic analysis of sequencing and packing heuristics. *Management Science*, 34:266–290, 1988.
8. M. Dell'Amico, M. Iori, and S. Martello. Heuristic algorithms and scatter search for the cardinality constrained $P||C_{\max}$ problem. *Journal of Heuristics*, 10: 169–204, 2004.
9. M. Dell'Amico, M. Iori, S. Martello, and M. Monaci. Lower bounds and heuristic algorithms for the k_i-partitioning problem. *European Journal of Operational Research*, 171:725–742, 2006.
10. M. Dell'Amico, M. Iori, S. Martello, and M. Monaci. Heuristic and exact algorithms for the identical parallel machine scheduling problem. *INFORMS Journal on Computing*, 2007 (to appear).
11. M. Dell'Amico and S. Martello. Optimal scheduling of tasks on identical parallel processors. *ORSA Journal on Computing*, 7:191–200, 1995.
12. M. Dell'Amico and S. Martello. Bounds for the cardinality constrained $P||C_{\max}$ problem. *Journal of Scheduling*, 4:123–138, 2001.
13. M. Dell'Amico and S. Martello. A note on exact algorithms for the identical parallel machine scheduling problem. *European Journal of Operational Research*, 160:576–578, 2005.
14. R. Martí (ed.). *Feature Cluster on Scatter Search Methods for Optimization*. *European Journal of Operational Research*, 169, 2, 2006.
15. G. Felinskas. *An investigation of heuristic methods and application to optimization of resource constrained project schedules*. PhD thesis, Vytautas Magnus University, Vilnius, Lithuania, 2007.

16. P.M. França, M. Gendreau, G. Laporte, and F.M. Müller. A composite heuristic for the identical parallel machine scheduling problem with minimum makespan objective. *Computers & Operations Research*, 21:205–210, 1994.
17. A. Frangioni, E. Necciari, and M. G. Scutellà. A multi-exchange neighborhood for minimum makespan machine scheduling problems. *Journal of Combinatorial Optimization*, 8:195–220, 2004.
18. F. Glover. Heuristic for integer programming using surrogate constraints. *Decision Sciences*, 8:156–166, 1977.
19. F. Glover. A template for scatter search and path relinking. In J. K. Hao, E. Lutton, E. Ronald, M. Schoenauer, and D. Snyers, editors, *Lecture Notes in Computer Science*, volume 1363, pages 1–45. Springer-Verlag, 1998.
20. F. Glover, M. Laguna, and R. Martí. Scatter search and path relinking: Foundations and advanced designs. In G. C. Onwubolu and B. V. Babu, editors, *New Optimization Techniques in Engineering*. Springer-Verlag, Heidelberg, 2004.
21. R.L. Graham. Bounds for certain multiprocessing anomalies. *Bell System Technical Journal*, 45:1563–1581, 1966.
22. R.L. Graham. Bounds on multiprocessing timing anomalies. *SIAM Journal on Applied Mathematics*, 17:416–429, 1969.
23. R.L. Graham, E.L. Lawler, J.K. Lenstra, and A.H.G. Rinnooy Kan. Optimization and approximation in deterministic sequencing and scheduling: a survey. *Annals of Discrete Mathematics*, 5:287–326, 1979.
24. M. Haouari, A. Gharbi, and M. Jemmali. Tight bounds for the identical parallel machine scheduling problem. *International Transactions in Operational Research*, 13:529–548, 2006.
25. M. S. Hillier and M. L. Brandeau. Optimal component assignment and board grouping in printed circuit board manufacturing. *Operations Research*, 46(5):675–689, 1998.
26. D. S. Hochbaum and D. B. Shmoys. Using dual approximation algorithms for scheduling problems: practical and theoretical results. *Journal of ACM*, 34: 144–162, 1987.
27. J.Y.T. Leung (ed.). *Handbook of Scheduling: Algorithms, Models, and Performance Analysis*. CRC Press, Boca Raton, FL, 2004.
28. R. Martí, M. Laguna, and F. Glover. Principles of scatter search. *European Journal of Operational Research*, 169:359–372, 2006.
29. E. Mokotoff. Parallel machine scheduling problems: a survey. *Asia-Pacific Journal of Operational Research*, 18:193–242, 2001.
30. E. Mokotoff. An exact algorithm for the identical parallel machine scheduling problem. *European Journal of Operational Research*, 152:758–769, 2004.
31. E. Mokotoff, J. J. Jimeno, and I. Gutiérrez. List scheduling algorithms to minimize the makespan on identical parallel machines. *TOP*, 9:243–269, 2001.

3

On the Effectiveness of Particle Swarm Optimization and Variable Neighborhood Descent for the Continuous Flow-Shop Scheduling Problem

Jens Czogalla and Andreas Fink

Helmut-Schmidt-University / UniBw Hamburg
Holstenhofweg 85, 22043 Hamburg, Germany
[czogalla,andreas.fink]@hsu-hamburg.de

Summary. Recently population-based meta-heuristics under the cover of swarm intelligence have gained prominence. This includes particle swarm optimization (PSO), where the search strategy draws ideas from the social behavior of organisms. While PSO has been reported as an effective search method in several papers, we are interested in the critical success factors of PSO for solving combinatorial optimization problems. In particular, we examine the application of PSO with different crossover operators and hybridization with variable neighborhood descent as an embedded local search procedure. Computational results are reported for the continuous (no-wait) flow-shop scheduling problem. The findings demonstrate the importance of local search as an element of the applied PSO procedures. We report new best solutions for a number of problem instances from the literature.

Key words: Flow-Shop Scheduling, Particle Swarm Optimization, Genetic Operators, Variable Neighborhood Descent, Hybridization.

3.1 Introduction

Swarm intelligence involves the design of heuristic methods for solving problems in a way that is inspired by the social behavior of organisms within swarm populations, where organisms interact locally with one another and with the environment. While there is no centralized control structure, interactions between the organisms may lead to the emergence of global behavior such as termite colonies building mounds, bird flocking, and fish schooling [7,37].

Recently population-based meta-heuristics under the cover of swarm intelligence have gained prominence. This includes particle swarm optimization (PSO), which was first introduced by Kennedy and Eberhart [14,29] for the optimization of continuous nonlinear functions. While PSO has been reported

J. Czogalla and A. Fink: *On the Effectiveness of Particle Swarm Optimization and Variable Neighborhood Descent for the Continuous Flow-Shop Scheduling Problem*, Studies in Computational Intelligence (SCI) **128**, 61–89 (2008)
www.springerlink.com

as an effective search method in several papers, we are interested in the critical success factors when applied to combinatorial optimization problems.

We consider the continuous (no-wait) flow-shop scheduling problem (CFSP) and analyze PSO hybridized with construction heuristics and variable neighborhood descent as an embedded local search procedure according to [42, 43]. Different crossover operators will be investigated and a combined crossover strategy will be derived. The analysis is based on extensive computational experiments for benchmark problem instances for the CFSP. We are able to significantly improve present best results from the literature. The findings show that local search may constitute the most important element of effective PSO approaches for combinatorial optimization problems.

In Section 3.2 we introduce the continuous flow-shop scheduling problem. In Section 3.3 we discuss PSO and extensions for combinatorial optimization problems; this includes the investigation of different crossover operators. Sections 3.4 and 3.5 are dealing with construction heuristics used to create the initial swarm population and with a local search heuristic, respectively. Computational results will be the topic of Section 3.6.

3.2 The Continuous Flow-Shop Scheduling Problem

In this section we review the continuous flow-shop scheduling problem (CFSP). First, we make a short excursion into the steel producing industry to show the motivation for the ongoing research on the continuous flow-shop scheduling problem; additionally we list applications in other branches.

3.2.1 How Iron Ore Becomes a Steel Plate

For the production of steel in a first step iron ore and coke are molted together (ironmaking). In a second step impurities (e.g., excess carbon) are removed and alloying materials (e.g., manganese, nickel) are added (steelmaking). In the further process, on the way to finished products such as steel plates and coils, the molted steel undergoes a series of operations [22] which are capital and energy extensive. Thus companies have been putting consistent emphasis on technology advances in the production process to increase productivity and to save energy [59].

Today steel production is typically done in integrated steel plants where the blast furnace maintains a steady flow of molten iron since the stopping of this process is enormously expensive. The molten iron then passes into the primary steel making process composed of basic oxygen, electric arc or induction furnace, ladle treatment facility, continuous caster (where the liquid steel is cast into slabs, since the liquid steel can not be stored) and hot strip mill. After passing through the primary steel making processes the steel coils may then pass through further finishing processes, such as pickling to remove surface oxides and annealing to improve mechanical properties [10]. The modern integrated process of steelmaking directly connects the steelmaking furnace, the continuous caster and the hot rolling mill with hot metal flow and makes a synchronized production [59]. Such a process has many advantages over the traditional cold charge process but brings new challenges for production planning and scheduling [59].

The short excursion demonstrates necessities for the application of the no-wait restriction to real-world problems [13]. First, the production technology may require the continuous processing of jobs. Second, a lack of storage capacity may force the production schedule to adapt the no-wait restriction. However, lack of intermediate storage capacity between consecutive machines does not necessarily impose the now-wait restriction, since jobs can wait on machines (while blocking these machines) until the next machine becomes available. Hall and Sriskandarajah [22] give a survey on machine scheduling problems with blocking and no-wait characteristics. Further applications that demand a no-wait scheduling are concrete ware production [20], food processing [22], pharmaceutical processing [48], chemical production [49], or more general just in time production and assembly lines.

3.2.2 Formal Description

A flow-shop scheduling problem consists of a set of n jobs which have to be processed in an identical order on m machines. Each machine can process exactly one job at a time. The processing time of job i on machine j is given as t_{ij} with $1 \leq i \leq n$ and $1 \leq j \leq m$. If a job does not need to be processed on a machine the corresponding processing time equals zero.

The continuous flow-shop scheduling problem includes no-wait restrictions for the processing of each job, i.e., once the processing of a job begins, there must not be any waiting times between the processing of any consecutive tasks of this job. Continuous processing leads to a delay $d_{ik}, 1 \leq i \leq n, 1 \leq k < n, i \neq k$ on the first machine between the start of jobs i and k when i and k are processed directly after each other. The delay can be computed as

$$d_{ik} = \max_{1 \leq j \leq m} \left\{ \sum_{h=1}^{j} t_{ih} - \sum_{h=2}^{j} t_{k,h-1} \right\}. \tag{3.1}$$

The processing order of the jobs is represented as a permutation $\Pi = < \pi_1, ..., \pi_n >$ where π_i is the job processed at the i-th position of a schedule. Note that the continuous flow-shop scheduling problem with makespan objective can be modeled as an asymmetric traveling salesman problem and thus can be solved by applying corresponding methods. In this chapter we consider the objective to minimize the total processing time (flow-time)

$$F(\Pi) = \sum_{i=2}^{n} (n + 1 - i)\, d_{\pi(i-1),\pi(i)} + \sum_{i=1}^{n} \sum_{j=1}^{m} t_{ij}. \tag{3.2}$$

The first part of Equation 3.2 sums the implied delays. Since the delay between two jobs affects all succeeding jobs the respective delay is multiplied with the number of following jobs. The second part of the formula is the constant total sum of processing times.

3.2.3 Literature Review

Since Johnson's seminal paper [28] on the two-machine flow-shop scheduling problem the literature on flow-shop scheduling has grown rapidly. It was reviewed by Day and Hottenstein in 1970 [12] and by Dudek et al. in 1992 [13]. Day and Hottenstein provide a classification of sequencing problems including flow-shop problems. Dudek et al. give a detailed overview on flow-shop sequencing research addressing several problem solving strategies and diverse optimization criterions. A statistical review of flow-shop literature is presented by Reisman et al. [52]. Ruiz and Maroto [53] provide a comprehensive review and evaluation of permutation flow-shop heuristics.

The continuous flow-shop scheduling problem with the objective to minimize total flow-time as posed by van Deman and Baker [63] is theoretically discussed by Gupta [21], Papadimitriou and Kanellakis [44], Szwarc [56], Adiri and Pohoryles [1], and van der Veen and van Dal [64]. Rajendran and Chauduri [50] provide a construction heuristic with two priority rules for the CFSP including computational experiments. Chen et al. [8] provide a genetic algorithm including computational results. Bertolissi [6] presents a construction heuristic evaluated by computational experiments. Fink and Voß [15] used several construction heuristics, tabu search, and simulated annealing and provided detailed computational results. Pan et al. [42] present extensive computational experiments of a discrete particle swarm optimization algorithm hybridized with a construction heuristic and variable neighborhood descent.

3.3 Particle Swarm Optimization

In this section we will describe PSO in its canonical form and extensions for combinatorial optimization problems. Merkle and Middendorf [37] provide an introduction to PSO and the related literature. Kennedy and Eberhart [31] present a more detailed introduction to PSO within the scope of swarm intelligence.

3.3.1 Standard Particle Swarm Optimization

The basic idea of PSO is that a particle i represents a search space location (solution) $x_i \in \mathbb{R}^n$. Depending on its velocity $v_i \in \mathbb{R}^n$ particles "fly" through the search space, thus exploring it and finding new solutions. The velocity of a particle is updated depending on locations where good solutions have already been found by the particle itself or other particles in the swarm. That is, there are two kinds of memory, individual and social, which direct the "reasoning" of the particle about the exploration of the search space.

The velocity of a particle i is canonically updated as follows:

$$v_i := w * v_i + c_1 * U(0,1) * (p_i - x_i) + c_2 * U(0,1) * (g - x_i) \qquad (3.3)$$

where p_i is the best previous position of the particle and g is the best found position within the swarm so far. The parameter w is called the *inertia weight* and represents the influence of the previous velocity. The parameters c_1 and c_2 are acceleration coefficients that determine the impact of p_i and g, i.e., the individual and social memory, respectively. Randomness is introduced by weighting the influence of the individual and social memory by random values uniformly drawn from [0,1]. After updating the velocity, the new position of the particle is calculated as:

$$x_i := x_i + v_i. \qquad (3.4)$$

Listing 3.1 shows the classic PSO procedure GBEST according to Eberhart and Kennedy [14]. After initialization of the parameters the initial population is randomly generated and evaluated using Equation 3.2. The best position of each particle and the best position of the whole swarm is updated. Then the velocity and the position of the particles are updated according to Equations 3.3 and 3.4. After evaluating all particles a new iteration is started. The loop is repeated until a termination criterion is met (e.g., concerning the elapsed time or the number of iterations).

Listing 3.1: Standard GBEST particle swarm algorithm.

```
initialize parameters
for all particles i do
    initialize position x_i and velocity v_i
    evaluate particle i
end for
do
    for all particles i do
        update personal best p_i
        update global best g
    end for
    for all particles i do
        update velocity v_i
        update position x_i
        evaluate particle i
    end for
while stop criterion not met
```

Premature convergence, causing local search instead of global search, is a general problem of PSO algorithms. In the literature different approaches are proposed in order to improve swarm behavior and the balance between convergence speed and convergence quality. Shi and Eberhart [55] investigate the influence of the maximum particle velocity and the inertia weight on the ability to escape from local optima. They provide detailed computational results and give suggestions for balancing those parameters. Trelea [62] analyzes PSO using standard results from dynamic system theory, discusses the tradeoff between exploration and exploitation, and presents guidelines for the selection of individual and social parameters. As swarm diversity (according to some measure of the differences between individuals of a swarm) decreases with swarm evolution, this results in a more thorough search in a restricted space while running into danger of getting stuck at local optima. Clerc [9] and Peram et al. [46] propose methods to measure and control diversity in order to prevent premature convergence.

In Kennedy and Mendes [32] and Mendes et al. [36] neighborhood relations between particles are introduced. Instead of using g, the best solution of the swarm so far, to update the velocity of particles, the best particle in some neighborhood is used. Janson and Middendorf [27] introduce a hierarchical neighborhood. Parsopoulos and Vrahatis [45] study PSO that makes use of the combination of different neighborhoods.

Current research is also concernd with the hybridization of PSO with other heuristic methods. For example, Gimmler et al. [16] analyze the influence of the type of local search hybridized with PSO for a number of standard test functions. For machine scheduling applications, Liu et al. [33, 34] incorporate different local search approaches and an adaptive meta-Lamarckian learning strategy into PSO. Tasgetiren et al. [60, 61] combine PSO with variable neighborhood descent.

3.3.2 Discrete Particle Swarm Optimization

Standard PSO is defined for a solution space $\mathbb{R}^n$. In order to apply the concept of PSO for combinatorial optimization problems one may define a transformation from $\mathbb{R}^n$ to a given problem-specific solution space. This allows keeping the canonical updating rules of PSO which are based on real-valued solutions (positions). In the case of the CFSP the solution space and the evaluation of the total processing time (Equation 3.2) is based on a permutation. Based on the random key representation [5] Tasgetiren et al. [60, 61] use the smallest position value (SPV) rule, stating that the job with the smallest position value is scheduled first, the job with the second smallest position value comes next, and so on. That is, the order of the jobs is derived from ascending position values. A similar transformation, called ranked-order-value (ROV), is used in Liu et al. [33, 34]. There are two problems with respect to such an approach for a combinatorial optimization problem such as the CFSP. First, the transformation of particle positions into discrete job permutations may be time consuming. Second, and more importantly, updating of particle velocities and positions is not done directly on the solution representation, thus loosing information about the search space during the transformation process.

Therefore, it may be indicated to adapt PSO for combinatorial optimization problems by means of problem-specific operators. This approach is followed, e.g., by [42] for the no-wait flow-shop scheduling problem. It is called discrete particle swarm optimization (DPSO). In contrast to the standard PSO updating of positions is done directly on job permutations. In this model X_i represents as position the job permutation of the i-th particle. Using a binary operator $\oplus$ with the meaning that the first operand defines the probability that the operator given as second operand is applied, the update is done by

$$X_i := c_2 \oplus F_3 \left(c_1 \oplus F_2 \left(w \oplus F_1 \left(X_i\right), P_i\right), G\right). \tag{3.5}$$

The first term of Equation 3.5 is $\lambda_i = w \oplus F_1 (X_i)$ which represents the "velocity" of the particle. F_1 is a swap operator which swaps two randomly chosen jobs within the permutation X_i of the i-th particle. The second part $\delta_i = c_1 \oplus F_2 (\lambda_i, P_i)$ represents the individual memory of the particle and F_2 is a crossover operator applying a one-cut crossover on λ_i and the best position of the particle so far. The social part is represented by $X_i = c_2 \oplus F_3 (\delta_i, G)$ with F_3 as a crossover operator using a two-cut crossover on δ_i and the best position of the swarm G. The parameters w, c_1, and c_2 determine the probabilities of the application of swap and crossover operators, respectively.

Listing 3.2 shows the pseudo code for the DPSO according to an adapted GBEST model from [14]. Since we are interested in the impact of local search on the solution quality there is an optional local search method included. Such a local search method may be employed to improve individual solutions (e.g., the best swarm solution G), which results in a hybridization of the concepts of swarm intelligence and local search.

Listing 3.2: Discrete particle swarm algorithm.

```
initialize parameters
for all particles i do
    initialize position X_i; evaluate particle i
end for do
    for all particles i do
        update personal best P_i
        update global best G
    end for
    for all particles i do
        update position X_i
        evaluate particle i
    end for
    local search (optional)
while stop criterion not met
```

Since the first DPSO algorithm proposed by Kennedy and Eberharth [30] there are several (recent) research papers which consider the design of DPSO procedures for different (scheduling) applications. Allahverdi and Al-Anzi [2] present a DPSO approach where particles are associated with job sequences and two velocities that correspond to probabilities of changing job positions. Results for the minimization of the maximum lateness for assembly flow-shop scheduling outperformed results obtained by tabu search and the earliest due date heuristic. Anghinolfi and Paolucci [3] investigate a DPSO approach for the single-machine total weighted tardiness scheduling problem with sequence-dependent setup times. Based on a discrete model for particle positions the velocity of a particle is interpreted as the difference in job positions of two particles. Pan et al. [43] apply DPSO to the single machine total earliness and tardiness problem with a common due date. A PSO approach for combinatorial optimization problems employing problem independent operators is introduced in Moraglio et al. [38].

The discrete particle swarm optimization concept, with the introduction of crossover and mutation operators, is similar to other methods that follow the paradigm of evolutionary computation [4]. Both genetic/evolutionary algorithms and PSO are based on populations formed of individuals representing solutions. The individuals are subject to some probabilistic operators such as recombination, mutation, and selection in order to evolve the population towards better fitness values. Scatter search [17] provides a complementary perspective on PSO. Scatter search generally operates on a relatively small number of solutions, called reference set. Some combination of two or more candidates from the reference set creates new solutions, which may be improved by means of local search, which is also a feature of memetic algorithms [25, 40]. Some of the obtained solutions may be inserted into the population according to some rule with the aim to guarantee both high solution quality and high diversity. A further generalization is possible when following the characterization of adaptive memory programming by Taillard [58]:

1. A set of solutions or a special data structure that aggregates the particularities of the solutions produced by the search is memorized;
2. A provisional solution is constructed using the data in memory;
3. The provisional solution is improved using a local search algorithm or a more sophisticated meta-heuristic;
4. The new solution is included in the memory or is used to update the data structure that memorizes the search history.

Nowadays most effective evolutionary methods use some kind of inherent hybridization with local search to improve individual solutions (in some cases termed as a mutation operator, which in its original meaning mainly introduces random effects). In Sections 3.4 and 3.5 we will discuss specific procedures for generating the initial swarm population and the hybridization of PSO with a local search algorithm, respectively. In the following we first consider the design of crossover operators for DPSO when applied for the CFSP.

3.3.3 Crossover Operators

In Equation 3.5 crossover operators are employed to create new particle positions represented by permutations. There are three major interpretations for permutations [4, 39]. Depending on the problem the relevant information contained in the permutation is the adjacency relation among the elements, the relative order of the elements, or the absolute positions of the elements in the permutation. Considering Equation 3.2 the adjacency relation between the jobs has the main impact on the objective function. In addition, the absolute positions of jobs or sequences of jobs are of importance since delays scheduled early are weighted with a larger factor then delays scheduled late. Besides those considerations the offspring created has to be a valid permutation or a repair mechanism must be applied. In the following we review some recombination methods developed for permutation problems where the adjacency relation among the jobs is mainly maintained.

The **order-based crossover** (OB, see MODIFIED-CROSSOVER in [11]) takes a part of a parent, broken at random, and orders the remaining jobs in accordance with their order in the second parent [11]. To use the order-based crossover as a two-cut crossover two crossover points are chosen randomly. The jobs between the crossover points are copied to the children. Starting from the second crossover site the jobs from the second parent are copied to the child if they are not already present in the child [54]. A different crossover operator is obtained when starting at the first position (OB').

As an example consider the two following parents and crossover sites:

```
Parent 1  = 0 1 2 | 3 4 5 | 6 7
Parent 2  = 3 7 4 | 6 0 1 | 2 5.
```

In the first step we keep jobs 3, 4, and 5 from parent 1 and jobs 6, 0, and 1 from parent 2:

```
Child 1   = x x x | 3 4 5 | x x
Child 2   = x x x | 6 0 1 | x x.
```

The empty spots are filled with the not yet used jobs, in the order they appear in the other parent, starting after the second crossover site:

```
Child 1   = 0 1 2 | 3 4 5 | 7 6
Child 2   = 4 5 7 | 6 0 1 | 2 3.
```

We get different offspring when starting at the first position:

```
Child 1   = 7 6 0 | 3 4 5 | 1 2
Child 2   = 2 3 4 | 6 0 1 | 5 7.
```

Like the order-based crossover the **partially matched crossover** (PMX, introduced in [19]) mainly preserves ordering within the permutations. Under PMX two crossing sites are picked at random which define a matching section. In this section jobs are exchanged with jobs corresponding to those mapped by the other parent. Consider the example from above:

```
Parent 1  = 0 1 2 | 3 4 5 | 6 7
Parent 2  = 3 7 4 | 6 0 1 | 2 5.
```

The first matching jobs are 3 in parent 1 and 6 in parent 2. Therefore jobs 3 and 6 are swapped. Similarly jobs 4 and 0, and 5 and 1 are exchanged:

```
Child 1   = 4 5 2 | 6 0 1 | 3 7
Child 2   = 6 7 0 | 3 4 5 | 2 1.
```

After the exchanges each child contains ordering information partially determined by each of its parents [18]. To obtain a one-cut crossover one of the crossing sites has to be the left or the right border.

The **PTL crossover** as proposed in [43] always produces a pair of different permutations even from identical parents. Two crossing sites are randomly chosen. The jobs between the crossing points of the first parent are either moved to the left or the right side of the children. The remaining spots are filled with the jobs not yet present, in the order they appear in the second parent.

```
Parent 1  = 0 1 2 | 3 4 5 | 6 7
Parent 2  = 3 7 4 | 6 0 1 | 2 5.
```

In our example we move the job sequence 345 to the left side to obtain child 1 and to the right side in order to get child 2:

```
Child 1   = 3 4 5 | x x x | x x
Child 2   = x x x | x x 3 | 4 5.
```

The missing jobs are filled according to their appearance in the second parent:

```
Child 1   = 3 4 5 | 7 6 0 | 1 2
Child 2   = 7 6 0 | 1 2 3 | 4 5.
```

3.4 Initial Swarm Population

The initial swarm population of a PSO implementation can be created randomly. More common is the employment of dedicated construction heuristics, which are often based on priority rules. With respect to the permution solution space, objects (jobs) are iteratively added to an incomplete schedule according to some preference relation.

The nearest neighbor heuristic (NNH), a well known construction heuristic related to the traveling salesman problem, can be modified to serve as construction heuristic for the CFSP. Starting with a randomly chosen job the partial schedule $\Pi = \langle \pi_1, ..., \pi_k \rangle$ is iteratively extended by adding an unscheduled job π_{k+1} with a minimal delay between π_k and π_{k+1}. This intuitive and fast construction heuristic seems to be appropriate for the CFSP since jobs scheduled earlier have a greater impact on the objective function.

Another variant of construction heuristics are simple insertion heuristics. In a first step the jobs are presorted in some way. Then the first two presorted jobs are selected and the best partial sequence for these two jobs is found considering the two possible partial schedules. The relative positions of these two jobs are fixed for the remaining steps of the algorithm. At each of the following iterations of the algorithm the next unscheduled job from the presorted sequence is selected and all possible insertion positions and the resulting partial schedules are examined.

In order to create the presorted sequence of jobs the NNH can be employed (NNNEH) as shown in [42]. Another approach is described in [41], where jobs with a larger total processing time get a higher priority then jobs with less total processing time (NEH). Therefore the jobs are ordered in descending sums of their processing times $T_i = \sum_{j=1}^{m} t_{i,j}$.

The effectiveness of the described construction heuristics generally depends on the initial job. So we repeat the construction heuristics with each job as the first job. The resulting solutions build up the initial swarm population (i.e., we fix the size of the swarm population as the number of jobs).

3.5 Local Search

Local search algorithms are widely used to tackle hard combinatorial optimization problems such as the CFSP. A straightforward local search algorithm starts at some point in the solution space and iteratively moves to a neighboring location depending on an underlying neighborhood structure and a move selection rule. In greedy local search the current solution is replaced if a better solution with respect to the objective function is found. The search is continued until no better solution can be found in the neighborhood of the current solution. In this section we examine neighborhood structures for the CFSP and review the variable neighborhood descent approach.

A neighborhood structure is defined by using an operator that transforms a given permutation Π into a new permutation Π^* [51]. We consider two alternative neighborhoods, both leading to a connected search space. The swap (or interchange) operator exchanges a pair of jobs π_{p_1} and π_{p_2} with $p_1 \neq p_2$, while all the other jobs in the permutation Π remain unchanged. The size of the swap neighborhood structure is $n * (n-1)/2$. The shift (or insertion) operator removes the job π_{p1} and inserts it behind the job π_{p2} with $p_1 \neq p_2$ and $p_1 \neq p_2 + 1$. The size of the shift neighborhood structure results as $(n-1)^2$.

Local search will be applied to the best position of the swarm G after each iteration of DPSO [42]. To prevent local search from getting stuck in a local minimum the search may be restarted from another starting point, in the simplest way using a random restart [35]. A more advanced approach is the variable neighborhood descent (VND) [24]. The VND algorithm is based on the fact that a local minimum with respect to one neighborhood structure is not necessarily a local minimum for another neighborhood structure [23]. Consequently, VND changes the neighborhood during the search [24]. Listing 3.3 shows the pseudo code for VND. Let $N_k, k = 1, ..., k_{max}$, be a set of neighborhood structures usually ordered by the computational complexity of their application [23]. In our case N_1 represents the swap neighborhood and N_2 the shift neighborhood. The VND algorithm performs local search until a local minimum with respect to the N_1 neighborhood is found. The search process is continued within the neighborhood N_2. If a better solution is found the search returns to the first neighborhood, otherwise the search is terminated.

The role of *perturbation* is to escape from local minima. Therefore the modifications produced by a *perturbation* should not be undone immediately by the following local search [26]. We follow the approach of [42] and use 3 to 5 random swap and 3 to 5 random shift moves as perturbation.

Listing 3.3: Variable neighborhood descent.

```
S := perturbation(G)
evaluate S
k := 1
do
    find best neighbor S* in neighborhood N_k of S
    evaluate S*
    if solution S* is better than solution S
        S := S*
        k := 1
    else
        increase k by 1
    end if
while k ≤ k_max
if solution S is better than solution G
    G := S
end if
```

3.6 Computational Results

Since we are interested in analyzing the factors contributing to the results reported in [42] we implemented DPSO for the CFSP in C#, pursuing an object-oriented approach, thus allowing easy combination of different heuristics. The DPSO was run on an Intel® Core^{TM} 2 Duo processor with 3.0 GHz and 2 GB RAM under Microsoft® Windows® XP[1]. We applied it to Taillard's benchmark instances[2] [57] treated as CFSP instances (as in [15]).

[1] In [42] DPSO was coded in Visual C++ and run on an Intel® Pentium® IV 2.4 GHz with 256 MB RAM.

[2] available at: http://mistic.heig-vd.ch/taillard/problemes.dir/problemes.html

In our computational experiments each configuration is evaluated by 10 runs for each problem instance. The quality of the obtained results is reported as the percentage relative deviation

$$\Delta_{avg} = \frac{solution - solution_{ref}}{solution_{ref}} * 100 \tag{3.6}$$

where *solution* is the total processing time generated by the variants of DPSO and $solution_{ref}$ is the best result reported in [15] (using simulated annealing, tabu search, or the pilot method). The relative deviation based on the results of [15] is also the measure used in [42]. The best objective function values reported in [15] and [42] are given in Table 3.8 (Appendix of this chapter) together with the results obtained in our own new computational experiments (denoted by CF). Note that we have been able to obtain new best results for all the larger problem instances (with $n = 100$ and $n = 200$) while we have at least re-obtained present best results for the smaller problem instances.

In [15] the 3-index formulation of Picard and Queyranne [47] was used to compute optimal solutions for the 20 jobs instances (ta001–ta030) and to obtain lower bounds provided by linear programming relaxation for the 50 jobs and 100 jobs instances (ta031–ta060 and ta061–ta090, respectively). For the smaller instances with 20 jobs the best results shown in Table 3.8 correspond to optimal solutions. The lower bounds for the instances ta031–ta060 and ta061–ta090 are given in Table 3.9 (Appendix of this chapter).

The parameter values of DPSO as described in Section 3.3.2 are taken from [42] in order to enable a comparison of results. The one-cut and the two-cut crossover probabilities are set to $c_1 = c_2 = 0.8$. The swap probability (w) was set to $w_{start} = 0.95$ and multiplied after each iteration with a decrement factor $\beta = 0.975$ but never decreased below $w_{min} = 0.40$. As mentioned in Section 3.4 the swarm size was fixed at the number of jobs.

Since in [42] there is no detailed information about the employed crossover operators we were at first interested in evaluating different crossover operators as described in Section 3.3.3. These experiments have been performed with a random construction of the initial swarm population and without using local search. The quality of the results obtained for different iteration numbers is shown in Table 3.1 (with the computation times given in CPU seconds). No particular crossover operator is generally superior. However, it seems that the PTL crossover operator is best suited for small problem instances while OB' and PMX work better on large instances.

In order to combine the advantages of the crossover operators we chose different crossover methods for the one-cut and the two-cut crossover; see Table 3.2. The notation PTL/PMX stands for the combination of PTL as one-cut crossover and PMX as two-cut crossover. Again there seems to be no superior combination of crossover operators. However, PTL/PMX performed best on the smaller instances while OB'/PMX provides the best results for the larger instances.

Table 3.1: Comparison of different crossover operators for DPSO.

jobs	iterations	PMX Δ_{avg}	PMX t	OB Δ_{avg}	OB t	OB' Δ_{avg}	OB' t	PTL Δ_{avg}	PTL t
20	1000	4.99	0.01	4.88	0.04	3.48	0.03	3.84	0.03
	2000	4.73	0.03	3.50	0.07	3.13	0.07	2.40	0.07
	10000	4.19	0.14	1.63	0.36	2.04	0.34	1.15	0.33
	50000	3.39	0.68	0.94	1.78	1.80	1.73	0.54	1.57
50	1000	9.26	0.05	17.15	0.14	8.74	0.14	15.52	0.13
	2000	8.29	0.10	13.70	0.28	6.71	0.27	12.34	0.26
	10000	7.72	0.49	7.52	1.46	4.71	1.31	6.60	1.24
	50000	7.16	2.47	4.35	7.15	4.05	6.93	3.94	6.19
100	1000	11.56	0.16	27.43	0.48	12.51	0.46	25.51	0.42
	2000	9.54	0.31	23.33	0.91	9.40	0,90	21.45	0.81
	10000	8.72	1.50	13.80	4.78	5.79	4.30	13.08	3.98
	50000	8.20	7.27	7.32	21.70	4.62	21.70	7.33	19.70
200	1000	14.43	0.61	38.03	1.94	17.88	1.92	36.43	1.66
	2000	11.51	1.19	34.20	3.82	13.43	3.88	32.51	3.28
	10000	9.08	5.88	22.79	20.44	7.09	18.14	22.40	15.91
	50000	8.63	26.52	11.86	97.98	4.71	95.25	12.86	76.13

Table 3.2: Comparison of the combination of crossover operators for DPSO.

jobs	iter.	PMX/OB' Δ_{avg}	PMX/OB' t	PMX/PTL Δ_{avg}	PMX/PTL t	OB'/PMX Δ_{avg}	OB'/PMX t	OB'/PTL Δ_{avg}	OB'/PTL t	PTL/PMX Δ_{avg}	PTL/PMX t	PTL/OB' Δ_{avg}	PTL/OB' t
20	1000	3.76	0.02	4.63	0.02	3.32	0.02	3.24	0.03	1.60	0.02	3.09	0.03
	2000	3.21	0.05	3.25	0.05	3.02	0.05	2.24	0.07	1.11	0.05	2.34	0.07
	10000	2.56	0.25	1.18	0.24	2.37	0.24	0.94	0.32	0.53	0.23	0.64	0.24
	50000	1.93	1.23	0.30	1.16	1.98	1.17	0.40	1.61	0.41	1.11	0.93	1.70
50	1000	8.33	0.10	18.75	0.09	7.88	0.09	13.57	0.13	7.69	0.09	9.81	0.14
	2000	6.64	0.20	17.27	0.18	6.43	0.18	10.84	0.25	5.72	0.19	7.81	0.27
	10000	5.11	0.94	11.01	0.92	4.71	0.92	6.24	1.26	2.82	0.91	4.58	1.37
	50000	4.45	4.85	6.68	4.40	4.26	4.50	3.67	6.22	1.59	4.34	3.05	6.56
100	1000	11.85	0.33	29.03	0.31	10.98	0.30	22.08	0.43	13.47	0.31	15.62	0.44
	2000	9.03	0.64	26.37	0.59	8.75	0.59	18.52	0.64	10.05	0.59	12.03	0.88
	10000	5.90	3.08	19.81	3.02	5.58	3.04	11.49	4.14	4.78	2.89	6.64	4.48
	50000	5.07	64.39	13.98	14.17	4.70	14.14	7.05	20.02	2.39	14.00	3.95	21.35
200	1000	17.39	1.40	39.29	1.24	14.91	1.35	32.48	1.82	20.35	1.26	23.41	1.84
	2000	12.57	2.80	36.38	2.47	11.58	2.74	28.25	3.60	15.51	2.51	17.55	3.75
	10000	6.83	13.21	29.66	12.33	6.34	13.41	19.67	17.78	7.27	12.01	9.32	18.62
	50000	5.07	64.39	23.18	57.43	4.55	62.92	12.30	86.98	3.17	57.70	5.25	87.57

Consequently, we used PMX as two-cut crossover and randomly selected from PTL and OB' as one-cut crossover for each particle. Table 3.3 shows the obtained results in comparison to the results reported by Pan et al. in [42] (notice that [42] do not give results for 10000 iterations). While Pan et al. report faster computation times, we implemented the methods using an object-oriented approach with the focus on an easy combination of the different algorithm elements without putting much value in the optimization of the code in terms of computation times. However, we mainly obtain significantly better results even when comparing on the basis of computation times.

The OB'+PTL/PMX crossover operator combination will be used throughout the remainder of this chapter and will be referred to as DPSO.

Table 3.3: Comparison of advanced crossover strategies for DPSO.

jobs	iterations	OB'/PMX Δ_{avg}	t	PTL/PMX Δ_{avg}	t	OB'+PTL/PMX Δ_{avg}	t	Pan et al. [42] Δ_{avg}	t
20	1000	3.32	0.02	1.60	0.02	1.89	0.02	2.15	0.01
	2000	3.02	0.05	1.11	0.05	1.35	0.05	1.55	0.02
	10000	2.37	0.24	0.53	0.23	0.64	0.24	-	-
	50000	1.98	1.17	0.41	1.11	0.48	1.12	0.40	0.18
50	1000	7.88	0.09	7.69	0.09	6.78	0.09	12.26	0.06
	2000	6.43	0.18	5.72	0.19	5.20	0.18	9.41	0.11
	10000	4.71	0.92	2.82	0.91	2.86	0.92	-	-
	50000	4.26	4.50	1.59	4.34	1.83	4.34	2.65	2.35
100	1000	10.98	0.30	13.47	0.31	10.84	0.30	24.11	0.19
	2000	8.75	0.59	10.05	0.59	8.06	0.59	19.99	0.38
	10000	5.58	3.04	4.78	2.89	4.14	2.97	-	-
	50000	4.70	14.14	2.39	14.00	2.44	13.97	6.30	8.98
200	1000	14.91	1.35	20.35	1.26	15.58	1.30	34.15	0.69
	2000	11.58	2.74	15.51	2.51	11.46	2.57	29.15	1.36
	10000	6.34	13.41	7.27	12.01	5.26	12.21	-	-
	50000	4.55	62.92	3.17	57.70	2.66	59.52	8.88	33.14

In accordance with [42] we also employ a dedicated construction heuristic to create the initial swarm population. The results presented in Table 3.4 are produced by evaluating the initial population without any iteration of DPSO performed. The insertion heuristics clearly outperform NN. Interestingly NEH, originally developed for the flow-shop scheduling problem with makespan criterion, performed slightly better than NNNEH. The results indicate that it may be advantageous to employ NEH in the initial phase of DPSO.

In order to improve the results we hybridized DPSO with VND as local search component as described in Section 3.5. To evaluate this combination we first show, in Table 3.5, the results of the application of VND as local

Table 3.4: Comparison of different construction heuristics.

jobs	random		NNH		NEH		NNNEH	
	Δ_{avg}	t	Δ_{avg}	t	Δ_{avg}	t	Δ_{avg}	t
20	28.54	0.00	19.04	0.00	1.71	0.00	2.10	0.00
50	47.37	0.00	25.77	0.00	4.04	0.01	4.92	0.01
100	57.29	0.00	25.65	0.01	4.62	0.04	5.58	0.04
200	65.28	0.01	19.62	0.05	3.56	0.34	4.15	0.35

search component within DPSO ($DPSO_{VND}$) for a randomly created initial swarm population.

Table 3.5: Results for $DPSO_{VND}$.

t		Δ_{avg}			
	jobs	20	50	100	200
0.02		0.07	2.65	-	-
0.05		0.01	1.41	4.23	-
0.09		0.00	0.98	3.18	-
0.16		0.00	0.75	2.49	-
0.50		0.00	0.46	1.11	3.79
1.00		0.00	0.28	0.65	2.38
2.00		0.00	0.18	0.27	1.43

In Table 3.6 DPSO is compared to different hybridized variants. The subscript VND denotes a possible local search component, which is applied after each iteration to the best solution of the swarm G; the superscript NEH denotes the application of a dedicated construction heuristic. The results show that DPSO is clearly outperformed by $DPSO_{VND}^{NEH}$. This demonstrates the pivotal effect of local search for the effectiveness of the applied PSO procedures. Comparing the results for the 50 jobs instances DPSO obtained Δ_{avg}=1.83 after 4.3 seconds while $DPSO_{VND}^{NEH}$ needed just 0.016 seconds to obtain the same solution quality. For the 100 and 200 jobs instances DPSO was not able to obtain results with a quality comparable to solutions obtained by $DPSO_{VND}^{NEH}$.

The effect of an increasing computational time on the solution quality, according to Table 3.6, is visualized in Figs. 3.1 – 3.4 (Appendix of this chapter). These figures show that hybridization of DPSO improves the search regardless of elapsed computation time.

In order to examine the contribution of DPSO mechanisms to the results obtained by the hybridized instances we run the algorithm with parameter values for inertia weight and acceleration coefficients set to zero. That is, after the initialization of the particle swarm neither crossover nor mutation

Table 3.6: Comparison of DPSO with its hybridized variants.

jobs	t	DPSO Δ_{avg}	DPSO^{NEH} Δ_{avg}	DPSO$_{VND}$ Δ_{avg}	DPSO$^{NEH}_{VND}$ Δ_{avg}
20	0.016	2.38	0.81	0.07	0.04
	0.031	1.68	0.66	0.02	0.02
	0.047	1.35	0.60	0.01	0.01
	0.094	1.07	0.53	0.01	0.00
	0.156	0.94	0.48	0.00	
	0.250	0.66	0.40		
	0.500	0.57	0.32		
	1.118	0.48	0.26		
50	0.016	15.42	3.82	2.65	1.83
	0.031	10.99	3.51	1.83	1.29
	0.047	9.30	3.30	1.41	1.03
	0.094	6.81	2.94	0.98	0.70
	0.156	5.54	2.67	0.75	0.58
	0.250	4.46	2.43	0.61	0.43
	0.500	3.62	2.17	0.46	0.32
	1.000	2.82	1.86	0.28	0.17
	2.000	2.14	1.59	0.18	0.09
100	0.047	23.72	4.62	-	-
	0.094	18.30	4.51	3.18	1.92
	0.156	14.57	4.34	2.49	1.39
	0.250	11.73	4.11	1.89	1.07
	0.500	8.72	3.64	1.11	0.59
	1.000	6.25	3.13	0.65	0.27
	2.000	4.76	2.70	0.27	-0.01
	4.000	3.78	2.38	-0.02	-0.20
	10.000	2.81	1.93	-0.25	0.12
200	0.500	23.12	3.55	3.79	-
	0.750	19.62	3.51	2.88	1.04
	1.000	17.57	3.48	2.38	0.78
	2.000	12.64	3.26	1.43	0.28
	5.000	8.15	2.77	0.41	-0.29
	10.000	5.94	2.41	-0.08	-0.62
	15.000	4.79	2.16	-0.35	-0.77
	60.000	2.66	1.46	-0.94	-1.22

is performed. Therefore, instead of evolving the swarm towards better solutions, just the best found solution after the initialization phase is subject to a repeated application of VND. The results are presented in Table 3.7 and Figs. 3.5 – 3.8 (Appendix of this chapter). The results indicate that the obtained high quality solutions are mainly caused by VND (and an effective construction of the initial population by NEH).

Table 3.7: Comparison of hybridized DPSO with VND.

jobs	t	DPSO_{VND} Δ_{avg}	$_{VND}$ Δ_{avg}	DPSO_{VND}^{NEH} Δ_{avg}	$^{NEH}_{VND}$ Δ_{avg}
20	0.016	0.07	0.00	0.04	0.03
	0.031	0.02	0.00	0.02	0.01
	0.047	0.01		0.01	0.01
	0.094	0.01		0.00	0.00
	0.156	0.00			
50	0.016	2.65	2.66	1.83	1.82
	0.031	1.83	1.87	1.29	1.28
	0.047	1.41	1.39	1.03	0.99
	0.094	0.98	0.98	0.70	0.70
	0.156	0.75	0.75	0.58	0.54
	0.250	0.61	0.60	0.43	0.42
	0.500	0.46	0.41	0.32	0.30
	1.000	0.28	0.28	0.17	0.18
	2.000	0.18	0.19	0.09	0.12
100	0.094	3.18	3.18	1.92	1.90
	0.156	2.49	2.50	1.39	1.38
	0.250	1.89	1.89	1.07	1.07
	0.500	1.11	1.12	0.59	0.58
	1.000	0.65	0.65	0.27	0.23
	2.000	0.27	0.27	-0.01	-0.05
	4.000	-0.02	0.00	-0.20	-0.25
	10.000	-0.25	-0.26	-0.42	-0.44
200	0.500	3.79	3.80	-	-
	0.750	2.88	2.88	1.04	1.04
	1.000	2.38	2.39	0.78	0.77
	2.000	1.43	1.44	0.28	0.26
	5.000	0.41	0.40	-0.29	-0.31
	10.000	-0.08	-0.09	-0.62	-0.63
	15.000	-0.35	-0.32	-0.77	-0.79
	60.000	-0.94	-0.94	-1.22	-1.23

To the best of our knowledge the results reported in [15] and [42] are the best solutions for Taillard's benchmark instances treated as CFSP instances.

During our experiments we were able to improve the solutions for 74 out of 80 unsolved instances using DPSO_{VND}^{NEH}. The obtained best results are listed in Table 3.8 (Appendix of this chapter). The best solutions are printed in bold letters (with preference to the prior source in case a best solution was found repeatedly).

3.7 Conclusions

We examined the application of DPSO in combination with different crossover strategies, construction heuristics, and variable neighborhood descent. In our computational experiments DPSO without local search achieved clearly worse results than classical meta-heuristics such as, e.g., tabu search and simulated annealing [15]. Based on the computational results we conclude that some good results recently reported in different papers for DPSO may be mainly caused by the employment of local search but not by the original concepts of PSO. While we were able to significantly improve present best results from the literature by investigating and applying hybridization of DPSO with local search components (i.e., new best solutions for 74 out of 80 unsolved instances), these results are not caused by the core concepts of PSO but by the embedded local search procedure. In particular, we have found that the embedded variable neighborhood descent procedure obtained the best results when particle interactions (i.e., the core concepts of PSO) were deactivated.

With the results at hand PSO and its modified variant DPSO may not be the number one choice for solving scheduling problems. However, the performance of DPSO may be improved by a sophisticated employment of crossover operators. Further research on crossover operators (see, e.g., geometric PSO [38]) and on the hybridization with local search may make DPSO more effective for solving combinatorial optimization problems.

References

1. I. Adiri and D. Pohoryles. Flowshop/no-idle or no-wait scheduling to minimize the sum of completion times. *Naval Research Logistics Quarterly*, 29:495–504, 1982.
2. A. Allahverdi and F. S. Al-Anzi. A PSO and a tabu search heuristics for the assembly scheduling problem of the two-stage distributed database application. *Computers & Operations Research*, 33:1056–1080, 2006.
3. D. Anghinolfi and M. Paolucci. A new discrete particle swarm optimization approach for the total tardiness scheduling scheduling problem with sequence-dependent setup times. *Working Paper*, 2007.
4. T. Bäck, D. B. Fogel, and Z. Michalewicz. *Evolutionary Computation 1: Basic Algorithms and Operators*. Institute of Physics Publishing, Bristol, 2000.
5. J. C. Bean. Genetic algorithms and random keys for sequencing and optimization. *ORSA Journal on Computing*, 6:154–160, 1994.

6. E. Bertolissi. Heuristic algorithm for scheduling in the no-wait flow-shop. *Journal of Materials Processing Technology*, 107:459–465, 2000.
7. E. Bonabeau, M. Dorigo, and G. Theraulaz. *Swarm Intelligence: From Natural to Artificial Systems*. Oxford University Press, New York, NY, 1999.
8. C.-L. Chen, R. V. Neppalli, and N. Aljaber. Genetic algorithms applied to the continuous flow shop problem. *Computers & Industrial Engineering*, 30:919–929, 1996.
9. M. Clerc. The swarm and the queen: Towards a deterministic and adaptive particle swarm optimization. *Proceedings of the 1999 Congress on Evolutionary Computation*, volume 3, pages 1951–1957, 1999.
10. P. Cowling and W. Rezig. Integration of continuous caster and hot strip mill planning for steel production. *Journal of Scheduling*, 3:185–208, 2000.
11. L. Davis. Applying adaptive algorithms to epistatic domains. *Proceedings of the 9th International Joint Conference on Artificial Intelligence*, 1:162–164, 1985.
12. J. E. Day and M. P. Hottenstein. Review of sequencing research. *Naval Research Logistics Quarterly*, 17:11–39, 1970.
13. R. A. Dudek, S. S. Panwalkar, and M. L. Smith. The lessons of flowshop scheduling research. *Operations Research*, 40:7–13, 1992.
14. R. Eberhart and J. Kennedy. A new optimizer using particle swarm theory. In *Sixth International Symposium on Micro Machine and Human Science*, pages 39–43, 1995.
15. A. Fink and S. Voß. Solving the continuous flow-shop scheduling problem by meta-heuristics. *European Journal of Operational Research*, 151:400–414, 2003.
16. J. Gimmler, T. Stützle, and T. E. Exner. Hybrid particle swarm optimization: An examination of the influence of iterative improvement algorithms on performance. In M. Dorigo, L. M. Gambardella, M. Birattari, A. Martinoli, R. Poli, and T. Stützle, editors, *Ant Colony Optimization and Swarm Intelligence, 5th International Workshop ANTS 2006*, volume 4150 of *Lecture Notes in Computer Science*, pages 436–443. Springer, Berlin, 2006.
17. F. Glover, M. Laguna, and R. Marti. Scatter search. In A. Ghosh and S. Tsutsui, editors, *Advances in Evolutionary Computing: Theory and Applications*, pages 519–538. Springer, Berlin, 2003.
18. D. E. Goldberg. *Genetic Algorithms in Search, Optimization, and Machine Learning*. Addison-Wesley, Reading, MA, 1989.
19. D. E. Goldberg and R. Lingle. Alleles, loci, and the traveling salesman problem. In J. Grefenstette, editor, *Proceedings of the International Conference on Genetic Algorithms and Their Applications*, pages 154–159. Lawrence Erlbaum Associates, Hillsdale, NJ, 1985.
20. J. Grabowski and J. Pempera. Sequencing of jobs in some production system. *European Journal of Operational Research*, 125:535–550, 2000.
21. J. N. D. Gupta. Optimal flowshop with no intermediate storage space. *Naval Research Logistics Quarterly*, 23:235–243, 1976.
22. N. G. Hall and C. Sriskandarajah. A survey of machine scheduling problems with blocking and no-wait in process. *Operations Research*, 44:510–525, 1996.
23. P. Hansen and N. Mladenović. Variable neighborhood search. In E. K. Burke and G. Kendall, editors, *Search Methodologies. Introductory Tutorials in Optimization and Decision Support Techniques*, pages 211–238. Springer, New York, NY, 2005.
24. P. Hansen, N. Mladenovic, and D. Perez-Brito. Variable neighborhood decomposition search. *Journal of Heuristics*, 7:335–350, 2001.

25. W. E. Hart, N. Krasnogor, and J. E. Smith. Memetic evolutionary algorithms. In W. E. Hart, N. Krasnogor, and J. E. Smith, editors, *Recent Advances in Memetic Algorithms*, volume 166 of *Studies in Fuzziness and Soft Computing*, pages 3–30. Springer, Berlin, 2005.
26. H. H. Hoos and T. Stützle. *Stochastic Local Search: Foundations and Applications*. Elsevier, Amsterdam, 2005.
27. S. Janson and M. Middendorf. A hierarchical particle swarm optimizer and its adaptive variant. *IEEE Transactions on Systems, Man, and Cybernetics. Part B: Cybernetics*, 35:1272–1282, 2005.
28. S. M. Johnson. Optimal two- and three-stage production schedules with setup times included. *Naval Research Logistics Quarterly*, 1:61–68, 1954.
29. J. Kennedy and R. C. Eberhart. Particle swarm optimization. In *Proceedings of the 1995 IEEE International Conference on Neural Networks*, pages 1942–1948, 1995.
30. J. Kennedy and R. C. Eberhart. A discrete binary version of the particle swarm algorithm. *Proceedings of IEEE International Conference on Systems, Man, and Cybernetics*, volume 5, pages 4104–4108, 1997.
31. J. Kennedy and R. C. Eberhart. *Swarm Intelligence*. Morgan Kaufmann Publishers, San Francisco, 2001.
32. J. Kennedy and R. Mendes. Population structure and particle swarm performance. *Proceedings of the 2002 Congress on Evolutionary Computation*, volume 2, pages 1671–1676, 2002.
33. B. Liu, L. Wang, and Y.-H. Jin. An effective hybrid particle swarm optimization for no-wait flow shop scheduling. *The International Journal of Advanced Manufacturing Technology*, 31:1001–1011, 2007.
34. B. Liu, L. Wang, and Y.-H. Jin. An effective PSO-based memetic algorithm for flow shop scheduling. *IEEE Transaction on Systems, Man, and Cybernetics. Part B: Cybernetics*, 37:18–27, 2007.
35. H. R. Lourenço, O. C. Martin, and T. Stützle. Iterated local search. In F. Glover and G. A. Kochenberger, editors, *Handbook on Meta-heuristics*, pages 321–353. Kluwer Academic Publishers, Norwell, MA, 2002.
36. R. Mendes, J. Kennedy, and J. Neves. The fully informed particle swarm: Simpler, maybe better. *IEEE Transactions on Evolutionary Computation*, 8: 204–210, 2004.
37. D. Merkle and M. Middendorf. Swarm intelligence. In E. K. Burke and G. Kendall, editors, *Search Methodologies. Introductory Tutorials in Optimization and Decision Support Techniques*, pages 401–435. Springer, New York, NY, 2005.
38. A. Moraglio, C. Di Chio, and R. Poli. Geometric particle swarm optimisation. In M. Ebner, M. O'Neill, A. Ekárt, L. Vanneschi, and A. I. Esparcia-Alcázar, editors, *Proceedings of the 10th European Conference on Genetic Programming*, volume 4445 of *Lecture Notes in Computer Science*, pages 125–136. Springer, Berlin, 2007.
39. A. Moraglio and R. Poli. Topological crossover for the permutation representation. In F. Rothlauf et al., editors, *Genetic and Evolutionary Computation Conference, GECCO 2005 Workshop Proceedings, 2005*, pages 332–338. ACM Press, New York, NY, 2005.
40. P. Moscato. On evolution, search, optimization, genetic algorithms and martial arts: Towards memetic algorithms. Technical Report Report 826, Caltech

Concurrent Computation Program, California Institute of Technology, Pasadena, 1989.
41. M. Nawaz, E. E. Enscore, jr., and I. Ham. A heuristic algorithm for the m-machine, n-job flow-shop sequencing problem. *Omega*, 11:91–95, 1983.
42. Q.-K. Pan, M. F. Tasgetiren, and Y.-C. Liang. A discrete particle swarm optimization algorithm for the no-wait flowshop scheduling problem. *Working Paper*, 2005.
43. Q.-K. Pan, M. Tasgetiren, and Y.-C. Liang. A discrete particle swarm optimization algorithm for single machine total earliness and tardiness problem with a common due date. In *IEEE Congress on Evolutionary Computation 2006*, pages 3281–3288, 2006.
44. C. H. Papadimitriou and P. C. Kanellakis. Flowshop scheduling with limited temporary storage. *Journal of the ACM*, 27:533–549, 1980.
45. K. E. Parsopoulos and M. N. Vrahatis. Studying the performance of unified particle swarm optimization on the single machine total weighted tardiness problem. In A. Sattar and B.-H. Kang, editors, *AI 2006: Advances in Artificial Intelligence*, volume 4304 of *Lecture Notes in Artificial Intelligence*, pages 760–769. Springer, Berlin, 2006.
46. T. Peram, K. Veeramachaneni, and K. M. Chilukuri. Fitness-distance-ratio based particle swarm optimization. *Proceedings of the 2003 IEEE Swarm Intelligence Symposium*, pages 174–181, 2003.
47. J.-C. Picard and M. Queyranne. The time-dependent traveling salesman problem and its application to the tardiness problem in one-machine scheduling. *Operations Research*, 26:86–110, 1978.
48. W. Raaymakers and J. Hoogeveen. Scheduling multipurpose batch process industries with no-wait restrictions by simulated annealing. *European Journal of Operational Research*, 126:131–151, 2000.
49. C. Rajendran. A no-wait flowshop scheduling heuristic to minimize makespan. *Journal of the Operational Research Society*, 45:472–478, 1994.
50. C. Rajendran and D. Chauduri. Heuristic algorithms for continuous flow-shop problem. *Naval Research Logistics Quarterly*, 37:695–705, 1990.
51. C. R. Reeves. Fitness landscapes. In E. K. Burke and G. Kendall, editors, *Search Methodologies: Introductory Tutorials in Optimization and Decision Support Techniques*, pages 587–610. Springer, Berlin, 2005.
52. A. Reisman, A. Kumar, and J. Motwani. Flowshop scheduling/sequencing research: A statistical review of the literature, 1952-1994. *IEEE Transactions on Engineering Management*, 44:316–329, 1997.
53. R. Ruiz and C. Maroto. A comprehensive review and evaluation of permutation flowshop heuristics. *European Journal of Operational Research*, 165:479–494, 2005.
54. K. Sastry, D. Goldberg, and G. Kendall. Genetic algorithms. In E. K. Burke and G. Kendall, editors, *Search Methodologies. Introductory Tutorials in Optimization and Decision Support Techniques*, pages 96–125. Springer, New York, NY, 2005.
55. Y. Shi and R. C. Eberhart. Parameter selection in particle swarm optimization. In V. W. Porto, N. Saravanan, D. Waagen, and A. E. Eiben, editors, *Evolutionary Programming VII: Proceedings of the 7th International Conference of Evolutionary Programming*, volume 1447 of *Lecture Notes in Computer Science*, pages 591–600. Springer, London, 1998.

56. W. Szwarc. A note on the flow-shop problem without interruptions in job processing. *Naval Research Logistics Quarterly*, 28:665–669, 1981.
57. E. Taillard. Benchmarks for basic scheduling problems. *European Journal of Operational Research*, 64:278–285, 1993.
58. E. D. Taillard, L. M. Gambardella, M. Gendreau, and J.-Y. Potvin. Adaptive memory programming: A unified view of meta-heuristics. *European Journal of Operational Research*, 135:1–16, 2001.
59. L. Tang, J. Liu, A. Rong, and Z. Yang. A review of planning and scheduling systems and methods for integrated steel production. *European Journal of Operational Research*, 133:1–20, 2001.
60. M. Tasgetiren, M. Sevkli, Y.-C. Liang, and G. Gencyilmaz. Particle swarm optimization algorithm for single machine total weighted tardiness problem. In *Proceedings of the 2004 Congress on Evolutionary Computation*, volume 2, pages 1412–1419, 2004.
61. M. F. Tasgetiren, Y.-C. Liang, M. Sevkli, and G. Gencyilmaz. A particle swarm optimization algorithm for makespan and total flowtime minimization in the permutation flowshop sequencing problem. *European Journal of Operational Research*, 177:1930–1947, 2007.
62. I. C. Trelea. The particle swarm optimization algorithm: convergence analysis and parameter selection. *Information Processing Letters*, 85:317–325, 2003.
63. J. M. van Deman and K. R. Baker. Minimizing mean flowtime in the flow shop with no intermediate queues. *AIIE Transactions*, 6:28–34, 1974.
64. J. A. A. van der Veen and R. van Dal. Solvable cases of the no-wait flow-shop scheduling problem. *Journal of the Operational Research Society*, 42:971–980, 1991.

Figures and Tables

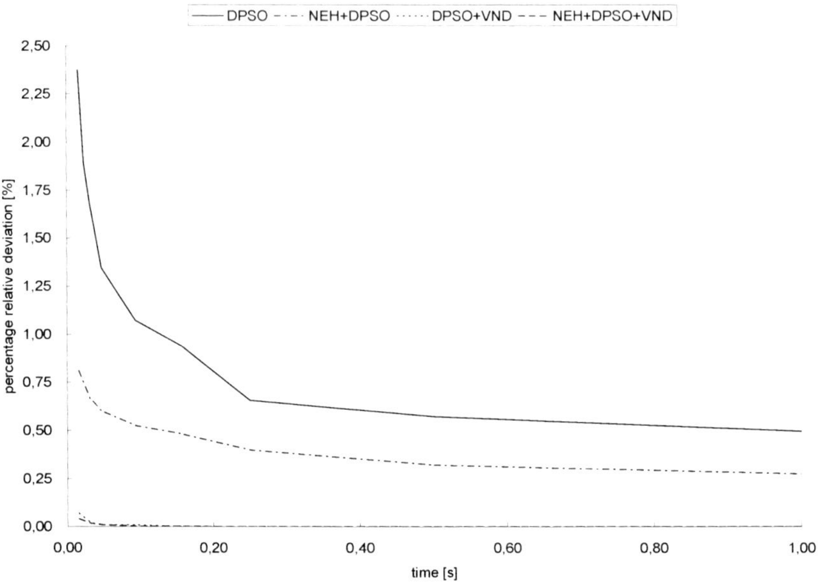

Fig. 3.1: Comparison of DPSO with its hybridized variants for 20 jobs instances.

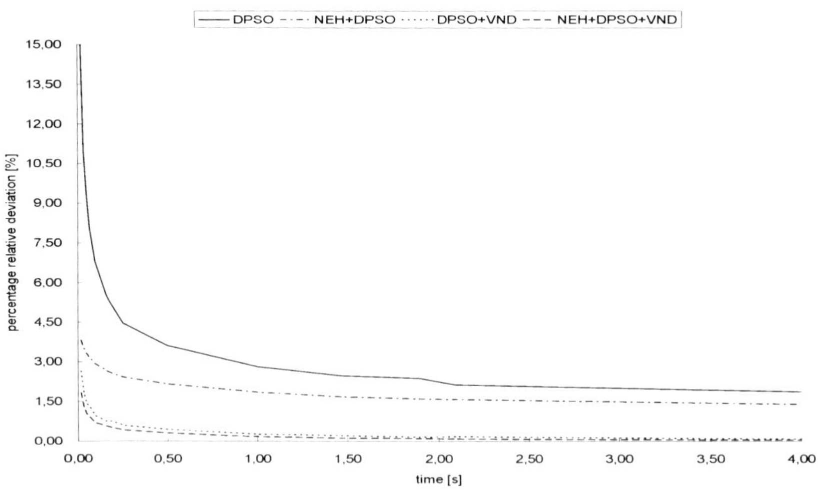

Fig. 3.2: Comparison of DPSO with its hybridized variants for 50 jobs instances.

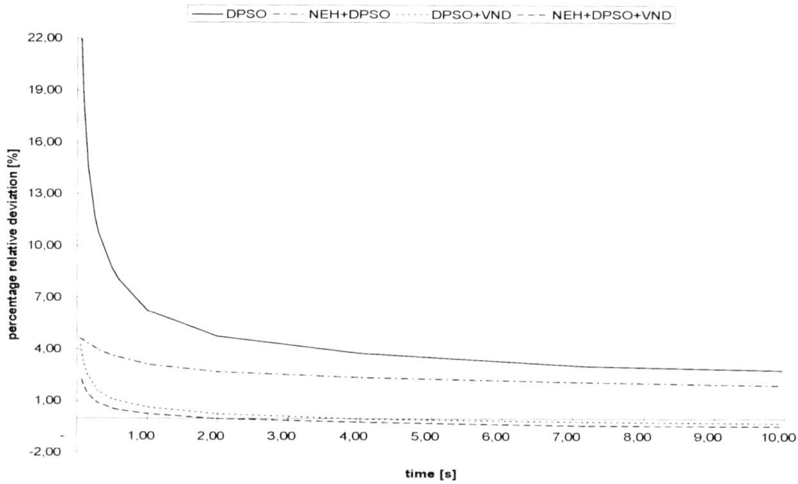

Fig. 3.3: Comparison of DPSO with its hybridized variants for 100 jobs instances.

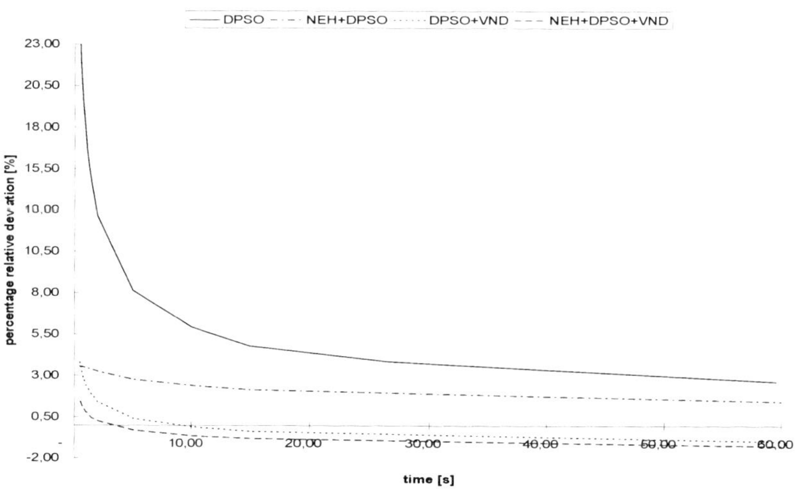

Fig. 3.4: Comparison of DPSO with its hybridized variants for 200 jobs instances.

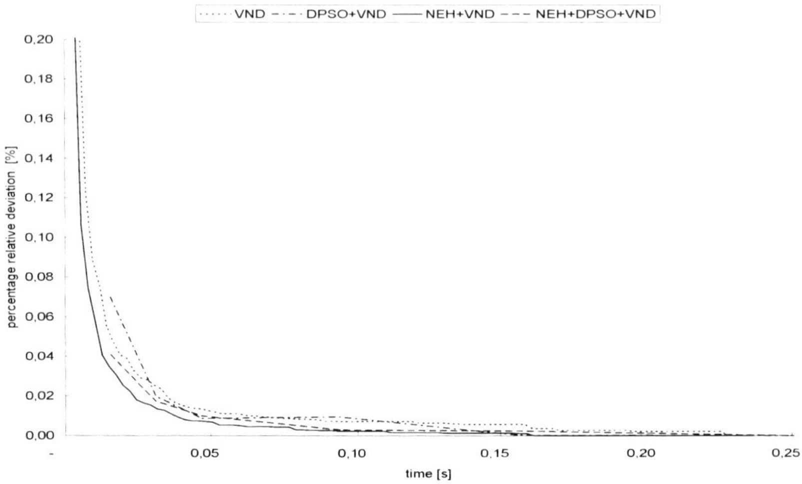

Fig. 3.5: Comparison of hybridized variants of DPSO with NEH & VND (and DPSO mechanisms turned off) for 20 jobs instances.

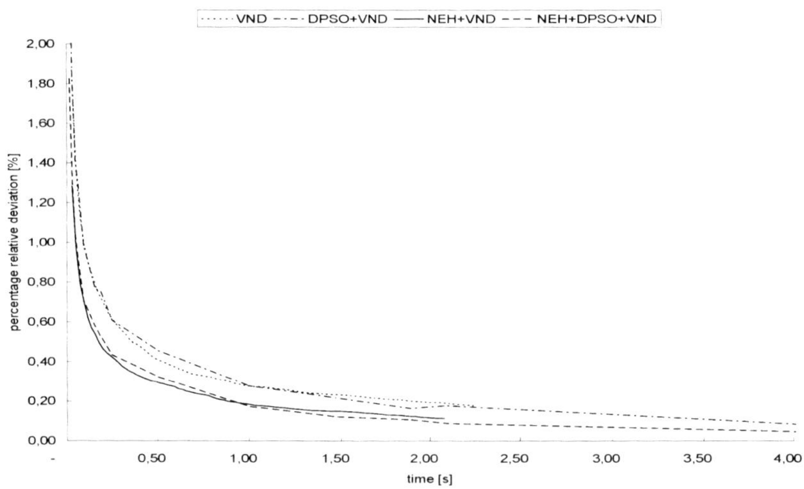

Fig. 3.6: Comparison of hybridized variants of DPSO with NEH & VND (and DPSO mechanisms turned off) for 50 jobs instances.

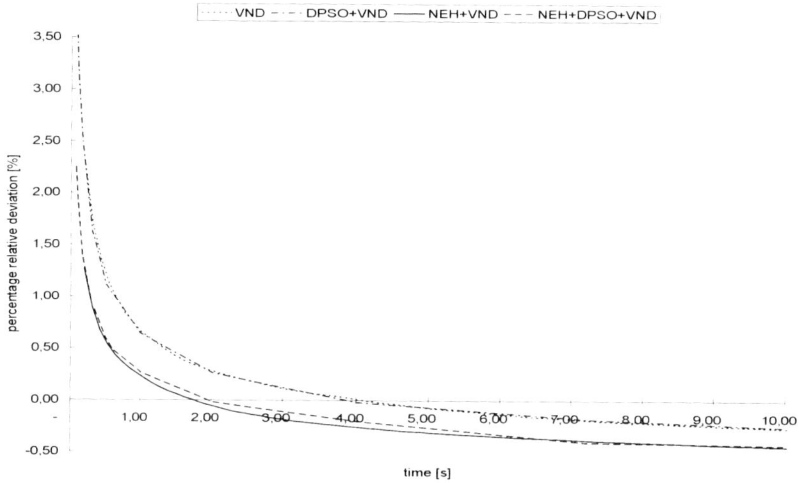

Fig. 3.7: Comparison of hybridized variants of DPSO with NEH & VND (and DPSO mechanisms turned off) for 100 jobs instances.

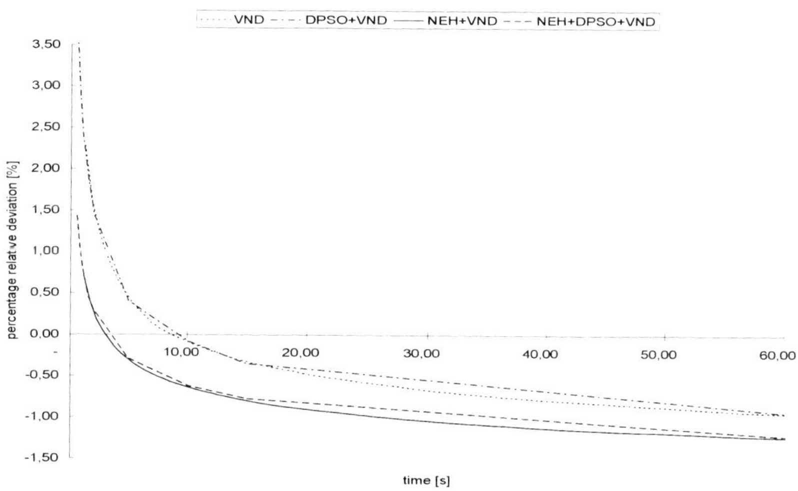

Fig. 3.8: Comparison of hybridized variants of DPSO with NEH & VND (and DPSO mechanisms turned off) for 200 jobs instances.

Table 3.8: Best known solutions for Taillard's benchmark instances.

Instance	CF	PTL [42]	FV [15]	Instance	CF	PTL [42]	FV [15]
Ta001	15674	15674	**15674**	Ta061	**303273**	304721	308052
Ta002	17250	17250	**17250**	Ta062	**297723**	297816	302386
Ta003	15821	15821	**15821**	Ta063	**291033**	292227	295239
Ta004	17970	17970	**17970**	Ta064	**276224**	276507	278811
Ta005	15317	15317	**15317**	Ta065	**289639**	289735	292757
Ta006	15501	15501	**15501**	Ta066	**286860**	287133	290819
Ta007	15693	15693	**15693**	Ta067	**297466**	298039	300068
Ta008	15955	15955	**15955**	Ta068	**286961**	287073	291859
Ta009	16385	16385	**16385**	Ta069	**303357**	303550	307650
Ta010	15329	15329	**15329**	Ta070	**297367**	297486	301942
Ta011	25205	25205	**25205**	Ta071	**408840**	409116	412700
Ta012	26342	26342	**26342**	Ta072	**389674**	391125	394562
Ta013	22910	22910	**22910**	Ta073	**402489**	403157	405878
Ta014	22243	22243	**22243**	Ta074	**418588**	419711	422301
Ta015	23150	23150	**23150**	Ta075	**398088**	398544	400175
Ta016	22011	22011	**22011**	Ta076	**388344**	388667	391359
Ta017	21939	21939	**21939**	Ta077	**390162**	390350	394179
Ta018	24158	24158	**24158**	Ta078	**399060**	399549	402025
Ta019	23501	23501	**23501**	Ta079	**413699**	413802	416833
Ta020	24597	24597	**24597**	Ta080	**408855**	409007	410372
Ta021	38597	38597	**38597**	Ta081	**556991**	559288	562150
Ta022	37571	37571	**37571**	Ta082	**561424**	563231	563923
Ta023	38312	38312	**38312**	Ta083	**558763**	558870	562404
Ta024	38802	38802	**38802**	Ta084	**558797**	560726	562918
Ta025	39012	39012	**39012**	Ta085	**552417**	552861	556311
Ta026	38562	38562	**38562**	Ta086	**559950**	560296	562253
Ta027	39663	39663	**39663**	Ta087	**569714**	571150	574102
Ta028	37000	37000	**37000**	Ta088	**573225**	573834	578119
Ta029	39228	39228	**39228**	Ta089	**559989**	560095	564803
Ta030	37931	37931	**37931**	Ta090	**568013**	570292	572798
Ta031	**75668**	75682	76016	Ta091	**1488029**	1495277	1521201
Ta032	82874	**82874**	83403	Ta092	**1474847**	1479484	1516009
Ta033	78103	**78103**	78282	Ta093	**1488869**	1495698	1515535
Ta034	**82413**	82533	82737	Ta094	**1458413**	1467327	1489457
Ta035	**83476**	83761	83901	Ta095	**1471096**	1471586	1513281
Ta036	**80671**	80682	80924	Ta096	**1468536**	1472890	1508331
Ta037	**78604**	78643	78791	Ta097	**1502101**	1512442	1541419
Ta038	**78726**	78821	79007	Ta098	**1493418**	1497303	1533397
Ta039	**75647**	75851	75842	Ta099	**1475304**	1480535	1507422
Ta040	**83430**	83619	83829	Ta100	**1486285**	1492115	1520800
Ta041	**114051**	114091	114398	Ta101	**1983177**	1991539	2012785
Ta042	**112116**	112180	112725	Ta102	**2024959**	2031167	2057409
Ta043	**105345**	105365	105433	Ta103	**2018601**	2019902	2050169
Ta044	**113206**	113427	113540	Ta104	**2006227**	2016685	2040946
Ta045	**115295**	115425	115441	Ta105	**2006481**	2012495	2027138
Ta046	**112477**	112871	112645	Ta106	**2014598**	2022120	2046542
Ta047	**116521**	116631	116560	Ta107	**2015856**	2020692	2045906
Ta048	**114944**	114984	115056	Ta108	**2020281**	2026184	2044218
Ta049	110367	**110367**	110482	Ta109	**2000386**	2010833	2037040
Ta050	113427	**113427**	113462	Ta110	**2018903**	2025832	2046966
Ta051	**172831**	172896	172845				
Ta052	**160888**	161029	161092				
Ta053	**160104**	160561	160213				
Ta054	**161492**	161690	161557				
Ta055	**167410**	167635	167640				
Ta056	**161589**	161784	161784				
Ta057	**167115**	167136	167233				
Ta058	167822	**167822**	168100				
Ta059	**165207**	165468	165292				
Ta060	168386	168386	**168386**				

Table 3.9: Lower bounds for Taillard's benchmark instances ($n = 50$, $n = 100$).

Instance	LB	Instance	LB
Ta031	75610.70	Ta061	302064.02
Ta032	82874.00	Ta062	295154.95
Ta033	78057.74	Ta063	289407.54
Ta034	82225.27	Ta064	274015.72
Ta035	83203.86	Ta065	287266.94
Ta036	80463.71	Ta066	284052.95
Ta037	78468.19	Ta067	294788.77
Ta038	78439.81	Ta068	284345.61
Ta039	75447.72	Ta069	301875.64
Ta040	83393.71	Ta070	295123.95
Ta041	113828.62	Ta071	406099.44
Ta042	111730.73	Ta072	388125.73
Ta043	105231.59	Ta073	400153.15
Ta044	112979.41	Ta074	415442.87
Ta045	115006.40	Ta075	394557.15
Ta046	112346.17	Ta076	384510.12
Ta047	115807.41	Ta077	388602.18
Ta048	114625.41	Ta078	395695.23
Ta049	110076.01	Ta079	409478.90
Ta050	113192.35	Ta080	405318.77
Ta051	171424.24	Ta081	553223.49
Ta052	158919.99	Ta082	553759.85
Ta053	159668.42	Ta083	553943.03
Ta054	161031.81	Ta084	554770.65
Ta055	166300.55	Ta085	546111.96
Ta056	160464.79	Ta086	554097.83
Ta057	166434.16	Ta087	565740.72
Ta058	166807.69	Ta088	568122.69
Ta059	163972.67	Ta089	555852.17
Ta060	167745.60	Ta090	562187.82

4

A Dynamical Ant Colony Optimization with Heuristics for Scheduling Jobs on a Single Machine with a Common Due Date

Zne-Jung Lee[1], Shih-Wei Lin[1], and Kuo-Ching Ying[2]

[1] Department of Information Management, Huafan University, No. 1, Huafan Rd. Shihding Township, Taipei County, 22301, Taiwan johnlee@hfu.edu.tw

[2] Department of Industrial Engineering and Management Information, Huafan University, No. 1, Huafan Rd., Shihding Township, Taipei County, 22301, Taiwan.

Summary. The problem of scheduling jobs on a single machine with a common due date is one of NP-complete problems. It is to minimize the total earliness and tardiness penalties. This chapter introduces a Dynamical Ant Colony Optimization (DACO) with heuristics for scheduling jobs on a single machine with a common due date. In the proposed algorithm, the parameter of heuristic information is dynamically adjusted. Furthermore, additional heuristics are embedded into DACO as local search to escape from local optima. Compared with other existing approaches in the literature, the proposed algorithm is very useful for scheduling jobs on a single machine with a common due date.

Key words: Scheduling, Single Machine, Dynamical Ant Colony Optimization, Heuristics.

4.1 Introduction

The scheduling problem with a common due date, known as NP-complete problem, has been investigated extensively [1–18]. This type of problem has become an attraction research with the advent of just-in-time (JIT) concept that an early or a tardy job completion is highly discouraged. To meet the JIT requirement, there is only one job can be completed exactly on the due date when scheduling jobs on a single machine with a common due date. All other jobs have to be completed either before or after the common due date. An early job completion results in an earliness penalty. On the other hand, a tardy job completion incurs a tardiness penalty. The objective of scheduling problem with a common due date is to find an optimal schedule that minimizes the sum of earliness and tardiness costs for all jobs.

Z.-J. Lee et al.: *A Dynamical Ant Colony Optimization with Heuristics for Scheduling Jobs on a Single Machine with a Common Due Date*, Studies in Computational Intelligence (SCI) **128**, 91–103 (2008)
www.springerlink.com

In the literature, many exact and heuristics algorithms have been proposed to solve the problem of scheduling jobs on a single machine with a common due date [1, 4, 11–13, 33]. Biskup and Feldmann [1] proposed a mixed integer programming model for this problem, and also designed a problem generator to solve 280 instances using two heuristics for identifying the upper bounds on the optimal function value. A comprehensive survey, applying polynomial or pseudo-polynomial time solvable algorithms on special cases, for the common due date assignment and scheduling problems can be found in [4]. In [11], Liaw proposed a branch-and-bound algorithm to find optimal solutions for problems that jobs have distinct due dates. Mondal and Sen [12] developed a dynamic programming for solving this problem. In [13], a sequential exchange approach is proposed for minimizing earliness-tardiness penalties of single-machine scheduling with a common due date. Due to the complexity of this problem, it is difficult for above approaches to obtain the optimal solution when the problem size is large [14, 15].

Recently, meta-heuristic approaches such as Simulated Annealing (SA), Genetic Algorithms (GAs) and Tabu Search (TS) have been proposed to find the near optimal solutions for the problem of scheduling jobs on a single machine with a common due date [6, 7, 9] and [16–18]. Feldmann and Biskup [5] applied five meta-heuristic approaches to obtain near-optimal solutions by solving 140 benchmarks. In [7], James developed the TS algorithm for solving the problem of scheduling jobs for general earliness and tardiness penalties with a common due date. Hino et al. [9] proposed a TS-based heuristic and a GA to minimize the sum of earliness and tardiness penalties of the jobs with 280 problems with up to 1000 jobs. Mittenthal et al. [16] proposed a hybrid algorithm, greedy approach and simulated annealing, for the V-shaped sequence of solution spaces. Lee and Kim [17] developed a parallel genetic algorithm for solving the problem of scheduling jobs for general earliness and tardiness penalties with a common due date. These approaches schedule their solutions with the first job starting at time zero, and may not find the optimal solutions. Liu and Wu [18] proposed a GA for the optimal common due date assignment and optimal policy in parallel machine earliness/tardiness scheduling problems. Pan et al. [33] also presented a discrete Particle Swarm Optimization algorithm for minimizing total earliness and tardiness penalties with a common due data on a single-machine. Even though these approaches could find the best solution in those test problems, the search performances seem not good enough. In this chapter, we propose a Dynamical Ant Colony Optimization (DACO) with heuristics for scheduling jobs on a single machine with a common due date.

The rest of this chapter is organized as follows. Section 4.2 describes the problem formulation. Section 4.3 introduces the basic ACO concepts. In Section 4.4, the structure of the proposed algorithm is discussed in detail. Section 4.5 reports the use of the proposed algorithm for test instances, and the effectiveness of the proposed algorithm is also illustrated. Concluding remarks are presented in Section 4.6.

4.2 Problem Formulation

The problem of scheduling jobs on a single machine with common due dates is to minimize the total earliness and tardiness penalties. There are n jobs available at time zero, each of which requires exactly one operation to be scheduled on a single machine with the common due date d. There is no preemption of jobs, and all jobs are sequence independent. For each job i, the processing time P_i, the penalty per unit time of earliness α_i, and the penalty per unit time of tardiness β_i are deterministic and known for $i = 1, \ldots, n$. Let C_i represent the completion time of job i. The earliness EA_i and tardiness TA_i can be obtained by $\max\{d - C_i, 0\}$ and $\max\{C_i - d, 0\}$, respectively. The penalty $\alpha_{i*} EA_i$ is incurred when job i is completed EA_i time units earlier than d, whereas the penalty $\beta_{i*} TA_i$ is incurred when it is completed TA_i time units later than d. The minimization of earliness and tardiness penalties of single-machine scheduling problems with a common due date can be formulated as follows [19–22].

$$ST_i + P_i + EA_i - TA_i = d, i = 1, \ldots, n \quad (4.1)$$

$$ST_i + P_i - ST_j - \gamma(1 - x_{ij}) \leq 0, i = 1, \ldots, n-1; j = i+1, \ldots, n \quad (4.2)$$

$$ST_j + P_j - ST_i - \gamma x_{ij} \leq 0, i = 1, \ldots, n-1; j = i+1, \ldots, n \quad (4.3)$$

$$ST_i, EA_i, TA_i \geq 0, i = 1, \ldots, n \quad (4.4)$$

where $x_{ij} \in \{0, 1\}$; x_{ij}=1 if job i precedes job j and x_{ij}=0, otherwise. n is the number of jobs, γ denotes a sufficiently large number and ST_i denotes the starting time of job i. Eq. 4.1 indicates each job is early or tardy. Eq. 4.2 represents that the starting time plus processing time of job i is earlier than or equal to the starting time of job j if job i precedes job j. Eq. 4.3 represents that the starting time plus processing time of job j is far ahead of the starting time of job j if job i lags job j. Eq. 4.4 ensures that the starting time, tardiness and earliness of jobs must be exceeding or equal to zero. Then, the objective function (F) for scheduling jobs on a single machine with a common due date is presented as follows.

$$F(S) = \sum_{i=1}^{n} (\alpha_i * EA_i + \beta_i * TA_i), \quad (4.5)$$

where S is the feasible schedule of the jobs. To efficiently obtain a better value for Eq. 4.5, three well-known theorems for scheduling jobs on a single machine with common due dates are shown below [13].

Theorem 4.1 *For scheduling jobs on a single machine with common due dates, there is an optimal schedule in which either the first job starts at time zero or one job is completed at the common due date d.*

Proof. The proof is shown in [10].

Theorem 4.2 *An optimal schedule exists if there is no idle time between any consecutive jobs for scheduling jobs on a single machine with common due dates.*

Proof. The proof is presented in [20].

Theorem 4.3 *For the optimal schedule, V-shaped property exists around the common due date. This means that jobs completed before or on common due date d are scheduled in non-increasing order of the ratios p_i/α_i and jobs starting on or after d are scheduled in non-decreasing order of the ratios p_i/β_i in an optimal schedule.*

Proof. The proof can be made by a job interchange argument [1], or can be followed from Smith's ratio rule [23].

By Theorem 4.1, the generated schedule follows that the starting time of the first job at time zero, or the completion time of a job coincides with the common due date [9, 13, 18]. By Theorem 4.2, the completion time for job j is calculated by adding the completion time of the previous job and the processing time of current job i when the schedule is generated [1, 9, 13, 18]. By Theorem 4.3, jobs can be classified into the subsets S_{EA} and S_{TA}, in which starting before or on/after the common due date [1, 9, 13, 18, 23]. It should be noted that above properties must be embedded into the proposed algorithm for obtaining the global solution and speeding up search performance.

In this chapter, the dynamical architecture of DACO is first derived to obtain feasible solutions. Furthermore, heuristics are used to ameliorate its performance and escape from local optima.

4.3 Overview of Ant Colony Optimization

The proposed algorithm is based on Ant Colony Optimization (ACO). In this section, the basic concept of ACO is introduced. ACO is also a class of meta-heuristic optimization algorithms inspired by the foraging behavior of real ants, and has been successively applied in many fields [24–32]. Real ants can explore and exploit pheromone information, which have been left on the traversed paths. The ACO algorithm is shown as follows [24]:

```
Procedure: ACO algorithm
        ScheduleActivities
            ConstructAntsSolutions
            UpdatePheromones
            DaemonActions
        end ScheduleActivities
end procedure
```

`ConstructAntsSolutions` decides a colony of ants that cooperatively and interactively visit next states by choosing from feasible neighbor nodes. They move by applying a stochastic local ant-decision policy that consists of pheromone trails and heuristic information. In this way, ants can construct solutions and find near-optimal solutions for the optimization problems. `UpdatePheromones` consists of pheromone evaporation and new pheromone deposition by which the pheromone trails are modified. Pheromone evaporation is a process of decreasing the intensity of pheromone trails. On the contrary, the trail's value can be increased as ants deposit pheromone on the traversed trails. Pheromone evaporation is a useful form of forgetting that ants can forage the promising area of the search space, and then can avoid trapping into local optima. The deposit of new pheromone can increase the probability that future ants will be directed to use a good solution again. `DaemonActions` is used to implement centralized actions such as local optimization procedure or the collection of global information that decides whether to deposit additional pheromone or not. `DaemonActions` cannot be performed by a single ant and are optional for ACO. The three above described procedures are managed by `ScheduleActivities`. `ScheduleActivities` construct but does not specify how these three procedures are scheduled and synchronized. In this chapter, we design a Dynamic ACO to specify the interaction between these three procedures for scheduling jobs on a single machine with common due dates.

4.4 The Proposed Algorithm

The ACO has shown its ability to find good solutions for NP-complete optimization problems. The problem of scheduling jobs on a single machine against the common due date with respect to earliness and tardiness penalties is also known as an optimization problem. It is promising that DACO is applied to solve this problem. In DACO, ants successively choose feasible jobs into subsequence to construct feasible solutions until all jobs are scheduled. For constructing solution, each ant decides that the l-th ant positioned on job r successively selects the next job $e$into subsequence at iteration$t$ with the ant-decision policy governed by

$$e = \begin{cases} \arg\{\max\limits_{u=allowed_l(t)} [\tau_{ru}(t)\ \eta_{ru}^{\varpi}]\}, & \text{when } q \leq q_0 \\ \\ E, & \text{otherwise;} \end{cases} \tag{4.6}$$

where $\tau_{ru}(t)$ is the pheromone trail, η_{ru} is the problem-specific heuristic information, and ϖ is a parameter representing the importance of heuristic information, q is a random number uniformly distributed in [0,1], q_0 is a prespecified parameter ($0\leq q_0 \leq 1$), and $allowed_l(t)$ is the set of feasible nodes currently not assigned by ant l at time t. In Eq. 4.6, q_0 is the probability of exploiting the learned knowledge when $q \leq q_0$. It indicates that ants will directly select next jobs by the product of learned pheromone trails and heuristic information. While $q > q_0$, it performs a biased exploration for the next job, and E is an index of node selected from $allowed_l(t)$ according to the probability distribution given by:

$$P_{re}^{l}(t) = \begin{cases} \frac{\tau_{re}(t)\eta_{re}^{\varpi}}{\sum\limits_{u \in allowed_l(t)} \tau_{ru}(t)\eta_{ru}^{\varpi}}, & \text{if e} \in allowed_l(t) \\ 0, & \text{otherwise;} \end{cases} \tag{4.7}$$

For η_{ru}, it is decided according to whether the next job positions in the S_{EA} or S_{TA} subset of the V-shaped. According to Theorem 4.3, η_{ru} is set to $\eta_{ru} = \frac{p_r}{\alpha_r} + \frac{p_u}{\alpha_u}$ if the next job is positioned in S_{EA} subset, otherwise η_{ru} is set to $\eta_{ru} = (\frac{p_r}{\beta_r} + \frac{p_u}{\beta_u})^{-1}$.

In DACO, the entropy information for estimating the variation of the pheromone trails is derived to adjust the parameter of heuristic information (ϖ). Each trail represents as a discrete random variable and the entropy (H) of the pheromone trails (Y) at the t-th iteration is defined as:

$$H(Y) = -\sum_{l=1}^{r} p(y_l) \log p(y_l) \tag{4.8}$$

where r represents the total number of pheromone trails. It is easy to show that the probability of initial pheromone trails is the same, H_t has the maximum value ($\log r$) [3]. Thereafter, the ratio value of H', $H' = H_t / \log r$, is used to dynamically adjust the value of heuristic information (ϖ) according to the rule given by

$$\varpi = \begin{cases} 4, & \Gamma < H' \leq 1 \\ 3, & \Pi < H' \leq \Gamma \\ 2, & \Omega < H' \leq \Pi \\ 1, & 0 < H' \leq \Omega \end{cases} \tag{4.9}$$

where the values of Γ, Π, and Ω could be predefined constants. The value of ϖ is set as the highest value in Eq. 4.9, because it guides the ant to increase the diversity search in the initial iteration. After constructing solutions, the amount of the pheromone trails will be more and more non-uniform, and the entropy will decrease gradually. Thus, a lower value of ϖ is used in Eq. 4.9.

In finding feasible solutions, ants perform online step-by-step pheromone updates as:

$$\tau_{ij}(t+1) = (1-\varphi)\tau_{ij}(t) + \varphi\tau_0, \tag{4.10}$$

where $0 < \varphi \leq 1$ is a constant, $\tau_0 = (m * \sum_{i=1}^{n} p_i)^{-1}$ is the initial value of pheromone trails and m is the number of ants. After all ants have constructed complete solutions, the global update is performed. Global update gives only the best solution to contribute to the pheromone trail update. The pheromone trail update rule is performed as:

$$\tau_{ij}(t+1) = (1-\rho)\tau_{ij}(t) + \rho\tau_{ij}(t), \tag{4.11}$$

where $0 < \rho \leq 1$ is a parameter governing the pheromone decay process, $\Delta\tau_{ij}(t) = 1/F^{best}$, and F^{best} is the objective function of the best solution obtained from the beginning of the search process.

After obtaining the best solution, additional heuristics are performed to escape form local optima and could also find the global optima. In the proposed algorithm, the idea of additional heuristics is to greedily swap jobs between the subsets of S_{EA} and S_{TA} for the best solution. There are 4 phases in the heuristics of the proposed algorithm. Firstly, the jobs in S_{EA} are successively selected from the first job to swap with all jobs in S_{TA} that could obtain better objective function than that of best solution. In phase 1, the i-th job in S_{EA} and $j-$th job in S_{TA} are swapped when this swapping causes the objective function improvement. Additively, the best solution is replaced by the swapped solution and the global update of Eq. 4.11 is performed. On the contrary, the best solution is not changed if this swapping of the i-th job in S_{EA} and $j-$th job in S_{TA} does not cause the any objective function improvement. Phase 1 continues until all jobs of the best solution in S_{EA} have been examined. In phase 2, the jobs in S_{TA} are successively selected from the first job to swap with the jobs in S_{EA} that could obtain better objective function than that of best solution. This phase continues until all jobs of the best solution in S_{TA} have been examined, and the swapping process is similar to phase 1 if a better objective function is obtained. In phase 3, a randomly selected $i-$th job in S_{EA} may be moved to S_{TA} if this move leads to the improvement of objective function. In phase 4, a randomly selected $j-$th job in S_{TA} may be moved to S_{EA} if this move leads to the improvement of objective. Additively, the best solution is also replaced by the new best solution if it is found in phase 3 and 4. It is noted that all jobs in subsets of S_{EA} and S_{TA} must follow the V-shaped property of Theorem 4.3.

4.5 Simulation Results

In simulation, we need to identify a set of parameters. The simulations with various values are performed, and the results are all similar. Experiments were conducted on PCs with PIV 3GHz processor. In the following simulations, we keep the following values as default: $\rho = 0.5$, $\psi = 0.1$, $q_0 = 0.8$, $\Gamma = 0.8$, $\Pi = 0.6$, $\Omega = 0.3$, and the number of ants $m = 20$. It is noted that the parameters of the proposed algorithm are set to the same values of ACO except for $\varpi = 2$ in ACO. For fair comparisons, these compared approaches are performed to see which approach can find the best solution after a fixed period of running without improving objective function [13]. To verify the effectiveness of the proposed algorithm, the problem sets are taken from [1,13]. The numbers of jobs n are set to 50, 100, 200, 500 and 1000, and four restrictive factors h = 0.2, 0.4, 0.6, and 0.8 are used to determine the common due date defined as $d = \lfloor h \sum P_i \rfloor$. For each combination of n and h, ten problems represented the k−th instance of a combination are used for testing. Tables 4.1 to 4.5 tabulate all simulation results for the proposed algorithm.

Table 4.1: Simulation results of n=50 for the proposed algorithm.

n=50	h=0.2	h=0.4	h=0.6	h=0.8
k=1	40,697	23,792	17,969	17,934
k=2	30,613	17,907	14,050	14,040
k=3	34,425	20,500	16,497	16,497
k=4	27,755	16,657	14,080	14,080
k=5	32,307	18,007	14,605	14,605
k=6	34,969	20,385	14,251	14,066
k=7	43,134	23,038	17,616	17,616
k=8	43,839	24,888	21,329	21,329
k=9	34,228	19,984	14,202	13,942
k=10	32,958	19,167	14,366	14,363

Table 4.2: Simulation results of n=100 for the proposed algorithm.

n=100	h=0.2	h=0.4	h=0.6	h=0.8
k=1	145,516	85,884	72,017	72,017
k=2	124,916	72,982	59,230	59,230
k=3	129,800	79,598	68,537	68,537
k=4	129,584	79,405	68,759	68,759
k=5	124,351	71,275	55,286	55,103
k=6	139,188	77,778	62,398	62,398
k=7	135,026	78,244	62,197	62,197
k=8	160,147	94,365	80,708	80,708
k=9	116,522	69,457	58,727	58,727
k=10	118,911	71,850	61,361	61,361

To show the superiority of the proposed algorithm, the percentage improvement of the obtained values (F^{OB}) for the proposed algorithm and various approaches were compared with regard to the benchmarks, provided by Biskup and Feldman (F^{BF}), which can be calculated as follows [1,13]:

$$Improvement\ rate\ (IR) = \frac{F^{BF} - F^{OB}}{F^{BF}} * 100\% \qquad (4.12)$$

Table 4.3: Simulation results of n=200 for the proposed algorithm.

n =200	h=0.2	h=0.4	h=0.6	h=0.8
k=1	498,653	295,684	254,259	254,259
k=2	541,180	319,199	266,002	266,002
k=3	488,665	293,886	254,476	254,476
k=4	586,257	353,034	297,109	297,109
k=5	513,217	304,662	260,278	260,278
k=6	478,019	279,920	235,702	235,702
k=7	454,757	275,017	246,307	246,307
k=8	494,276	279,172	225,215	225,215
k=9	529,275	310,400	254,637	254,637
k=10	538,332	323,077	268,353	268,353

Table 4.4: Simulation results of n=500 for the proposed algorithm.

n=500	h=0.2	h=0.4	h=0.6	h=0.8
k=1	2,954,852	1,787,698	1,579,031	1,579,031
k=2	3,365,830	1,994,788	1,712,195	1,712,195
k=3	3,102,561	1,864,365	1,641,438	1,641,438
k=4	3,221,011	1,887,284	1,640,783	1,640,783
k=5	3,114,759	1,806,978	1,468,231	1,468,231
k=6	2,792,231	1,610,015	1,411,830	1,411,830
k=7	3,172,398	1,902,617	1,634,330	1,634,330
k=8	3,122,267	1,819,185	1,540,377	1,540,377
k=9	3,364,310	1,973,635	1,680,187	1,680,187
k=10	3,120,383	1,837,336	1,519,181	1,519,181

Table 4.5: Simulation results of n=1000 for the proposed algorithm.

n=1000	h=0.2	h=0.4	h=0.6	h=0.8
k=1	14,054,929	8,110,906	6,410,875	6,410,875
k=2	12,295,998	7,271,371	6,110,091	6,110,091
k=3	11,967,290	6,986,816	5,983,303	5,983,303
k=4	11,796,599	7,024,050	6,085,846	6,085,849
k=5	12,449,588	7,364,810	6,341,477	6,341,477
k=6	11,644,121	6,927,585	6,078,373	6,078,375
k=7	13,277,006	7,861,297	6,574,297	6,574,297
k=8	12,274,736	7,222,137	6,067,312	6,067,312
k=9	11,757,063	7,058,786	6,185,321	6,185,321
k=10	12,427,441	7,275,945	6,145,737	6,145,737

The averaged improvement rate of the proposed algorithm and various approaches over the benchmarks are shown in Table 4.6. In Table 4.6, each cell represents the averaged value of ten instances ($k = 1, 2, \ldots, 10$), and the best results in the literature are reported in bold. It shows that the proposed algorithm has the best performance among these compared approaches except for n=200 and h=0.4. We noted that the proposed algorithm is also superior to ACO as shown in Table 4.6.

Table 4.6: The averaged improvement rate of the proposed algorithm and various approaches.

n	h	Meta-heuristics [6]	TS [9]	GA [9]	HTG [9]	HGT [9]	SEA [13]	DPSO [33]	ACO	Our algorithm
50	0.2	5.65	**5.70**	5.68	5.70	5.70	5.58	5.68	**5.70**	**5.70**
	0.4	4.64	4.66	4.60	**4.66**	4.66	4.42	**4.66**	**4.66**	**4.66**
	0.6	0	0.32	0.31	0.27	0.31	0.31	**0.34**	0.33	**0.34**
	0.8	0	**0.24**	0.19	0.23	0.23	**0.24**	**0.24**	**0.24**	**0.24**
100	0.2	6.18	**6.19**	6.17	**6.19**	**6.19**	6.12	**6.19**	**6.19**	**6.19**
	0.4	**4.94**	4.93	4.91	4.93	4.93	4.85	**4.94**	4.92	**4.94**
	0.6	0	0.01	0.12	-0.08	-0.04	0.14	**0.15**	**0.15**	**0.15**
	0.8	0	0.15	0.12	0.08	0.11	0.17	**0.18**	0.17	**0.18**
200	0.2	5.73	5.76	5.74	5.76	5.76	5.75	**5.78**	5.76	**5.78**
	0.4	**3.79**	3.74	3.75	3.75	3.75	3.72	3.74	3.75	3.75
	0.6	0	0.01	0.13	-0.37	-0.07	**0.15**	**0.15**	**0.15**	**0.15**
	0.8	0	0.04	0.14	-0.26	-0.07	**0.15**	**0.15**	**0.15**	**0.15**
500	0.2	6.40	6.41	6.41	6.41	6.41	6.42	6.42	6.42	**6.43**
	0.4	3.52	3.57	**3.58**	**3.58**	**3.58**	3.56	3.56	3.57	**3.58**
	0.6	0	-0.25	**0.11**	-0.73	-0.15	**0.11**	**0.11**	**0.11**	**0.11**
	0.8	0	-0.21	**0.11**	-0.73	-0.13	**0.11**	**0.11**	**0.11**	**0.11**
1000	0.2	6.72	6.73	6.75	6.74	6.74	**6.77**	6.76	6.76	**6.77**
	0.4	4.30	4.39	**4.40**	4.39	4.39	4.39	4.38	4.38	**4.40**
	0.6	0	-1.01	0.05	-1.28	-0.42	0.05	**0.06**	**0.06**	**0.06**
	0.8	0	-1.13	0.05	-1.28	-0.40	0.05	**0.06**	**0.06**	**0.06**

4.6 Conclusions

In this chapter, we propose a Dynamical Ant Colony Optimization with heuristics for scheduling jobs on a single machine with a common due date. In the proposed algorithm, the entropy information is used to estimate the variation of the pheromone trails and then to dynamically adjust the parameter of heuristic information. Furthermore, heuristics are embedded into the proposed algorithm to ameliorate its search performance. We use the benchmarks, provided by Biskup and Feldman, to test the performance of the proposed algorithm. From simulation results, it indicates that the proposed algorithm outperforms original ACO and other approaches.

References

1. Biskup, D. and Feldmann, M. (2001) Benchmarks for scheduling on a single machine against restrictive and unrestrictive common due dates, Computers and Operations Research: 28(8) 787-801.
2. Garey, M. and Johnson, D. (1979). Computers and Intractability: A Guide to the Theory of NP Completeness, W. H. Freeman and Company, San Francisco, California.
3. Bose, R. (2002) Information theory, coding, and cryptography, McGraw Hill.
4. Gordon, V., Proth, J.-M. and Chu, C. (2002) A survey of the state-of-art of common due date assignment and scheduling research, European Journal of Operational Research: 139(1) 1-25.
5. Gupta, J. N. D., Lauff, V. and Wernerm F. (2004) Two-machine flow shop problems with nonregular criteria, Journal of Mathematical Modelling and Algorithms: 3 123-151.
6. Feldmann, M. and Biskup D. (2003) Single-machine scheduling for minimizing earliness and tardiness penalties by meta-heuristic approaches, Computers and Industrial Engineering: 44(2) 307-323.
7. James, R. J. W. (1997) Using tabu search to solve the common due date early/tardy machine scheduling problem, Computers and Operations Research: 24(3) 199-208.
8. Hall, N. G., Kubiak, W. and Sethi, S. P. (1991) Earliness-tardiness scheduling problem, II: deviation of completion times about a restrictive common due date, Operations Research: 39(5) 847-856.
9. Hino, C. M., Ronconi, D. P., and Mendes, A. B. (2005) Minimizing earliness and tardiness penalties in a single-machine problem with a common due date, European Journal of Operational Research: 160(1) 190-201.
10. Hoogeveen, J. A. and Velde Van De S. L. (1991) Scheduling around a small common due date, European Journal and Operational Research: 55(2) 237-242.
11. Liaw, C.-F. (1999) A Branch-and-Bound Algorithm for the Single Machine Earliness and Tardiness Scheduling Problem, Computers and Operations Research: 26 679-693.
12. Mondal, S. A. and Sen, A. K. (2001) Single machine weighted earliness–tardiness penalty problem with a common due date, Computers and Operations Research: 28 649-669.
13. Lin, S.-W., Chou, S.-Y., and Ying, K.-C. (2007) A sequential exchange approach for minimizing earliness-tardiness penalties of single-machine scheduling with a common due date, European Journal of Operational Search: 177 1294-1301.
14. Ibarraki T. and Katoh N. (1988) Resource Allocation Problems: The MIT Press: Cambridge, Massachusetts.

15. Lee, Z.-J., Lee, C.-Y. (2005) A Hybrid Search Algorithm with Heuristics for Resource Allocation Problem, Information sciences: 173 155-167.
16. Mittenthal, J., M. Raghavachari, and A. I. Rana. (1993) A hybrid simulated annealing approach for single machine scheduling problems with non-regular penalty functions, Computers and Operations Research: 20 103-111.
17. Lee, C. Y. and Kim, S. J. (1995) Parallel genetic algorithms for the earliness-tardiness job scheduling problem with general penalty weights, Computers and Industrial Engineering: 28(2) 231-243.
18. Liu, M. and Wu, M. (2006) Genetic algorithm for the optimal common due date assignment and the optimal policy in parallel machine earliness/tardiness scheduling problems, Robotics and Computer-Integrated Manufacturing: 22 279-287.
19. Jaynes, E. T. (1982) On the rationale of the maximum entropy methods. Proceedings of the IEEE: 70(9) 939-952.
20. Kahlbacher, H. G. (1993) Scheduling with monotonous earliness and tardiness penalties, European Journal of Operational Research: 64(2) 258-277.
21. Raghavachari, M. (1988) Scheduling problems with non-regular penalty functions: a review, Operations Research: 25 144-164.
22. Szwarc, W. (1989) Single-machine scheduling to minimize absolute deviation of completion times from a common due date, Naval Research Logistics: 36 663-673.
23. Smith, W. E. (1956) Various optimizers for single-stage production, Naval Research Logistics Quarterly: 3 59-66.
24. Dorigo M. and Stützle T. (2004). Ant Colony Optimization: The MIT Press.
25. Lee, C.-Y., Lee, Z.-J. and Su, S.-F. (2005) Ant Colonies With Cooperative Process Applied To Resource Allocation Problem, Journal of the Chinese Institute of Engineers: 28 879-885.
26. Lee, Z.-J., Lee, C.-Y. and Su, S.-F. (2002) An Immunity Based Ant Colony Optimization Algorithm for Solving Weapon-Target Assignment Problem, Applied Soft Computing 2(1) 39-47.
27. Lee, Z.-J. and Lee, W.-L. (2003) A Hybrid Search Algorithm of Ant Colony Optimization and Genetic Algorithm Applied to Weapon-Target Assignment Problems, Lecture Notes in Computer Science 2690: 278-285.
28. Bauer, A. et al. (1999) An ant colony optimization approach for the single machine total tardiness problem, Proceedings of the 1999 Congress on Evolutionary Computation: 2 1445-1450.
29. Varela, G. N. and Sinclair, M. C. (1999) Ant colony optimization for virtual wavelength path routing and wavelength allocation, Proceedings of the 1999 Congress on Evolutionary Computation: 3 1809-1816.
30. Dicaro, G. and Dorigo, M. (1998) Mobile agents for adaptive routing, Proceedings of the Thirty-First Hawaii International Conference on System Sciences: 7 74-83.
31. Yu, I. K., Chou, C. S. and Song,Y. H. (1998) Application of the ant colony search algorithm to short-term generation scheduling problem of thermal units, Proceedings of the 1998 International Conference on Power System Technology: 1 552-556.
32. Nemes L. and Roska, T. (1995) A CNN model of oscillation and chaos in ant colonies: a case study, IEEE Transactions on Circuits and Systems I, Fundamental Theory and Applications: 42 (10) 741-745.

33. Pan Q.-K., Tasgetiren M. F. and Liang Y.-C. (2006) Minimizing total earliness and tardiness penalties with a common due date on a single-machine using a discrete particle swarm optimization algorithm, Lecture Notes in Computer Science: 4150 460-467.

5

Deterministic Search Algorithm for Sequencing and Scheduling

Seamus M. McGovern[1] and Surendra M. Gupta[2]

[1] U.S. DOT National Transportation Systems Center, Cambridge, MA 02142, (617)-494-2054, mcgoverns@volpe.dot.gov
[2] Northeastern University, Boston, MA 02115, (617)-373-4846, gupta@neu.edu

Summary. Many sequencing and scheduling problems are recognized as being NP-complete combinatorial problems. This class of problems often necessitates the use of near-optimal solution techniques including heuristics and meta-heuristics. The recently developed H-K general purpose heuristic has been successfully demonstrated on members of this class. This chapter proposes the use of this deterministic heuristic for use on complex scheduling problems. Specifically, the heuristic is applied to a newly defined problem from the field of environmentally conscious manufacturing, the goal of which is to determine a product's part removal schedule. This schedule provides the sequence of parts to be removed from a product at its end-of-life on a paced reverse-manufacturing line. It seeks not only to determine a sequence that is feasible (due to precedence constraints) but also to minimize the number of workers on the line, equalize the time level-of-effort of each, remove environmentally hazardous and high-demand parts early on, and to schedule the removal of parts with similar removal directions adjacently. In addition, the problem used in this chapter is shown to be similar to the Multiprocessor Scheduling Problem, a comparison that is further carried through the application of a complexity proof. Finally, a scheduling application of the H-K heuristic is demonstrated using an electronic product case study from the literature.

Key words: Sequencing, Scheduling, Deterministic Heuristics, Multiprocessor Scheduling, Reverse-manufacturing, H-K heuristic.

5.1 Introduction

Scheduling problems can be found in a variety of fields, while their complexity often necessitates the use of sub-optimal solutions through the use of heuristics or meta-heuristics. A well-studied scheduling problem is the Simple Assembly Line Balancing problem type I (SALB-I) and type II (SALB-II) [2]. SALB-I seeks to determine the minimum number of workstations necessary to maintain a production rate while observing any precedence constraints (i.e., typically parts cannot be installed in any sequence; order is essential).

S.M. McGovern and S.M. Gupta: *Deterministic Search Algorithm for Sequencing and Scheduling*, Studies in Computational Intelligence (SCI) **128**, 105–124 (2008)
www.springerlink.com

These problems have recently been extended to the field of environmentally conscious manufacturing, specifically reverse manufacturing.

More and more manufacturers are acting to recycle and remanufacture their post-consumer products due to new and more rigid environmental legislation, increased public awareness, and extended manufacturer responsibility. A crucial first step is disassembly. Disassembly is defined as the methodical extraction of valuable parts, subassemblies, and materials from discarded products through a series of operations. A disassembly system faces many unique challenges; for example, it has significant inventory problems because of the disparity between the demands for certain parts or subassemblies and their yield from disassembly. The flow process is also different. As opposed to the normal "convergent" flow in regular assembly environment, in disassembly the flow process is "divergent" (a single product is broken down into many subassemblies and parts). There is also a high degree of uncertainty in the structure and the quality of the returned products. The conditions of the products received are usually unknown and the reliability of the components is suspect. In addition, some parts of or materials in the product may cause pollution or may be hazardous. Unintentional release of these materials or damage to these parts is a risk during disassembly; in addition these parts or materials may require special handling, all of which can influence the utilization of the disassembly workstations. For example, an automobile slated for disassembly contains a variety of parts and materials that are dangerous to remove and/or present a hazard to the environment such as the battery, airbags, fuel, and oil. Various demand sources may also lead to complications in disassembly sequencing. The reusability of parts creates a demand for them, however, the demands and availability of the reusable parts is significantly less predicable than what is found in the assembly process. Most products contain parts that are installed (and must be removed) in different attitudes, from different areas of the main structure, or in different directions. Since any required directional changes increase the setup time for the disassembly process, it is desirable to minimize the number of directional changes in the chosen disassembly sequence. Finally, balancing the disassembly line is critical in minimizing the use of valuable resources (such as time and money) invested in disassembly and maximizing the level of automation of the disassembly process and the quality of the parts or materials recovered.

The Disassembly Line Balancing Problem (DLBP) seeks a sequence of parts for removal from an end-of-life product on a paced reverse-manufacturing line that minimizes the resources required for disassembly as well as maximizes the automation of the process in the interest of ensuring the quality of the parts or materials recovered. This chapter first mathematically models the multi-criteria DLBP, which is shown here to be similar to the familiar Multi-processor Scheduling Problem. The DLBP is then proven to be NP-complete in the strong sense, necessitating use of specialized solution techniques such as those from combinatorial optimization. Combinatorial optimization is a field that combines techniques from applied mathematics, operations research, and computer science to solve optimization problems - such as scheduling problems - over discrete structures. From this field, a newly developed deterministic search algorithm, the Hunter-Killer (H-K) general purpose heuristic, holds promise for application to scheduling problems. A case study instance of DLBP from the recent literature is solved using the H-K combinatorial optimization method.

5.2 Literature Review

Key to addressing any scheduling problem is to understand how complex or easy it is, what it shares with similar problems, and appropriate methods to obtain reasonable solutions. For these reasons, a background in optimization and algorithms is valuable. Tovey [31] provides a well-structured review of complexity, NP-hardness, NP-hardness proofs, typical NP-hard problems, the techniques of specialization, forcing, padding, and gadgets, mathematical programming versus heuristics, and other complexity classifications. Rosen [28] provides a useful text in the general area of discrete mathematics including set theory, logic, algorithms, graph theory, counting, set theory, and proofs. Papadimitriou and Steiglitz [26] is the de-facto text on combinatorial optimization, as is Garey and Johnson [4] in the area of NP-completeness. Osman and Laporte [25] provide a well-researched paper on all forms of metaheuristics, the basic concepts of each, and references to applications. A follow-on paper by Osman [24] is more compact and also more current.

A major part of scheduling manufacturing and assembly operations, the assembly line is a production line where material moves continuously at a uniform rate through a sequence of workstations where assembly work is performed. With research papers going back to the 1950s, the Assembly Line Balancing problem is well defined and fairly well understood. While possessing some significant differences from assembly line balancing, the recent development of DLBP requires that related problems be fully investigated and understood in order to better define DLBP and to obtain guidance in the search for appropriate methodologies to solve it. Gutjahr and Nemhauser [12] first described a solution to the Assembly Line Balancing problem, while Erel and Gokcen [3] developed a modified version by allowing for mixed-model lines (assembly lines used to assemble different models of the same product). Suresh *et al.* [29] first presented a genetic algorithm to provide a near-optimal solution to the Assembly Line Balancing problem. Tabu search is used in balancing assembly lines in Lapierre *et al.* [17] using SALB-I with instances from the literature (Arcus 1 and 2) and a case study from industry. Hackman *et al.* [13] proposed a branch-and-bound heuristic for the SALB-I problem. Ponnambalam *et al.* [27] compared line-balancing heuristics with a quantitative evaluation of six assembly line balancing techniques.

Many papers have discussed the different aspects of product recovery. Brennan *et al.* [1] and Gupta and Taleb [11] investigated the problems associated with disassembly planning and scheduling. Torres *et al.* [30] reported a study for non-destructive automatic disassembly of personal computers. Güngör and Gupta [6], [7], [9] presented the first introduction to disassembly line balancing, and then developed an algorithm for solving the DLBP in the presence of failures with the goal of assigning tasks to workstations in a way that probabilistically minimizes the cost of defective parts [8]. McGovern *et al.* [23] first proposed combinatorial optimization techniques for the DLBP. Güngör and Gupta [5] provide a review of environmentally conscious manufacturing and product recovery. Lambert [15] and Lambert and Gupta [16] detail a comprehensive review of disassembly sequencing.

5.3 Formulating The Reverse Production Scheduling Problem

The desired solution to a DLBP instance consists of an ordered sequence (i.e., n-tuple) of work elements (also referred to as tasks, components, or parts). For example, if a solution consisted of the eight-tuple $\langle 5, 2, 8, 1, 4, 7, 6, 3 \rangle$ then the reverse manufacturing schedule would consist of removing component 5 first (with a corresponding part removal time, or PRT_k, of PRT_5), followed by component 2, then component 8, and so on.

While different authors use a variety of definitions for the term "balanced" in reference to assembly [2] and disassembly lines, we propose the following definition [23], [20] that considers the total number of workstations NWS and the station times (i.e., the total processing time requirement in workstation j) ST_j; this definition will be used consistently throughout this chapter:

Definition 5.1 *A disassembly line is optimally balanced when the fewest possible number of workstations is needed and the variation in idle times between all workstations is minimized, while observing all constraints. This is mathematically described by*

$$\textit{Minimize } NWS$$

then

$$\textit{Minimize } [\max(ST_x) - \min(ST_y)] \forall\ x,\ y \in \{1, 2, \ldots, NWS\}.$$

The scheduling solution methodology demonstrated here addresses the multiple-criteria aspects (e.g., hazardous parts, high-demand parts, etc.) of DLBP as follows. Since line balance is the primary consideration in this chapter, additional objectives are only considered subsequently; that is, the H-K heuristic first seeks to select the best-balanced solution schedule; solutions equal in balance measure are then evaluated for hazardous part removal positions; equal balance and hazard measure solutions are evaluated for high-demand part removal positions; and equal balance, hazard measure, and high-demand part removal position solutions are evaluated for the number of direction changes. Similar to *preemptive goal programming*, this priority ranking approach was selected over a weighting scheme for its simplicity, ease in re-ranking the priorities, ease in expanding or reducing the number of priorities, due to the fact that other weighting methods can be readily addressed at a later time, and primarily to enable unencumbered *efficacy* (a method's effectiveness in finding good solutions) analysis of the combinatorial optimization methodology and problem data instance under consideration here.

Given the cycle time (the maximum time available at each workstation on the paced line) CT, a minimum value of the sum of the workstation idle times (which is also indicative of a minimum total number of workstations) I is desired and can be described by

$$I = \sum_{j=1}^{NWS} (CT - ST_j) \tag{5.1}$$

When perfect balance (i.e., all idle times equal to zero) is not achievable, either Line Efficiency (LE) or the Smoothness Index (SI) is often used as a performance evaluation tool [2]. SI rewards similar idle times at each workstation, but at the expense of allowing for a large (sub-optimal) number of workstations. This is because SI compares workstation elapsed times to the largest ST_j instead of to CT. (SI is very similar in format to the sample standard deviation from the field of statistics, but using $\max(ST_j) \mid j \in \{1, 2, ..., NWS\}$ rather than the mean of the station times.) LE rewards the minimum number of workstations but allows unlimited variance in idle times between workstations because no comparison is made between ST_js. The balancing method developed by McGovern *et al.* [23], [20] seeks to simultaneously minimize the number of workstations while ensuring that idle times at each workstation are similar, though at the expense of the generation of a nonlinear objective function. A resulting minimum numerical value is the more desirable solution, indicating both a low number of workstations and similar idle times across all workstations. The measure of balance F is represented as

$$F = \sum_{j=1}^{NWS} (CT - ST_j)^2 \tag{5.2}$$

Note that mathematically, Eq. 5.2 effectively makes Eq. 5.1 redundant due to the fact that it addresses both the variation in the idle times at each workstation as well as the total number of workstations.

Hazard, demand, and direction measures have also been developed for use in measuring the performance of a heuristic-generated solution schedule (where n represents the number of parts for removal and PS_k identifies the

k^{th} part in the solution sequence; e.g., for solution $\langle 3, 1, 2 \rangle$, $PS_2 = 1$). Hazard measure H (with binary value h_{PS_k} equal to one if part k is known to be hazardous, else zero), demand measure D (with value d_{PS_k} equal to the quantity of part k requested), and part removal direction measure R (with value r_{PS_k} corresponding to the kth part's removal direction and those directions being represented by integers) can be summarized as

$$H = \sum_{k=1}^{n} (k \cdot h_{PS_k}) \qquad h_{PS_k} = \begin{cases} 1 \text{ , hazardous} \\ 0 \text{ , otherwise} \end{cases} \tag{5.3}$$

$$D = \sum_{k=1}^{n} (k \cdot d_{PS_k}) \qquad d_{PS_k} \in \mathbf{N} \; \forall \; PS_k \tag{5.4}$$

$$R = \sum_{k=1}^{n-1} R_k \qquad R_k = \begin{cases} 1 \text{ , } r_{PS_k} \neq r_{PS_{k+1}} \\ 0 \text{ , otherwise} \end{cases} \tag{5.5}$$

McGovern and Gupta provide further explanation and details of Formulae (1) through (5) including upper and lower theoretical bounds [21].

5.4 Modeling The Problem As A Multiprocessor Scheduling Problem And Proof As Unary NP-Complete

The theory of NP-completeness is applied only to *decision problems*. While an *optimization problem* asks for a structure of a certain type that has a minimum (or maximum) cost among all such structures, associating a numerical threshold as an additional parameter and asking whether there exists a structure of the required type having cost no more (less) than that threshold creates the desired decision problem. The decision version of DLBP is NP-complete in the strong sense (i.e., unary NP-complete). This can be shown through a proof by restriction to the Multiprocessor Scheduling Problem [4]. The *Multiprocessor Scheduling Problem* (note that some authors define the problem differently; see Papadimitriou and Steiglitz [26]) is described by:

INSTANCE: A finite set A of tasks, a length $l(a) \in \mathbf{Z}^+$ for each $a \in A$, a number $m \in \mathbf{Z}^+$ of processors, and a deadline $B \in \mathbf{Z}^+$.

QUESTION: Is there a partition $A = A_1 \cup A_2 \cup ... \cup A_m$ of A into m disjoint sets such that $\max \left\{ \sum_{a \in A_i} l(a) : 1 \leq i \leq m \right\} \leq B$?

In DLBP, NWS is equivalent to m in the Multiprocessor Scheduling Problem, the set P of part removal tasks is equivalent to A, while CT is equivalent to B (note that a *polynomial time reduction* is represented by $\leq_P$ and implies that a function exists that can map the solution space of one problem to another in polynomial time).

Theorem 5.1 *The decision version of DLBP is NP-complete in the strong sense.*

INSTANCE: A finite set P of tasks; partial order $\prec$ on P; task time $PRT_k \in \mathbf{Z}^+$, hazardous part binary value $h_k \in \{0,1\}$, part demand $d_k \in \mathbf{N}$, and part removal direction $r_k \in \mathbf{Z}$ for each $k \in P$; workstation capacity $CT \in \mathbf{Z}^+$; number $NWS \in \mathbf{Z}^+$ of workstations; difference between largest and smallest idle time $V \in \mathbf{N}$; hazard measure $H \in \mathbf{N}$; demand measure $D \in \mathbf{N}$; and direction change measure $R \in \mathbf{N}$.

QUESTION: Is there a partition of P into disjoint sets $P_A, P_B, ..., P_{NWS}$ such that the sum of the sizes of the tasks in each P_X is CT or less, the difference between largest and smallest idle times is V or less, the sum of the hazardous part binary values multiplied by their sequence position is H or less, the sum of the demanded part values multiplied by their sequence position is D or less, the sum of the number of part removal direction changes is R or less, and it obeys the precedence constraints?

Proof. DLBP $\in$ NP. Given an instance, it can be verified in polynomial time if the answer is "yes" by: counting the number of disjoint sets and showing that they are NWS or less, summing the sizes of the tasks in each disjoint set and showing that they are CT or less, examining each of the disjoint sets and noting the largest and the smallest idle times then subtracting the smallest from the largest and showing that this value is V or less, summing the hazardous part binary values multiplied by their sequence position and showing this is H or less, summing the demanded part values multiplied by their sequence position and showing that this is D or less, summing the number of changes in part removal direction and showing that this is R or less, and checking that each task has no predecessors listed after it in the sequence.

Multiprocessor Scheduling $\leq_P$ DLBP. Restrict to Multiprocessor Scheduling by allowing only instances in which $V = CT$, $\prec$ is empty, and $h_x = h_y$, $d_x = d_y$, $r_x = r_y$ $\forall x, y \in P$.

Therefore, the decision version of DLBP is NP-complete in the strong sense. □

It is easy to see that the Bin-Packing problem is similar to DLBP. The Bin-Packing problem is NP-hard in the strong sense, which indicates that there is little hope in finding even a pseudo-polynomial time optimization algorithm for it.

There are normally considered to be two general categories of techniques for addressing NP-complete problems [4]. The first includes approaches that attempt to improve upon exhaustive search as much as possible with, for example, commercial math-programming software. The second allows for sub-optimal solutions in a reasonable amount of time through the use of *heuristics* (intuitive rules-of-thumb). Here, a promising deterministic search heuristic is demonstrated using a newly developed DLBP case study instance.

5.5 H-K Heuristic

5.5.1 Heuristic Search Background

Exhaustive search techniques (e.g., pure depth-first or pure breadth-first) will fail to find a solution to any but the smallest instances within any practical length of time. *Blind search*, *weak search*, *naive search*, and *uninformed search* are all terms used to refer to algorithms that use the simplest, most intuitive method of searching through a search space, whereas *informed search* algorithms use heuristics to apply knowledge about the structure of the search space. An uninformed search algorithm is one that does not take into account the specific nature of the problem. This allows uninformed searches to be implemented in general, with the same implementation able to be used in a wide range of problems. Uninformed searches include breadth-first search, depth-first search, iterative deepening, and H-K. H-K seeks to take advantage of the benefits of uninformed search while addressing the exhaustive search drawbacks of runtime growth with instance size. The independent-agent nature of this search heuristic also lends itself to solution using a grid computer system, while the heuristic's deterministic nature lends itself to use as part of a hybrid.

5.5.2 Heuristic Motivation and Introduction

Exhaustive search is optimal because it looks at every possible answer. While an optimal solution can be found, this technique is impractical for all but the simplest combinatorial problems due to the explosive growth in search time. In many physical search applications (e.g., antisubmarine warfare, search and rescue) exhaustive search is not possible due to time or sensor limitations. In these cases, it becomes practical to sample the search space and operate under the assumption that, for example, the highest point of land found during the conduct of a limited search is either is the highest point in a given search area or is reasonably near the highest point. The search technique [18] in this chapter works by sampling the exhaustive solution set; that is, search the solution space in a method similar to an exhaustive search but in a pattern that skips solutions (conceptually similar to the STEP functionality in a FOR loop as found in computer programming) to significantly minimize the search space (Fig. 5.1; the shading indicates solutions visited, the border represents the search space).

This pattern is analogous to the radar acquisition search pattern known as "spiral scan," the search and rescue pattern of the "expanding square," or the antisubmarine warfare aircraft "magnetic anomaly detector hunting circle." Once the solution is generated, the space can be further searched with additional applications of the H-K heuristic (with modifications from the previous H-K) or the best-to-date solution can be further refined by performing subsequent local searches (such as 2-opt or smaller, localized H-K searches).

Fig. 5.1: Exhaustive search space and the H-K search space and methodology.

Depending on the application, H-K can be run once, multiple times on subsequent solutions, multiple times from the same starting point using different skip measure (potentially as a multiprocessor application using parallel algorithms or as a grid computing application), multiple times from a different starting point using the same skip measure (again, potentially as a multiprocessor or grid computing application), or followed up with an H-K or another, differing local search on the best or several of the best sub-optimal solutions generated. While termination normally takes place after all sequences are generated for a given skip size, termination can also be effected based on time elapsed or once finding a solution that is within a predetermined bound. H-K can also be used as the first phase of a hybrid algorithm or to hot start another methodology (e.g., to provide the initial population in a genetic algorithm). One interesting use for H-K is application to the unusual problem where quantifying a small improvement (i.e., a greedy decision, such as would be found in ant colony optimization where the ant agents build a solution incrementally and, therefore, need to know which of the available solution elements reflects an improvement) is not possible or is not understood, or where the incremental greedy improvements may not lead to a global optima. Finally, H-K would also be useful in quickly gathering a sampling of the solution space to allow for a statistical or other study of the data (e.g., H-K could enable the determination of the approximate worst-case and best-case solutions as well as search space statistics, e.g., mean, median, and mode).

The *skip size* ψ, or more generally ψ_k (the kth element's skip measure; i.e., for the solution's third element, visit every 2nd possible task given $\psi_3 = 2$) can be as small as $\psi = 1$ or as large as $\psi = n$. Since $\psi = 1$ is equivalent to exhaustive search and $\psi = n$ generates a trivial solution (it returns only one solution, that being the data in the same sequence as it is given to H-K, that is, $PS_k = \langle 1, 2, 3, ..., n \rangle$; also, this solution is already considered by any H-K search having the data presented at least in forward order, regardless of ψ), in general all skip values can be further constrained as

$$2 \leq \psi_k \leq n - 1 \tag{5.6}$$

Depending on structural decisions, H-K can take on a variety of forms, from a classical optimization algorithm in its most basic form, to a general evolutionary algorithm with the use of multiple H-K processes, to a *biological* or *natural process algorithm* by electing random functionality. In order to demonstrate the method, in this chapter the most basic form of the H-K heuristic is used: one process (visiting the data once, in forward order), constant starting point of $PS_k = \langle 1, 1, 1, ..., 1\rangle$ (since the solution set is a permutation, there are no repeated items; therefore, the starting point is effectively $PS_k = \langle 1, 2, 3, ..., n\rangle$), constant skip type (i.e., each element in the solution sequence is skipped in the same way), constant skip size (i.e., each element in the sequence is skipped using identical ψ values), and no follow-on solution refinement.

5.5.3 The H-K Process and DLBP Application

As far as the H-K process itself, since it is a modified exhaustive search allowing for solution sampling, it searches for solutions similar to depth-first search, iteratively seeking the next permutation iteration - allowing for skips in the sequence - in lexicographic order. In the basic H-K and with, for example $\psi = 2$ and $n = 4$, the first element in the first solution would be 1, the next element position would first consider 1, but since 1 is already in the solution (in element position one PS_1), element position two would be incremented and 2 would be considered and be acceptable. This is repeated for all of the elements until the first solution (i.e., $PS_k = \langle 1, 2, 3, ..., n\rangle$) is generated. For the next solution visited, the rightmost element that is able to be incremented by ψ (equal to 2 in this example) while not exceeding n would be incremented by ψ and the remaining element positions to the right would be filled lexicographically (smallest to biggest, left to right). This process is continued to the left until the first element position is reached. The part under consideration would then be $PS_1 = 1$ which would be incremented by $\psi = 2$ and, therefore, 3 would be considered and inserted as the first element position value ($PS_1 = 3$). Since part 1 is not yet in the sequence, it would be placed in the second position ($PS_2 = 1$), part 2 in the third ($PS_3 = 2$), etc., and the process is repeated. For example, with $n = 4$, $P = \{1, 2, 3, 4\}$, and no precedence constraints, instead of considering the $4! = 24$ possible permutations, only five are considered by the single-phase H-K with $\psi = 2$ and using forward-only data: $PS_k = \langle 1, 2, 3, 4\rangle$, $PS_k = \langle 1, 4, 2, 3\rangle$, $PS_k = \langle 3, 1, 2, 4\rangle$, $PS_k = \langle 3, 1, 4, 2\rangle$, and $PS_k = \langle 3, 4, 1, 2\rangle$. With $n = 5$, $P = \{1, 2, 3, 4, 5\}$, and no precedence constraints, instead of considering the $5! = 120$ possible permutations, only 16 are considered by the single-phase H-K with $\psi = 2$ and using forward-only data as demonstrated in Fig. 5.2.

For DLBP H-K this is further modified to test the proposed sequence part addition for precedence constraints. If all possible parts for a given solution position fail these checks, the remainder of the positions are not further inspected, the procedure falls back to the previously successful solution addition, increments it by one, and continues. These processes are repeated until all allowed items have been visited in the first solution position (and by default, due to the nested nature of the search, all subsequent solution positions). All of

$$PS_k = \langle 1, 2, 3, 4, 5 \rangle$$
$$PS_k = \langle 1, 2, 5, 3, 4 \rangle$$
$$PS_k = \langle 1, 4, 2, 3, 5 \rangle$$
$$PS_k = \langle 1, 4, 5, 2, 3 \rangle$$
$$PS_k = \langle 1, 4, 5, 3, 2 \rangle$$
$$PS_k = \langle 3, 1, 2, 4, 5 \rangle$$
$$PS_k = \langle 3, 1, 4, 2, 5 \rangle$$
$$PS_k = \langle 3, 1, 4, 5, 2 \rangle$$
$$PS_k = \langle 3, 4, 1, 2, 5 \rangle$$
$$PS_k = \langle 3, 4, 1, 5, 2 \rangle$$
$$PS_k = \langle 3, 4, 5, 1, 2 \rangle$$
$$PS_k = \langle 5, 1, 2, 3, 4 \rangle$$
$$PS_k = \langle 5, 1, 4, 2, 3 \rangle$$
$$PS_k = \langle 5, 3, 1, 2, 4 \rangle$$
$$PS_k = \langle 5, 3, 1, 4, 2 \rangle$$
$$PS_k = \langle 5, 3, 4, 1, 2 \rangle$$

Fig. 5.2: DLBP H-K results at $n = 5$ and $\psi = 2$.

the parts are maintained in a tabu-type list in order to maintain each solution as a permutation (i.e., no repeats). Each iteration of the DLBP H-K generated solution is considered for feasibility. If it is ultimately feasible in its entirety, DLBP H-K then looks at each element in the solution and places that element using the Next-Fit (NF) rule (from the Bin-Packing problem application; once a bin has no space for a given item attempted to be packed into it, that bin is never used again even though a later, smaller item may appear in the list and could fit in the bin, see Hu and Shing [14]). DLBP H-K puts the element under consideration into the current workstation if it fits. If it does not fit, a new workstation is assigned and previous workstations are never again considered. Although NF does not perform as well as First-Fit, Best Fit, First-Fit-Decreasing, or Best-Fit-Decreasing when used in the general Bin-Packing problem, it is the only one of these rules that will work with a DLBP solution sequence due to the existence of precedence constraints (see McGovern and Gupta [19] for a DLBP implementation of First-Fit-Decreasing). When all of the work elements have been assigned to a workstation, the process is complete and the balance, hazard, demand, and direction measures are calculated. The best of all of the inspected solution sequences is then saved as the problem solution. Although the actual software implementation for this study consisted of a very compact recursive algorithm, in the interest of clarity, the general DLBP H-K procedure is presented here as a series of nested loops (Fig. 5.3, where ISS_k is the binary flag representing the tabu-type list; set to one if part k is in the solution sequence, and where FS is the feasible sequence binary flag; set to one if the sequence is feasible). It should also be noted that this general representation of the heuristic allows for a different skip value to be used at each solution element position (as denoted by, for example, ψ_1)

while the heuristic implemented in this chapter used the same skip value at each solution element position (i.e., $\psi_x = \psi_y \; \forall x, y \in P$).

```
Procedure DLBP_H-K {
    SET ISS_k := 0∀ k∈ P
    SET FS := 1
    FOR PS_1 := 1 to n, STEP ψ_1
        SET ISS_PS1 := 1
            FOR PS_2 := 1 to n, STEP ψ_2
                WHILE       (ISS_PS2 = = 1∨
                             PRECEDENCE_FAIL∧
                             not at n)
                        Increment PS_2 by 1
                IF      ISS_PS2 = = 1
                THEN SET FS := 0
                ELSE SET ISS_PS2 := 1
                        :
                        :
                IF      FS = =1
                            FOR PS_n := 1 to n STEP ψ_n
                                WHILE       (ISS_PSn = =1∨
                                             PRECEDENCE_FAIL∧
                                             not at n)
                                        Increment PS_n by 1

                                IF      ISS_PSn = = 0
                                THEN evaluate solution PS
                        :
                        :
            IF      FS = = 1
            THEN SET ISS_PS2 := 0
            ELSE SET FS := 1
        SET ISS_PS1 := 0
        SET FS := 1
}
```

Fig. 5.3: The DLBP H-K algorithm.

Skip size affects various measures including any efficacy measures and time complexity. The general form of the skip size-to-problem size relationship is formulated as

$$\psi_k = n - \Delta\psi_k \tag{5.7}$$

where $\Delta\psi_k$ represents the kth element's delta skip measure; difference between problem size n and skip size ψ_k (i.e., for $\Delta\psi_k = 10$ and $n = 80$, $\psi_k = 70$).

Early tests of time complexity growth with skip size suggest another technique to be used as part of H-K search. Since any values of ψ_k that are larger than the chosen skip value for a given H-K instance take significantly less processing time, considering all larger skip values should also be considered in order to increase the search space at the expense of a minimal increase in search time. In other words, H-K can be run repeatedly on a given instance using all skip values from a smallest ψ_k (selected based upon time complexity considerations) to the largest (i.e., $n-1$ per Formula (6)) without a significant time penalty. In this case, any ψ_k would be constrained as

$$n - \Delta\psi_k \leq \psi_k \leq n - 1 \text{ where } 1 \leq \Delta\psi_k \leq n - 1 \tag{5.8}$$

If this technique is used it should also be noted that multiples of ψ_k visit the same solutions; for example, for $n = 12$ and $2 \leq \psi \leq 10$, the four solutions considered by $\psi = 10$ are also visited by $\psi = 2$ and $\psi = 5$.

In terms of time complexity, the rate of growth has been observed to be exponential in the inverse of ψ. The average-case time complexity of H-K is then listed as $O(b^b)$ in skip size, where $b = 1/\psi$. Due to the nature of H-K, the number of commands executed in the software do not vary based on precedence constraints, data sequence, greedy or probabilistic decision making, improved solutions nearby, etc., so the worst case is also $O(b^b)$, as is the best case (big-Omega of $\Omega(b^b)$), and, therefore, a tight bound exists, which is $\Theta(b^b)$. When the maximum skip size is fixed, time complexity can be seen as a function of problem size n. For example, using forward and reverse data, and $1 \leq \Delta\psi \leq 10$ (and resulting skip sizes of $n-10 \leq \psi \leq n-1$), the average-case time complexity of DLBP H-K is empirically seen to be $O(n^2)$ or polynomial complexity. The deterministic, single iteration nature of H-K (as used in this chapter) also indicates that the process would be no faster than this, so it is expected that the time complexity lower bound is $\Omega(n^2)$ and, therefore, H-K appears to have an asymptotically tight bound of $\Theta(n^2)$ (when configured as described).

With any heuristic it is important to understand how well it performs when compared to other solution-generating techniques. Analysis by the authors [22] shows that the heuristic compares favorably to some familiar search methodologies, including those from the areas of probabilistic distributed intelligent agent meta-heuristics, purely deterministic searches, and hybrid techniques. Table 5.1 shows comparisons with a genetic algorithm (GA), an ant colony optimization (ACO) meta-heuristic, a greedy algorithm, a greedy/adjacent element hill climbing hybrid, and a greedy/2-optimal (2-Opt) hybrid, as well as with exhaustive search. The table shows each of the combinatorial optimization methodologies' calculated number of workstations, balance, hazard, demand, and part removal direction *efficacy indices* (EI_x – where x is some metric under consideration e.g., H – is the ratio of the difference between a calculated measure and its worst-case measure, to the difference between the best-case measure and the worst-case measure) sample means; regression model average-case experimentally determined time complexity $T(n)$; and associated asymptotic upper bound (experimentally determined average case using a DLBP benchmark data set [22]).

The shading provides a quick reference to performance, with darker shades indicating worsening performance. While exhaustive search is included in the table, none of its row elements are considered for shading since its purpose in

Table 5.1: Summary of quantitative measures for various methodologies when compared with the H-K general purpose heuristic.

	$\overline{EI}_{NWS}$	$\overline{EI}_{F(norm)}$	$\overline{EI}_H$	$\overline{EI}_D$	$\overline{EI}_R$	$T(n)$	big-Oh
Exhaustive	100%	100%	100%	100%	100%	$1.199n!$	$O(n!)$
GA	97%	94%	84%	78%	49%	$0.033n$	$O(n)$
ACO	95%	90%	74%	51%	13%	$0.001n^3$	$O(n^3)$
Greedy	95%	83%	100%	64%	16%	$7\times10^{-8}n^2$	$O(n^2)$
Greedy/AEHC	95%	87%	100%	65%	48%	$3\times10^{-7}n^2$	$O(n^2)$
Greedy/2-Opt	96%	93%	100%	74%	56%	$5\times10^{-5}n^2$	$O(n^2)$
H-K	96%	92%	90%	49%	20%	$0.003n^2$	$O(n^2)$

the table is as the benchmark. Note that the balance measure sample mean efficacy index is based on a *normalized* balance (using the square root of F) in the interest of providing a more appropriate, comparable, and consistent scale. The time complexity regression model column only contains the highest-order base in the polynomial and its coefficient in the interest of space and since these are the portions of the regression model that result in the greatest runtime growth with instance size.

H-K's runtime performance can be improved by increasing skip size, while the opposite (i.e., decreasing skip size) and running different versions of the data (including running the data in forward and reverse order), would be expected to increase solution efficacy. Note that, as a recently developed methodology, other techniques as proposed in Section 5.5.2 may decrease runtime and/or increase solution efficacy of the heuristic as well.

5.6 Electronic Product Instance

As a point of reference, an automobile assembly plant has a rather short cycle time of $CT = 60$ seconds while at the other extreme is a general aviation aircraft manufacturer that provides $3\frac{1}{2}$ hours of cycle time. The number of components can run into the thousands with even an efficient contemporary automobile design (e.g., 2005 Ford Mustang) running over 4,000 individual parts and taking a total of 20 hours to build.

Since DLBP is a recent problem, very few problem instances exist to study the performance of different heuristic solutions. One data set includes the cellular telephone problem instance, which as developed, includes a list of parts and associated part removal times, hazardous content, demand, and removal direction, as well as precedence relationships.

Gupta *et al.* [10] provided the basis for the cellular telephone DLBP instance. The growth of cellular telephone use, and rapid changes in technology and features, has prompted the entry of new models on a regular basis while, according to a 2005 Collective Good International report, one-hundred-million cellular telephones are discarded each year. Unwanted cell phones typically end up in landfills and usually contain numerous hazardous parts that may contain mercury, cadmium, arsenic, zinc, nickel, lead, gallium arsenide, and beryllium, any of which can pose a threat to the environment. As reported in 2005, companies that refurbished cellular telephones typically paid consumers \$2 to \$20 per phone for the more popular Motorola and Nokia cellular telephones while cellular telephone recyclers (that may only extract precious metals such as gold from the circuit boards) paid between \$1 and \$6 per phone.

Gupta *et al.* [10] selected a 2001 model year Samsung SCH-3500 cell phone for disassembly analysis. The result is an appropriate, real-world instance consisting of $n = 25$ components having several precedence relationships. The associated search space is massive, with 25! (1.55×10^{25}) possible solutions. The data set includes a paced disassembly line operating at a speed which allows $CT = 18$ seconds per workstation. Collected data on the SCH-3500 is listed in Table 5.2. Demand was estimated based on part value and/or recycling value; part removal times and precedence relationships (Fig. 5.4) were determined experimentally (part removal times were repeatedly collected until a consistent part removal performance was attained).

1 → none
2 → none
3 → 1 & 2
4 → none
5 → none
6 → 2
7 → 2
8 → 2
9 → 3
10 → 4 & 5
11 → 10
12 → 11
13 → 6 & 7 & 8 & 9
14 → 6 & 7 & 8 & 9
15 → 6 & 7 & 8 & 9
16 → 6 & 7 & 8 & 9
17 → 13 & 14
18 → 15
19 → 13 & 14 & 16 & 18
20 → 17
21 → 17
22 → 21
23 → 16 & 22
24 → 19 & 23
25 → 21

Fig. 5.4: Cellular telephone precedence relationships (numbers represent task identification; arrow interpreted as "requires completion of").

5.7 Numerical Analysis

For consistency in software architecture and data structures, the authors wrote all of the search algorithm code; no off-the-shelf software was used. The computer program was written in C++ and run on a 1.6GHz PM x86-family workstation. After engineering, the program was first investigated on a variety of test cases for verification and validation purposes.

Table 5.2: Knowledge base of the cellular telephone instance.

Task	Part Removal Description	Time	Hazardous	Demand	Direction
1	Antenna	3	Yes	4	+y
2	Battery	2	Yes	7	-y
3	Antenna guide	3	No	1	-z
4	Bolt (type 1) a	10	No	1	-z
5	Bolt (type 1) b	10	No	1	-z
6	Bolt (type 2) 1	15	No	1	-z
7	Bolt (type 2) 2	15	No	1	-z
8	Bolt (type 2) 3	15	No	1	-z
9	Bolt (type 2) 4	15	No	1	-z
10	Clip	2	No	2	+z
11	Rubber seal	2	No	1	+z
12	Speaker	2	Yes	4	+z
13	White cable	2	No	1	-z
14	Red/blue cable	2	No	1	+y
15	Orange cable	2	No	1	+x
16	Metal top	2	No	1	+y
17	Front cover	2	No	2	+z
18	Back cover	3	No	2	-z
19	Circuit board	18	Yes	8	-z
20	Plastic screen	5	No	1	+z
21	Keyboard	1	No	4	+z
22	LCD	5	No	6	+z
23	Sub-keyboard	15	Yes	7	+z
24	Internal IC	2	No	1	+z
25	Microphone	2	Yes	4	+z

The instance was run (forward-only data; i.e., in the order presented in Table 5.2) with $\Delta\psi$ varying from 1 to 10 (e.g., at $n = 25$, skip would vary as $15 \leq \psi \leq 24$) with the best solution from these searches kept. Note that for other, smaller data sets the software was set up so that it would not attempt any skip size smaller than $\psi = 3$ to avoid exhaustive or near-exhaustive searches (which would result in unrealistically large search times on these small data sets). This data set was not run with the data in reverse order. Although H-K is deterministic (and hence always makes the same number of calculations when run on a given instance), the H-K combinatorial optimization computer software was run three times to obtain an average of the computation time in order to accommodate variation in computer operating system overhead (i.e., background) processing.

The H-K combinatorial optimization technique quickly provided a solution to this complex disassembly scheduling case study instance and the results are shown in Table 5.3 (shading is used to distinguish between workstations). This solution results in $NWS = 11$, $F = 399$, $H = 82$, $D = 940$, and $R = 10$ and took 0.01 seconds on average. While the optimal solution for this problem has not yet been determined (primarily due to the search space size of 10^{25}) a heuristic designed specifically for this instance was able to obtain $NWS = 9$ and $F = 9$ [10].

Table 5.3: H-K solution using the cellular telephone instance and $15 \leq \psi \leq 24$ (forward only).

Part ID	1	2	3	4	5	6	7	8	9	10	11	12	13	14	15	16	17	18	19	20	21	22	23	24	25
PRT	3	2	3	10	10	15	15	15	15	2	2	2	2	2	2	2	2	3	18	5	1	5	15	2	2
Workstation	1	1	1	1	2	3	4	5	6	6	7	7	7	7	7	7	7	7	8	9	9	9	10	10	11
Hazardous	1	1	0	0	0	0	0	0	0	0	0	1	0	0	0	0	0	0	1	0	0	0	1	0	1
Demand	4	7	1	1	1	1	1	1	1	2	1	4	1	1	1	1	2	2	8	1	4	6	7	1	4
Direction	2	3	5	5	5	5	5	5	5	4	4	4	5	2	0	2	4	5	5	4	4	4	4	4	4

While H-K is used here with a single instance in the interest of demonstrating the implementation of this method for use in sequencing and scheduling problems, the application of H-K to other instances can be seen in references [18] and [22].

5.8 Future Research

Directions for future research can be described as follows:

- It may be of interest to vary the multi-criteria ordering of the objectives; two possibilities include: a re-ordering of the objectives based on expert domain knowledge, and a comparison of all permutations of the objectives in search of patterns or unexpected performance improvements or decreases.
- While the multiple-criteria decision making approach used here made use of preemptive goal programming, many other methodologies are available and should be considered to decrease processing time, for an overall or selective efficacy improvement, or to examine other promising methods including weighting schemes.
- Multiple versions of the H-K heuristic can be run separately on stand-alone machines searching different areas of an instance's search space; this is especially timely with the advent of grid computing and as such, a study of the use of grid computing (or some similar hardware and software technology) using H-K may be in order.
- It may be of interest to make use of the promise of H-K in generating uninformed solutions from throughout a search space through the use of H-K as a first phase in a hybrid search (e.g., with follow-on solution refinement provided by a hill climbing search [19]) or to hot start a genetic algorithm.

This concludes some suggestions for future research. In addition to the items listed above, any further developments and applications of recent methodologies to the H-K algorithm or the Disassembly Line Balancing Problem that will help to extend the research in these areas may be appropriate.

5.9 Conclusions

An uninformed deterministic search approach to NP-complete scheduling problems and an application to the multiple-objective DLBP was presented in this chapter. The H-K general-purpose heuristic rapidly provides a feasible solution schedule for the DLBP using a search technique that samples the entire solution space. The DLBP H-K heuristic provides a near-optimal minimum number of workstations, with the level of optimality increasing with the number of constraints. It generates a feasible sequence with optimal or near-optimal balance, hazardous materials, demand, and part removal direction measures. The H-K general-purpose heuristic appears well suited to the scheduling problem format as well as for the solution of problems with nonlinear objectives. Also, H-K provides a natural application to grid or cluster computing where the many networked computers would each work on a portion of the farmed-out search space and with each exponential increase in processing speed (attributed to some combination of increased computer clock speed and/or increases in the number of computers in the grid) the skip size could be decremented, resulting in improvements in solutions with no increase in perceived runtime.

References

1. Brennan L., Gupta S. M., and Taleb K. N. (1994) Operations Planning Issues in an Assembly/Disassembly Environment. International Journal of Operations and Production Management, 14(9): 57–67
2. Elsayed E. A. and Boucher T. O. (1994) Analysis and Control of Production Systems. Prentice Hall, Upper Saddle River, New Jersey
3. Erel E. and Gokcen H. (1964) Shortest-Route Formulation of Mixed-Model Assembly Line Balancing Problem. Management Science, 11(2), 308-315
4. Garey M. and Johnson D. (1979) Computers and Intractability: A Guide to the Theory of NP Completeness. W. H. Freeman and Company, San Francisco, California
5. Gungor A. and Gupta S. M. (1999) A Systematic Solution Approach to the Disassembly Line Balancing Problem. In Proceedings of the 25th International Conference on Computers and Industrial Engineering, New Orleans, Louisiana, 70–73
6. Gungor A. and Gupta S. M. (1999) Disassembly Line Balancing. In Proceedings of the 1999 Annual Meeting of the Northeast Decision Sciences Institute, Newport, Rhode Island, 193–195

7. Gungor A. and Gupta S. M. (1999) Issues in Environmentally Conscious Manufacturing and Product Recovery: A Survey. Computers and Industrial Engineering, 36(4), 811–853
8. Gungor A. and Gupta S. M. (2001) A Solution Approach to the Disassembly Line Problem in the Presence of Task Failures. International Journal of Production Research, 39(7), 1427–1467
9. Gungor A. and Gupta S. M. (2002) Disassembly Line in Product Recovery. International Journal of Production Research, 40(11), 2569–2589
10. Gupta S. M., Evren E., and McGovern S. M. (2004) Disassembly Sequencing Problem: A Case Study of a Cell Phone. In Proceedings of the 2004 SPIE International Conference on Environmentally Conscious Manufacturing IV, Philadelphia, Pennsylvania, 43–52
11. Gupta S. M. and Taleb K. (1994) Scheduling Disassembly. International Journal of Production Research, 32(8), 1857–1866
12. Gutjahr A. L. and Nemhauser G. L. (1964) An Algorithm for the Line Balancing Problem. Management Science, 11(2), 308–315
13. Hackman S. T., Magazine, M. J., and Wee, T. S. (1989) Fast, Effective Algorithms for Simple Assembly Line Balancing Problems. Operations Research, 37(6), 916–924
14. Hu T. C. and Shing, M. T. (2002) Combinatorial Algorithms, Dover Publications, Mineola, New York
15. Lambert, A. J. D. (2003) Disassembly Sequencing: A Survey. International Journal of Production Research, 41(16), 3721–3759
16. Lambert, A. J. D. and Gupta, S. M. (2005) Disassembly Modeling for Assembly, Maintenance, Reuse, and Recycling, CRC Press (Taylor & Francis), Boca Raton, Florida
17. Lapierre S. D., Ruiz, A., and Soriano, P. (2006) Balancing Assembly Lines with Tabu Search. European Journal of Operational Research, 168(3), 826–837
18. McGovern, S. M. and Gupta, S. M. (2004) Demanufacturing Strategy Based Upon Meta-heuristics. In Proceedings of the 2004 Industrial Engineering Research Conference, Houston, Texas, CD–ROM
19. McGovern, S. M. and Gupta, S. M. (2005) Local Search Heuristics and Greedy Algorithm for Balancing the Disassembly Line. The International Journal of Operations and Quantitative Management, 11(2), 91–114
20. McGovern, S. M. and Gupta, S. M. (2006) Ant Colony Optimization for Disassembly Sequencing with Multiple Objectives. The International Journal of Advanced Manufacturing Technology, 30(5–6), 481–496
21. McGovern, S. M. and Gupta, S. M. (2007) A Balancing Method and Genetic Algorithm for Disassembly Line Balancing. European Journal of Operational Research, 179(3), 692–708
22. McGovern, S. M. and Gupta, S. M. (2007) Combinatorial Optimization Analysis of the Unary NP–Complete Disassembly Line Balancing Problem. International Journal of Production Research, 45(18–19), 4485–4511
23. McGovern, S. M., Gupta, S. M., and Kamarthi, S. V. (2003) Solving Disassembly Sequence Planning Problems Using Combinatorial Optimization. In Proceedings of the 2003 Northeast Decision Sciences Institute Conference, Providence, Rhode Island, 178–180
24. Osman, I. H. (2004) Meta-heuristics: Models, Design and Analysis. In Proceedings of the Fifth Asia Pacific Industrial Engineering and Management Systems Conference, Gold Coast, Australia, 1.2.1–1.2.16

25. Osman, I. H. and Laporte, G. (1996) Meta-heuristics: A Bibliography. Annals of Operations Research, 63, 513–623
26. Papadimitriou, C. H. and Steiglitz, K. (1998) Combinatorial Optimization: Algorithms and Complexity, Dover Publications, Mineola, New York
27. Ponnambalam, S. G., Aravindan, P., and Naidu, G. M. (1999) A Comparative Evaluation of Assembly Line Balancing Heuristics. The International Journal of Advanced Manufacturing Technology, 15, 577–586
28. Rosen, K. H. Discrete Mathematics and Its Applications, McGraw-Hill, Boston, Massachusetts (1999)
29. Suresh, G., Vinod, V. V., and Sahu, S. (1996) A Genetic Algorithm for Assembly Line Balancing. Production Planning and Control, 7(1), 38–46
30. Torres, F., Gil, P., Puente, S. T., Pomares, J., and Aracil, R. (2004) Automatic PC Disassembly for Component Recovery. International Journal of Advanced Manufacturing Technology, 23(1–2), 39–46
31. Tovey, C. A. (2002) Tutorial on Computational Complexity, Interfaces, 32(3), 30–61

6

Sequential and Parallel Variable Neighborhood Search Algorithms for Job Shop Scheduling

Mehmet E. Aydin[1] and Mehmet Sevkli[2]

[1] University of Bedfordshire, Dept. of Computing and Information Systems, Luton, UK mehmet.aydin@beds.ac.uk

[2] Fatih University, Dept. of Industrial Engineering, Buyukcekmece, Istanbul, Turkey msevkli@fatih.edu.tr

Summary. Variable Neighborhood Search (VNS) is a recently invented meta-heuristic to use in solving combinatorial optimization problems in which a systematic change of neighborhood with a local search is carried out. However, as happens with other meta-heuristics, it sometimes takes long time to reach useful solutions whilst solving some sort of hard and large scale combinatorial problems such as job shop scheduling. One of the most considerable way out to overcome this shortcoming is to parallelize VNS implementations. In this chapter, firstly, a number of variable neighborhood search algorithms are examined for Job Shop Scheduling (JSS) problems and then four different parallelization policies are tackled as part of efficiency investigation for parallel VNS algorithms. The experimentation reveals the performance of various VNS algorithms and the efficiency of policies to follow in parallelization. In the end, a policy based on unidirectional-ring topology is found most efficient.

Key words: Variable Neighborhood Search, Job Shop Scheduling, Parallelization, Unidirectional-ring Topology.

6.1 Introduction

Meta-heuristics are general frameworks for designing search-based procedures to solve optimization problems. The heuristic search procedures are carried out via neighborhood structures (NS), which generally transform solutions from one state to another throughout the solution space. Variable neighborhood search (VNS) is a recent meta-heuristic for solving combinatorial and global optimization problems whose basic idea is to systematically change the neighborhood by simply switching one NS to another whilst conducting the search. The idea is based upon a simple principle: change the neighborhood structure when the search is trapped on a local minimum, which is very likely in most of combinatorial and/or multi-model numerical optimization problems. Especially, as search space grows fast with growing problem sizes, the

M.E. Aydin and M. Sevkli: *Sequential and Parallel Variable Neighborhood Search Algorithms for Job Shop Scheduling*, Studies in Computational Intelligence (SCI) **128**, 125–144 (2008)
www.springerlink.com

likelihood of being trapped in local minima becomes inevitable. The main focus of the research in this field is to recover trapped search or to put effort for preventing on-line. VNS offers a multiple neighborhood structure with which one recovers the solutions trapped via the others. The main idea here is to choose heuristics/neighbourhood structures complementary to each other.

The main limitation with VNS implementations arises in recruiting the neighborhood structures, which restrain the search with spinning in some particular regions of the space. The perturbation facility does not sometimes help to overcome this matter since jumping to another region of the search space sometimes becomes almost impossible. The most effective healing option appears to be an efficient parallelism with rigorously developed VNS implementations since a parallel meta-heuristic allows running concurrent multiple searches over the domain of the problem; each to be carried out at different regions. That accelerates the search to cut down the computational time and provides with a better solution quality. Two main parallelization approaches have appeared within the literature tackling p-Median problem. Garcia-Lopez et al. [19] proposed a centrally coordinated synchronous policy while Crainic et al. [11] implemented an asynchronous policy to coordinate the processors. Apart form these two methods, we have not come across any other publications discussing different approaches. The main gap appeared to be whether or not a non-centrally coordinated parallelism would relieve to improve the performance with respect to both solution quality and computational time. For this purpose, we carried out this study examining two above-mentioned central parallelization approaches and two non-central methods.

Job Shop Scheduling (JSS), which is an NP-hard [20] problem, is the other part of this research. It is clear that small size instances of the JSS problems can be solved in reasonable computational time by exact algorithms such as branch-and-bound approach [4, 12]. However, when the problem size increases, the computational time of the exact methods grows exponentially. Therefore, the recent research on JSS problems is focused on heuristic algorithms such as Simulated Annealing (SA) [5, 26, 31], Genetic Algorithms (GA) [9, 16, 21, 22], Taboo Search (TS) [15, 28, 30], Ant Colony Optimization (ACO) [10, 14], Shifting Bottleneck Procedure [1, 24], Guided Local Search [7], Parallel Greedy Randomized Adaptive Search Procedure (GRASP) [2] and Constraint Propagation [17]. A comprehensive survey of the JSS problem can be found in [25].

In this chapter, we examine four parallelization policies with a particularly fine-tuned implementation of VNS for JSS problems. The main idea here is to test two previously studied central and two non-central parallelization methods/policies for JSS problems. The algorithms have been tested with several hard benchmark instances of JSS problems. The rest of the chapter is organized as follows. The second section provides background information of variable neighborhood search, while the third section presents parallelization policies for variable neighborhood search. The fourth section is about JSS problems, the way how to represent the problems and the neighborhood structures employed. The last subsection of the fourth section provides the implemented VNS variations. The extensively carried out experiments are reported and discussed in the fifth section while the sixth section provides with the conclusions.

6.2 Variable Neighborhood Search

Variable neighborhood search (VNS) is a recent meta-heuristic approach invented for problem solving in an easier way. It is one of the very well-known local search methods [27], takes more attention day-by-day, because of its ease of use and success in solving combinatorial optimization problems [18, 23]. Basically, a local search algorithm carries out exploration within a limited region of the whole search space. That only facilitates to find better solutions without going further investigation. The VNS is a simple and effective search procedure that proceeds to a systematic change of neighborhood. An ordinary VNS algorithm starts with an initial solution, xϵ S, where S is the whole set of search space, and manipulates it through a two-nested loop in which the core one alters and explores via two main functions so called *shake* and *local search*. The outer loop works as a refresher reiterating the inner loop, while the inner loop carries out the major search. *local search* explores for an improved solution within the local neighborhood, whilst shake diversifies the solution by switching to another local neighborhood. The inner loop iterates as long as it keeps improving the solutions, where an integer, k, controls the length of the loop. Once an inner loop is completed, the outer loop reiterates until the termination condition is met. Since the complementariness of neighborhood functions is the key idea behind VNS, the neighborhood structure should be chosen very rigorously so as to achieve an efficient VNS.

In order to develop an effective VNS algorithm, one needs two kinds of neighborhood functions; $N_k^s(x)$ and $N_l^{LS}(x)$ resulting each with a particular neighborhood structure, where $N_k^s(x)$ and $N_l^{LS}(x)$ denote neighborhood functions for *shake* and *local search* functions, respectively. The neighborhood structures used may be more than one for each function (shake and local search) so as to achieve a valuable neighborhood change. For that purpose, the indices, k and l, are to be used for shake and local search functions, respectively, in order to ease switching from one to another neighborhood. Obviously, both indices have upper boundaries, which are denoted with k_{max} and l_{max}. Hence, $1 \leq k \leq k_{max}$ and $1 \leq l \leq l_{max}$ are the ranges identified for each indices.

The VNS comprises the following main steps:

1. **Initialization:** Find an initial solution x.
2. **Repeat** the following steps until the stopping condition is met:

   (a) **Shake Procedure:** Generate solution $x' \epsilon N_k^s(x)$? shake function.
   (b) **Local Search:** Apply local search to solution x' to produce $x" \epsilon N_l^{LS}(x')$?

(c) **Improve or not:** If x" is better than x, do $x \leftarrow x$"

Once all the parameters initialized, the algorithm starts with a randomly selected initial state and then repeats the following steps. In step 2(a), the *shake* function generates new states, and passes it to the *local search* function in step 2(b). Once it completes, the result is evaluated in step 2(c), where the result is promoted if it is better, otherwise the *shake* function generates another state for the next iteration.

The *shake* function works to switch to another region of the search space so as to carry out a new local search there, as *shake* functions are to diversify the exploration. In this study, the *shake* function does not work in the sense of variable neighborhood descent (VSD) algorithm, but is designed to conduct a couple of successive random moves with aforementioned NSs. For instance, given a state of x^* is operated with NS_1 to obtain x', which is to be operated with NS_2 consecutively. Finally, NS_1 re-operates on the outcome of NS_2, say x", to obtain x.

6.3 Parallelization of VNS

The main motivation behind parallelism is to improve the performance of the algorithms with respect to primarily the computational time and secondarily the efficiency for the quality of solution. Similar to the other parallel meta-heuristics, VNS can be parallelized mainly in two ways; one is by decomposing the domain (data and/or the problem space) if it is separable, and the other is by parallelizing the algorithm itself in which the unnecessarily sequential parts of the procedures can be appropriately organized to be executed on parallel computational resources. The former way of parallelization is not preferable as due to the restrictions while the latter one has been examined several times as reported in the literature [4,6]. Besides the parallelism by data, the easiest way to parallelize VNS is to execute multiple independent runs of the same algorithm concurrently, and collect the results in the end. However, such parallelism does not usually provide with desirable performance. Hence, a better way of parallelism for VNS could be hidden in more complex methods in which the *shake* and *local search* stages of the algorithm can be re-arranged appropriately and work either synchronously or asynchronously. We came across with mainly two parallel VNS algorithms in the literature. Garcia-Lopez et al. [19] proposed with a synchronously organized client-server style of parallelism, where the clients run identical algorithms packed of *shake* and *local search* procedures and collect the best of the whole system every iteration. Crainic et al. [11] discusses a parallelism of an asynchronously organized client-server system. In this chapter, we examined 4 approaches including the two approaches acknowledged in [11,19] tackling with job shop scheduling problems. The main idea is to compare the performance of each way of parallelization so as to identify the strengths and weaknesses subject to the conditions undertaken.

VNS algorithms can be parallelized in a way that an identical local search is assigned to each processor and the intermediate results are collected for comparison by a central unit to select the solution to go for further iterations. Another way is to coordinate the processors in peer-to-peer communication fashions. Below is the parallelization methods considered in this study. All considered policies are based on multiple independent runs, where each processor runs a single iteration of a sequential VNS procedure, which comprises a pair of *shake* and *local search* functions. We call a completed execution of this procedure on a processor as a "run" and one completed set of runs across all processors as "generation" in the rest of this chapter.

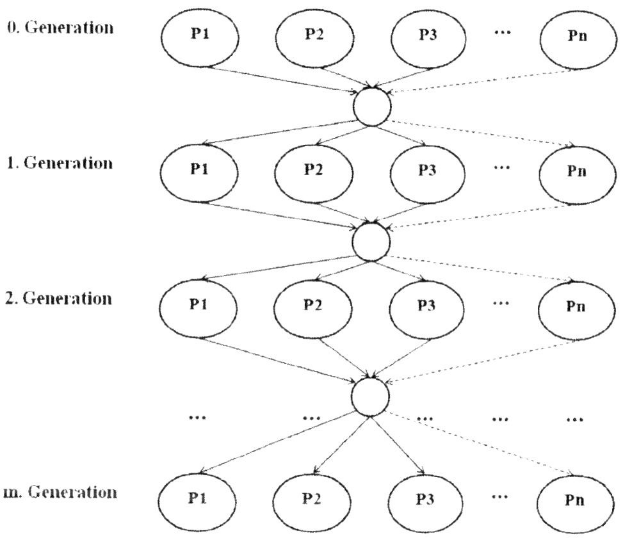

Fig. 6.1: Synchronous parallelization policy.

1. Synchronized parallelism is the framework proposed by Garcia-Lopez et al. [19], which implies centralism in coordinating the whole system, where each processor is equipped with a pair of *shake* and *local search* heuristics and a central unit serves for collaboration among the processors. Each particular processor executes the algorithm in which the solution supplied is shaken first and then passed into the local search algorithm to complete one run. Once each processor completes a run, it reports the result found to the central unit, and then the central unit compares the results collected from all processors for the best one to be considered as the initial solution of the next runs on the processors. Once the central unit identifies the best of the processors, and then assigns it to every processor for the next generation. This policy has been sketched in Fig. 6.1, where each iteration

ends up with reporting their best to the central unit, and the central unit identifies the best of the system and feeds the processors with it for the next generation.

2. Asynchronous parallelism is another centrally coordinated method proposed mainly by Crainic et al. [11]. The only difference in between this approach and the previous one is the way of coordination, where this approach does not propose a synchronous coordination while the previous one does. This approach does not require collecting the results allows each processor reporting the result and choosing the best of the time being for the next run regardless of whether the all processors reported or not. Once the central unit receives a new solution from any processor, it compares the best known and the new solution to update the new best for further generations. That makes this approach advantageous over the synchronous one as it allows starting with various initial solutions during intermediate generations, while the synchronous approach allows only one solution for every processor in every generation. Therefore, this method diversifies more. The asynchronous idea is presented in Fig. 6.2, where every processor is allowed to receive the best solutions from every others with an asynchronous scheme in which once a processor completes a generational search, then reports its best to the central unit, and requests the best of the system at that time regardless of waiting all others.

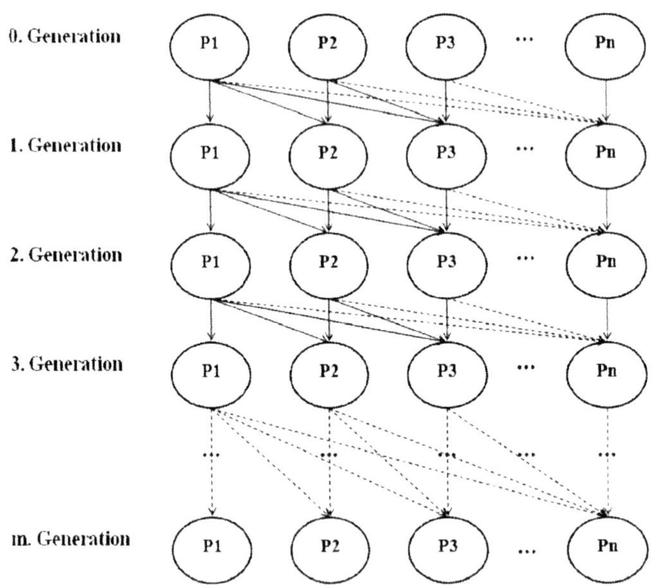

Fig. 6.2: Asynchronous parallelization policy.

3. Non-central parallelism via unidirectional-ring topology is the third method in which VNS is parallelized in this work. It implies that a set of processors are organized in a unidirectional-ring topology, where the processors named with descending numbers are meant to be organized in a scheme of unidirectional-ring order in which the succeeding numbers are to be adjacent of each other and the first processor is adopted to be the succeeding processor of the last one. The idea is to feed a particular processor for the next generation with the outcome of the previous processor while the first processor is fed by the last processor. Obviously, there is no central unit employed in this method. A particular processor launches the execution, and collects the final result in the end. Apart from that, there is no need for coordination, either, as the method imposes self-coordination. This method always provides with a number of different ongoing runs with different initial solutions, and diversifying with exchanging intermediate states. This idea is shown in Fig. 6.3, where one processor is allowed to receive the best of only one and to feed only one for the next generation.

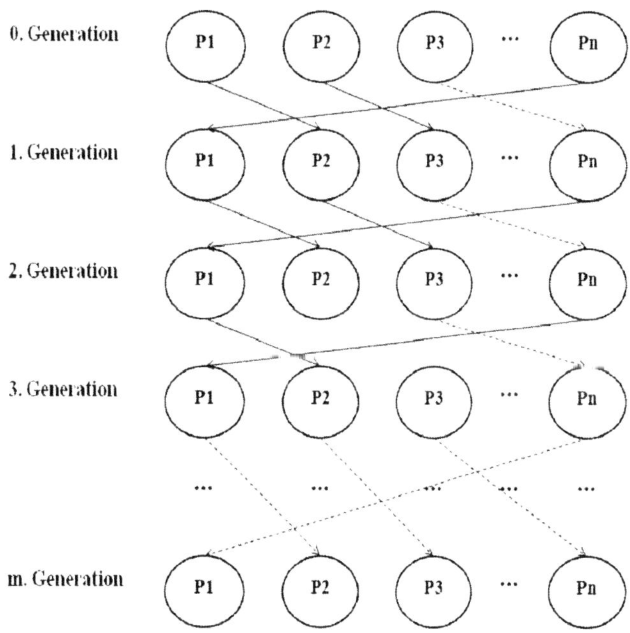

Fig. 6.3: Uni-directional ring policy for parallelization.

4. Non-central parallelism via mesh topology is another peer-to-peer organization of processors. The processors are labelled with the same scheme as the unidirectional-ring topology is done through. Each particular processor receives three different solutions resulted from the processor itself

as well as the previous and the next adjacent ones. It can be described as bi-directional ring topology since each processor receives messages and/or correspondence from both of its neighbors. Therefore, the ring organization works in both directions. The processor selects the best of three solutions for the next generation. There is no need for central coordination in this method, either. The mesh policy is indicated in Fig. 6.4.

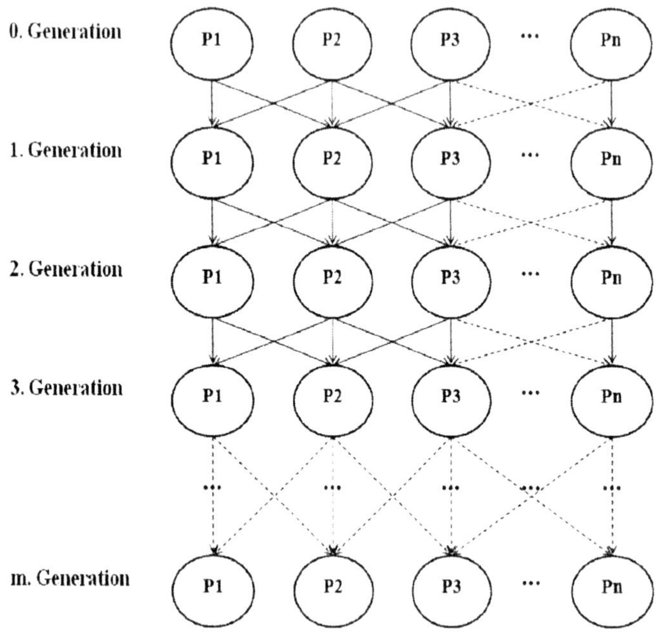

Fig. 6.4: Mesh policy for parallelization.

As expected, the parallelism helps saving computational time and diversifies the solutions for further search steps. The first aforementioned policy for parallelism enforces all processors to go with the same initial solutions with a sole expectation that the randomness in the search may diversify the solutions. Therefore, the performance of such a parallelism is expected to help with respect to computational time. The other three policies mentioned above have more fruitfulness for diversification of solutions besides saving in computational time as they offer running multiple independent runs with different initial solutions. The asynchronous central approach has the lowest likelihood for various initial solutions while the unidirectional-ring topology has the highest likelihood. The mesh topology looks in an intermediate position. Hence, the unidirectional-ring topology is expected to provide with a better quality of solution besides saving computational time since it carries out a simultaneous search in multiple regions of the search space, while the other methods conducts search in fewer regions concurrently. All methods are set to complete the same number of iterations (generation and moves). Note that there is no selection rule applicable in exchanging intermediate solutions. That means the new solutions received by processors are always replaced with the older ones regardless of their level of solution quality.

6.4 VNS Algorithms for Job Shop Scheduling

The implementation of VNS is provided in the following subsections, after introducing fundamentals of job shop scheduling problems, the representation adopted for this study and the alteration (neighborhood) structure for search.

6.4.1 Job shop scheduling problems

Job Shop Scheduling (JSS) problems have been studied for a long time. Due to their NP-Hard nature, this type of problem has never been dropped from scientific research and kept becoming a popular testbed for meta-heuristics.

The problem is comprised a set of jobs (J) to be processed on a set of machines (M) subject to a number of technological constraints. Each job consists of m operations, $O_j = \{o_{1j}, o_{2j}, ..., o_{mj}\}$, each operation must be processed on a particular machine, and there is only one operation of each job to be processed on each machine. There is a predefined order of the operations of each particular job in which each operation has to be processed after its predecessor (PJ_j) and before its successor (SJ_j). In the end of the whole schedule, each machine completes processing n operations in an order that is determined during the scheduling time, although there is no such order initially. Therefore, each operation processed on machine M_i has a predecessor (PM_i) and a successor (SM_i). A machine can process only one operation at a time. There are no set-up times, no release dates and no due dates.

Each operation has a processing time (p_{ij}) on related machine starting at the time of r_{ij}. The completion time of operation c_{ij} is therefore: $c_{ij} = r_{ij} + p_{ij}$, where $i = (1, 2, ..., m), j = (1, 2, ..., n)$ and $r_{ij} = max(c_{iPJ_j}, c_{PM_ij})$. Machines and jobs have particular completion times, which are denoted and identified as: $c_{Mi} = c_{in}$ and $c_{Jj} = c_{in}$ where c_{in} and c_{jm} are the completion time of the last (n^{th}) operation on i^{th} machine and the completion time of the last (m^{th}) operation of jth job, respectively. The overall objective is to minimize the completion time of the whole schedule (makespan), which is the maximum of machines' completion times, $C_{max} = max(CM_1, ..., CM_m)$. The representation is done via a disjunctive graph, as it is widely used.

6.4.2 Problem representation

Schedules are represented in a set of integers, where each stands for a particular operation of a problem of n jobs, m machines. The integers are job labels, and repeated m times within a particular state, where each represent the last non-completed operation of corresponding job. This way of representation prevents infeasibility, and always provide with a feasible active schedule. For instance, we are given a state of [2 1 2 2 1 3 1 3 3], where 1, 2, 3 represents j_1, j_2, j_3 respectively. Obviously, there are totally 9 operations, but, 3 different integers, each is repeated 3 times. For instance, the first integer, 2, represents the first operation of the second job, o_{21}, to be processed first on corresponding machine. Likewise, the second integer, 1, represents the first operation of the first job, o_{11}. Thus, the set of [2 1 2 2 1 3 1 3 3] is understood as $[o_{21}, o_{11}, o_{22}, o_{23}, o_{12}, o_{31}, o_{13}, o_{32}, o_{33}]$ where o_{ij}, stands for the ith, operation of jth job. More details on this representation method can be found in [13].

6.4.3 Neighborhood Structure

The neighborhood structure with which the neighboring solutions are determined to move to is one of the key elements of meta-heuristics that use neighborhood structures. That is why, the performance of those meta-heuristic algorithms significantly depends on the efficiency of the neighborhood structure. The following two neighborhood structures are employed in this study:

a) **Exchange** is a function used to explore in which any two randomly selected operations are simply swapped. For instance, suppose that we are given a state of [2 1 2 2 1 3 1 3 3] and the two random numbers generated are 2 and 8. After applying *Exchange*, the new state will be [2 1 3 2 1 3 1 3 2]. Obviously, the 2nd and 8th integers on the state were 2 and 3, respectively and turned to 3 and 2 with applying *Exchange* function.
b) **Insert** is another fine-tuning function that implies inserting a randomly chosen integer in front of another randomly chosen one. For instance, we are given the same state as before. ([2 1 2 2 1 3 1 3 3]). In order to apply *Insert*, we also need to derive two random numbers; one is for determining the integer to be inserted and the other is for the place where to insert in. Let us say those number are 2 and 5, where 2nd integer is 2 and the 5th one is 3. Consecutively, the new state will be [2 1 3 2 1 2 1 3 3].

Although there are many other, even more efficient, neighborhood structures reported in the literature, we preferred these two due to simplicity and ease of use alongside a reasonable efficiency. The others such as critical path based-functions, provide more efficiency, but definitely require much more computational time and experience and hard working.

6.4.4 VNS algorithms for JSS

In this chapter, we developed a number of VNS sequential algorithms based on particular shake and local search functions, which are formed up with the NS functions identified in previous sub section. The *local search* function is a simple hill climbing algorithm based on both aforementioned neighborhood structures (NS). As indicated in the following pseudo code, the aforementioned NSs are used complementary to each other in the way that the functioning NS keeps iterating as long as better moves are resulted. It switches to the other NS once the result produced is not better and the algorithm stops if the number of moves, n, meets a predefined number, n_{max}. The change of NS is organized with a binary integer variable, $l \epsilon (0, 1)$, in which the value of l is changed by using an absolute function denoted in $|\cdot|$ norm at the second part of Step 3(b) of the pseudo code. The local search procedure is mainly as follows.

1. **Get initial solution**, $x' \epsilon S$
2. **Set** n← 0, and $l \leftarrow 1$
3. **while** $n \leq n_{max}$ **do**
(a) if $(l = 1)$ then $x" \epsilon S \leftarrow Exchange(x')$;
else if $(l = 0)$ then $x" \epsilon S \leftarrow Insert(x')$
(b) if $f(x") \leq f(x')$ then $x' \leftarrow x"$; else $l \leftarrow |l - 1|$
(c) $n \leftarrow n + 1$

We have examined a number of VNS implementations in [32]; each differs from the others with the configuration of shake and local search functions by using the neighborhood structures. The idea is to develop efficient implementation, although the pseudo-code provided above remains as the main back-bone. Similarly, shake functions have been varying. Following is a list of 3 VNS algorithms provided with their functional configurations. Note that VNS-3 is described as no Shake function employed explicitly, but there is already one shaking procedure embedded into the Local search. The aim here is to understand whether there is a significant impact of starting with a particular function on the results.

VNS-1 : **Shake** ← *Exchange* + *Insert* + *Exchange*,
Local Search ← *Exchange* + *Insert*
VNS-2 : **Shake** ← *Exchange* + *Insert* + *Exchange*,
Local Search ← *Exchange*
VNS-3 : **No Explicit Shake**
Local Search ← *Exchange* + *Insert*

6.5 Experimental Study

In this chapter, we report the performance of four different parallel VNS implementations for job shop scheduling problems with a wide range of experimentation. First, we provided experimental results for various VNS algorithms in order to develop an efficient implementation. Then, a wide range of experimental result with relevant discussion on the parallelization approaches followed and the performance levels gained. The measures considered in this study are mainly about the solution quality and computational time. The success of the algorithms regarding the quality of solution has mainly been accounted with respect to the relative percentage of error (RPE) index, which is calculated as follows:

$$RPE = \frac{bf - opt}{opt} \times 100 \quad (6.1)$$

where bf is the best makespan found and opt is either the optimum or the lowest boundary known for unknown optimum values. Obviously, RPE is calculated based on the best value found, and also it can be measured benchmark-by-benchmark. In order to review the results in a broader view, we developed a second index based on the latter RPE calculation averaged over the number of repetitions (experiments are repeated 30 times). That is called ARPE standing for averaged relative percentage of error. The third index used is the hitting-ratio (HR) being calculated as the ratio between number of times optimum hit and the total number of repetitions. This is needed since the other indexes may not build a sufficient level of confidence with the results. The experimentation has been carried out on a local area network of PCs equipped with Intel Pentium IV 2.6 GHz processor and 256MB memory. The software coded in Java SDK 1.4 and the communications are conducted via TCP/IP protocol. The processor means, along this chapter, a particular PC over the network. The JSS benchmark problems, which are very well known within the field, were picked up from OR-Library [8].

6.5.1 Experimentation with VNS algorithms

Experimental results are obtained with examination of all 3 VNS variations described above. Table 6.1 presents the experimental results gained by each VNS algorithm with respect to the quality of solution measured in ARPE and HR indexes, where former (APRE) is minimized and the latter (HR) is maximized. The experiments have been conducted over 30 benchmarks tackled; some are known moderately hard but some are very hard. The accomplishments of the algorithms are clearly reflected. The APRE and HR indexes provide very consistently, which proves that measuring the performances reflects the real achievement.

The table is two part; each is made of 8 columns, where first two columns of each part are reserved for identifying the benchmarks and the rest are coupled two by two for each VNS variation. Here, one column of the two shows the achievement with respect to APRE and the other with HR. Obviously, VNS-1 is the best with APRE of 0.74 % and HR of 33 % and VNS-3 is the worst with respect to both measures. The difference between VNS-1 and VNS-2 with respect to the configuration is obviously the use of both neighborhood structures (NS) in both parts of the algorithm, where VNS-1 uses both NSs in both parts, but VNS-2 uses both in Shake but, only one in Local search.

As a result, we observe the significant change with a local search of double NS versions. The version with the worst performance, VNS-3, does not have an explicit shake function, though it is implicitly available Local search.

A statistical t test has been carried out over 30 repetitions to check whether or not the differences are significant. Since the differences between HRs are variable and, hence, the standard deviations are close to each other, the best is to compare APREs instead. The differences between performance of VNS-1 and other two is significant with 95% and 99.95%, respectively, where VNS-2 remains better than VNS-3, significantly, too.

Table 6.1: Results obtained from VNS algorithms with respect to 2 indexes for quality of solution.

		VNS-1		VNS-2		VNS-3				VNS-1		VNS-2		VNS-3	
Bench.	Opt.	(%)	HR	(%)	HR	(%)	HR	Bench.	Opt.	(%)	HR	(%)	HR	(%)	HR
ft10	930	0.55	0.60	1.11	0.40	4.79	0.00	ft20	1165	0.54	0.50	0.95	0.10	1.08	0.03
abz05	1234	0.10	0.60	0.22	0.40	1.43	0.20	la16	945	0.30	0.50	0.46	0.40	3.04	0.07
abz06	943	0.00	1.00	0.00	1.00	1.53	0.13	la19	842	0.09	0.93	0.15	0.87	2.66	0.03
abz07	656	2.01	0.00	2.70	0.00	3.11	0.00	la21	1046	0.62	0.03	1.30	0.00	2.74	0.00
abz08	665	1.99	0.00	2.86	0.00	3.76	0.00	la22	927	0.24	0.57	0.67	0.13	1.28	0.03
orb01	1059	1.47	0.13	2.62	0.03	5.21	0.00	la24	935	0.60	0.03	1.29	0.00	2.74	0.00
orb02	888	0.33	0.00	0.38	0.00	2.60	0.00	la25	977	0.64	0.07	0.99	0.00	2.27	0.03
orb03	1005	2.60	0.10	2.72	0.10	7.52	0.00	la27	1235	0.91	0.00	1.80	0.00	1.65	0.03
orb04	1005	0.62	0.40	0.93	0.13	2.69	0.00	la28	1216	0.03	0.87	0.42	0.17	0.30	0.53
orb05	887	0.31	0.17	0.76	0.00	4.07	0.00	la29	1152	1.80	0.00	3.04	0.00	3.40	0.00
orb06	1010	0.89	0.10	1.21	0.00	5.91	0.00	la36	1268	0.49	0.27	1.07	0.00	1.98	0.00
orb07	397	0.27	0.80	0.48	0.63	2.49	0.17	la37	1397	0.73	0.37	0.98	0.20	2.41	0.07
orb08	899	1.63	0.23	1.28	0.50	5.58	0.03	la38	1196	0.97	0.07	1.85	0.00	4.19	0.00
orb09	934	0.78	0.13	0.74	0.17	2.34	0.00	la39	1233	0.43	0.30	1.07	0.00	2.58	0.03
orb10	944	0.00	1.00	0.08	0.87	4.86	0.07	la40	1222	0.42	0.00	0.84	0.00	2.25	0.00
Average										**0.74**	**0.33**	**1.17**	**0.21**	**3.08**	**0.05**

6.5.2 Parallel VNS algorithms

As mentioned in Section 4, there are 4 parallelization policies followed in this study. The first two propose a central control over the communication among the processors while the last two imply non-central control policies. The first two approaches are synchronously and asynchronously control the messages sent by the processors and evaluate the results forwarded with respect to the quality of solution. The synchronously controlled approach broadcasts the best of all processors to all for the next generation while the asynchronously controlled approach offers the best of the time when any processor contacts the central unit. On the other hand, the non-centrally controlled approaches communicate in peer-to-peer way and exchange the results in a particular fashion. The third approach works following a unidirectional-ring topology and offers the outcome of any processor to the next adjacent for the next generation while the mesh topology suggests to collect the outcomes of two adjacent processors and selects the best of three including its own result for next generation. Since it is observed as the best VNS implementation, we kept track on parallelizing VNS-1 with the methods described above. The algorithm remains the same and the processors are equipped with a pair of *shake* and *local search* algorithms identical to each other. The only thing to change is the way in which processors are organized. As mentioned before, the motivation is to gain benefits for the quality of solution besides the computational time. In order to observe the benefit over the change of the numbers of processors, we examined 4 configurations for parallelism; 2, 5, 10, and 20-processor and kept the number of generations fixed at 200 in total. Hence, it becomes 100, 40, 20 and 10 when we examine 2, 5, 10 and 20 processors, respectively. The pairs of *number-of-processors* and *number-of-generations* complete 200 generations at the end of each trail of experiment. This is also important to build a fair comparison on computational time [32].

Table 6.2: The progress gained with parallelization methods with respect to the quality of solution.

	Synchronous		**Asynchronous**		**Ring Topology**		**Mesh Topology**	
	%	HR	%	HR	%	HR	%	HR
1p-200g	0.79	0.32	0.79	0.32	0.79	0.32	0.79	0.32
2p-100g	0.76	0.33	0.74	0.34	0.74	0.35	0.75	0.32
5p-40g	0.75	0.32	0.76	0.32	0.65	0.33	0.66	0.34
10p-20g	0.79	0.31	0.76	0.32	0.60	0.36	0.61	0.34
20p-10g	0.80	0.30	0.76	0.31	0.65	0.33	0.55	0.38
Average	**0.78**	**0.31**	**0.76**	**0.32**	**0.69**	**0.33**	**0.67**	**0.34**

Investigating the impact of these 4 parallelization approaches, we realized that the best results provided by mesh topology with respect to the quality of solutions while the performances remain the same in terms of the computational time. Table 6.2 presents corresponding information with respect to the quality of solution in APRE and HR indexes averaged over all problems. As can be observed from Table 6.2, the synchronous policy provides with almost the same results with respect to APRE of 0.75 and 0.80 percent for minimum and maximum values, respectively. The HR scores are very steady as well.

It seems the configuration of 5p-40g provides slightly better. In the asynchronous case, the slightly better results come with the configuration of 2p-100g. The other cases of configurations are nearly the same. On the other hand, non-central policies, provides slightly better than the central ones, as the HR values are getting higher and ARPE values are getting lower. In fact, the unidirectional-ring topology provides with 0.60 as the minimum in ARPE and 0.36 in HR with the configuration of 10p-20g while mesh topology offers 0.55 minimum of APRE and 0.38 maximum of HR with the configuration of 20p-10g. As can be observed, there is neither significant difference between the performances of synchronous central policy and asynchronous one nor between ring and mesh topology policies, but there is quite difference between the performances of central and non-central policies. Although the APRE index differs in between the central and non-central policies, still it is ignorable as the values of the index is within the range of 0.80 and 0.55 percent, which is not very considerable difference. However, the difference in terms of HR can be significant, as the best of central policies is 34 % while the best non-central ones is 38 %. This figure can be significant when one has worked with larger amount of data and larger problems.

The Experiments done so far, does not provide with a satisfactory level of confidence for understanding the characterization of the behaviors of the policies examined. We carried out a set of further experiments with different configurations. We set up 4 configurations: (i) 20 processors each to work for 100 generations, (ii) 20 processors each to work for 50 generations, (iii) 10 processors each to work for 100 generations and (iv) 10 processors each to work for 50 generations. As each cycle of VNS is counted as a generation, the total number of VNS cycles for each configuration will be 2000, 1000, 1000 and 500, respectively. All 4 parallelization policies have been examined with these 4 configurations and the results are shown in Fig. 6.5 and Fig. 6.6, where the performance measures regarded are APRE and HR indexes, respectively. There exist 4 groups of 4 bars, where each bar signifies a particular policy. As mentioned above, ARPE and HR are to be minimized and maximized, respectively. Thus the lower the APRE value and the higher the HR value the more desirable.

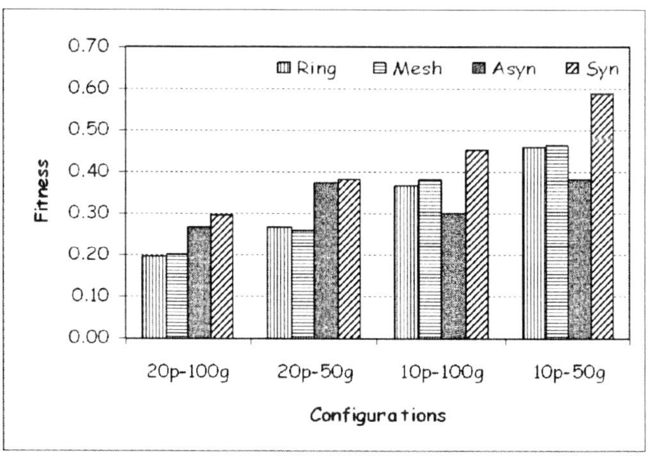

Fig. 6.5: The performance of parallel algorithms with respect to APRE index.

As can be seen from Fig. 6.5 and Fig. 6.6, the best performance with respect to both indexes and all policies is obtained with 20p-100g configuration

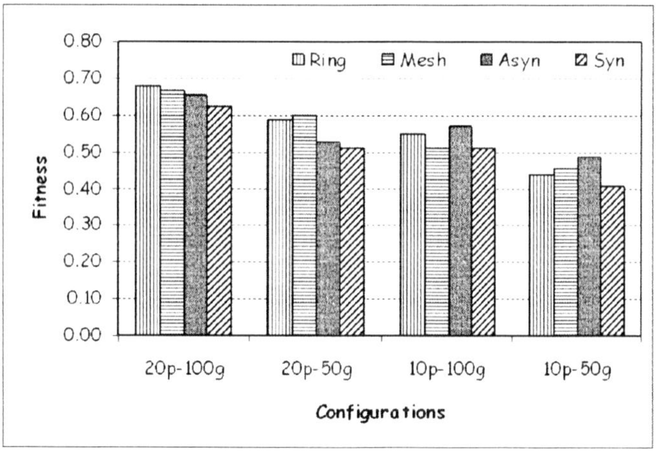

Fig. 6.6: The performance of parallel algorithms with respect to HR index.

as it has the highest number of generations while the poorest one is obtained with the configuration of 10p-50g since it runs for the lowest number of generations. On the other hand, the best performance observed is with the policy of unidirectional-ring topology and the weakest is with the synchronous one. This can be observed in all variations presented in both Fig. 6.5 and Fig. 6.6. This is because the unidirectional-ring topology implies to run the search in various regions of the space simultaneously while the synchronous policy offers a search with the multiple execution of the same neighborhood structure within the same region of the space. That affects the performance of the search significantly. The central policies (sync and asyn) are usually weaker than the non-central ones (ring and mesh). That means the non-central approaches diversify the search much more than the central ones. Although the unidirectional-ring topology looks usually better than mash, the difference between the performances is not that significant. Therefore, they can be considered competitive with respect to the search level carried out. The results reveal some other things that the experimentation with 10 processors seems more unstable than the ones with 20 processors. For instance, although the total number of generations is 1000 in both case of 10p-100g and 20p-50g, the performance of 20p-50g looks more consistent than 10p-100g referring to the performance gained with 20p-100g. That is perhaps because of the diversity of the search, which is double of 10p-100g in 20p-50g. The most stable results obtained with unidirectional-ring topology and the synchronous policy, while other two provide with less stability. The reason behind this is still the diversity as the unidirectional-ring topology provides the highest level and the synchronous one does the lowest while other two provide with varying diversities, which sometimes approaches to the highest level sometimes to the lowest. That is why, the performance looks quite varying.

In general, the performances with respect to APRE look not significant as the differences are below 1%, but it is significant with respect to HR, especially, the difference between the best and the worst ones are about 6%, which means the search carried out with unidirectional-ring topology provides 69% while the synchronous one does 63%. That is the rate of hitting the optima of the benchmarks through the whole experimentation. This difference will definitely get larger whilst solving larger problems.

6.5.3 Related works

As explained earlier, the relevance of this work should be exercised in terms of two issues; parallel VNS algorithms and job shop scheduling problems. The related works has been discussed earlier in introduction section and Section 3. The theme of this subsection is to build a level of confidence with bringing forward a comparison between our results and the results gained by some other works related to job shop scheduling recently published. Alongside our serial and parallel VNS algorithms, Table 6.3 also presents results provided with various meta-heuristics recently published with respect to the solution quality in PRE index. The reason to switch to RPE index is due to the difficulty of calculating APRE index with the results provided in the related literature. The benchmarks those are considered very hard are chosen from OR-Library. These algorithms taken into account are listed as follows:

- Distributed evolutionary simulated annealing algorithm (dESA) by Aydin and Fogarty [5].
- Ant colony optimization algorithm (ACO) by Blum and Sampels [10].
- A Tabu Search Method (TSSB) by Pezzella and Merelli [30].

Table 6.3: A comparison among the meta-heuristics recently published with respect to the quality of the solution in RPE index.

		Related results			**Serial VNS**	**Parallel VNS**		
Bench.	Opt.	dESA	ACO	TSSB	VNS-1	5p-40g	10p-20g	20p-10g
abz07	656	2.44	2.74	1.52	0.46	0.91	0.46	0.91
abz08	665	2.41	3.61	1.95	0.60	0.75	0.89	0.75
abz09	679	2.95	3.39	2.06	0.15	0.29	0.59	0.29
la21	1046	0.00	0.10	0.00	0.00	0.00	0.00	0.00
la24	935	0.32	0.96	0.32	0.00	0.00	0.32	0.00
la25	977	0.00	0.00	0.20	0.00	0.00	0.00	0.00
la27	1235	0.40	0.65	0.00	0.08	0.00	0.00	0.00
la29	1152	2.08	1.39	1.39	0.95	1.03	0.86	1.03
la38	1196	0.42	2.59	0.42	0.00	0.00	0.00	0.00
la40	1222	0.49	0.49	0.90	0.16	0.16	0.16	0.16
Average		**1.15**	**1.59**	**0.88**	**0.24**	**0.31**	**0.33**	**0.31**

Since the best of our VNS algorithms in Table 6.1 is VNS-1 with respect to RPE index, we put the results obtained with that variant and its parallel versions providing them in the last four columns of Table 6.3. The parallel versions of VNS have been picked up from the unidirectional-ring topology as it is found as the best overall policy. We can observe that the serial and parallel versions of VNS outperform the other algorithms with respect to RPE index, (not APRE), as the lowest value provided by TSSB is about 0.88 while VNS provides with 0.33 at most. A t test has been carried out to measure the significance of differences between any particular parallel and serial VNS variants and the others, which resulted that there is no significant difference between any two VNS variants with a probability higher than 75%. However, the difference between any other work and any VNS is 99.95% significant. That means parallel and serial VNS algorithms remain competitive among themselves at this level of configuration, as one parallel variant provides better for one benchmark but worse for another. VNS-1, which is a serial variant, provides definitely better than other algorithms.

6.6 Conclusions

In this chapter, we examined a number of serial and parallel VNS algorithms for job shop scheduling problems, which has been studied for far long time. Because of its hardness and being representative for planning problems, many methods have been tested with this family of problems. In this chapter, the VNS implementations have been tested with respect to the efficiency over classical job shop problems. The best of the implementations, VNS-1, has been exploited in investigation of a better parallelism for VNS algorithms. For the purpose of parallelism, four approaches have been considered through the whole experimental investigation; synchronous, asynchronous, unidirectional-ring and mesh policies. The first two approaches imply central coordination among the processors while the last two requires a non-central coordination, but rather a peer-to-peer approach.

The main focus steers for a higher level of efficiency via parallelism to obtain a higher quality of solution within far shorter time. The experimental investigation reveals that central coordination does not help as much as non-central coordination does. Since the synchronous policy offers a rigorous simultaneous search within a particular region of the space, the weakest performance took place with this approach. On the other hand, unidirectional-ring topology proposes a concurrent search in various region of the space, and it outperforms the others with respect to the solution quality. It has been shown that the VNS implementation parallelized via unidirectional-ring topology policy has done well and outperformed a number of meta-heuristics recently published.

References

1. Adams J, Balas E, Zawack D (1988) The Shifting Bottleneck Procedure for Job Shop Scheduling, Management Science 34:391–401
2. Aiex R M, Binato S, Resende (2003) Parallel GRASP with Path-Relinking for Job Shop Scheduling, Parallel Computing 29:393–430
3. Alba E, Tomassini M (2002) Parallelism and evolutionary algorithms, IEEE Transactions on Evolutionary Computation, 6(5):443–462
4. Applegate D, Cook W (1991) A Computational Study of Job-Shop Scheduling, ORSA Journal on Computing 3(2):149–156
5. Aydin ME, Fogarty, TC (2004) A Distributed Evolutionary Simulated Annealing Algorithm for Combinatorial Optimisation Problems, Journal of Heuristics 10:269–292
6. Aydin ME, Yigit V (2005) Parallel simulated annealing. In: Parallel Meta-Heuristics, E. Alba (Ed), pp. 267-288, Wiley.
7. Balas E, Vazacopoulos A (1998) Guided Local Search with Shifting Bottleneck for Job Shop Scheduling. Management Science 44: 262–275
8. Beasley JE Obtaining Test Problems via Internet, Journal of Global Optimisation 8:429-433, http://people.brunel.ac.uk/ mastjjb/jeb/info.html.
9. Bierwith C (1995) A Generalized Permutation Approach to Job Shop Scheduling with Genetic Algorithms. OR Spektrum 17:87–92
10. Blum C, Sampels M (2004) An Ant Colony Optimization Algorithm for Shop Scheduling Problems. Journal of Mathematical Modelling and Algorithms 3:285–308
11. Crainic TG, Gendreau M, Hansen P, Mladenovic N, (2004) Cooperative Parallel Variable Neighborhood Search for the p-Median, Journal of Heuristics, 10: 293–314
12. Carlier J, Pison E (1989) An Algorithm for Solving the Job-Shop Problem, Management Science 35: 164–176
13. Cheng R, Gen M, Tsujimura Y (1996) A Tutorial Survey of Job Shop Scheduling Problems Using genetic Algorithms-I. Representation. Journal of Computers and Industrial Engineering 30(4):983–997
14. Colorni A, Dorigo M, Maniezzo V, Trubian M (1994) Ant System for Job-Shop Scheduling. Belgian Journal of Operations Research, Statistics and Computer Science (JORBEL) 34(1):39–53
15. Dell'Amico M, Trubian M (1993) Applying Tabu-Search to the Job-Shop Scheduling Problem. Annals of Operations Research 4:231–252
16. Dorndorf U, Pesch E (1995) Evolution Based Learning in a Job Shop Scheduling Environment, Computers and Operations Research 22:25–44
17. Dorndorf U, Pesch E, Phan-Huy T (2002) Constraint Propagation and Problem Decomposition: A Preprocessing Procedure for the Job Shop Problem, Annals of Operations Research 115:125–145
18. Fleszar K, Hindi KS (2002) New heuristics for one-dimensional bin-packing. Computers and Operations Research, 29:821–839
19. Garcia-Lopez F, Melian-Batista B, Moreno-Perez JA, Moreno-Vega M (2002) The parallel variable neighbourhood search for the p-Median problem, Journal of Heuristics, 8(3):375–388
20. Garey M, Johnson D, Sethy R (1976) The Complexity of Flow Shop and Job Shop Scheduling. Mathematics of Operations Research 1: 117–129

21. Croce FD, Tadei R, Volta G (1995) A Genetic Algorithm for the Job Shop Problem. Computers and Operations Research 22: 15–24
22. Goncalves JF, Mendes JM, Resende M (2004) A hybrid genetic algorithm for the job shop scheduling problem, European Journal of Operations Research 167(1):77–95
23. Hansen P, Mladenovic N, Dragan U (2004) Variable neighborhood search for the maximum clique Discrete Applied Mathematics, 145(1):117–125
24. Huang W, Yin A (2004) An Improved Shifting Bottleneck Procedure for the Job Shop Scheduling Problem. Computers and Operations Research 31:2093–2110
25. Jain A, Meeran S (1999) Deterministic Job-Shop Scheduling: Past, Present and Future. European Journal of Operational Research 113:390–434
26. Kolonko M, (1999) Some New Results on Simulated Annealing Applied to the Job Shop Scheduling Problem. European Journal of Operational Research 113:123–136.
27. Mladenovic N, Hansen P (1997) Variable Neighborhood Search. Computers and Operations Research 24:1097–1100
28. Nowicki E, Smutnicki C (1996) A Fast Taboo Search Algorithm for the Job Shop Problem. Management Science 42: 797–813
29. Nowicki E, Smutnicki C (2005) An advanced tabu search algorithm for the job shop problem. Journal of Scheduling 8:145–159
30. Pezzella F, Merelli E (2000) A Tabu Search Method Guided by Shifting Bottleneck for the Job Shop Scheduling Problem. European Journal of Operational Research 120:297–310
31. Satake T, Morikawa K, Takahashi K, Nakamura N (1999) Simulated Annealing Approach for Minimizing the Makespan of the General Job- Shop, International Journal of Production Economics 60:515-522
32. Sevkli M, Aydin M. E., (2007) Parallel variable neighbourhood search algorithms for job- shop scheduling problems, IMA Journal of Management Mathematics, 18:117-133

7

Solving Scheduling Problems by Evolutionary Algorithms for Graph Coloring Problem

Pawel B. Myszkowski

Institute of Applied Informatics, Wroclaw University of Technology (Poland)
pawel.myszkowski@pwr.wroc.pl

Summary. This chapter presents a new evolutionary approach to the Graph Coloring Problem (GCP) as a generalization of some scheduling problems: timetabling, scheduling, multiprocessor scheduling task and other assignment problems. The proposed evolutionary approach to the Graph Coloring Problem utilizes information about the conflict localization in a given coloring. In this context a partial fitness function (*pff*) and its usage to specialize genetic operators (IBIS and BCX) and phenotypic measure of diversity in population are described. The particular attention is given to the practical usage of GCP. The performance of the proposed algorithm is verified by computer experiments on the set of benchmark graphs instances (DIMACS). Additional experiments were done on benchmark graph for timetabling problem.

Key words: Evolutionary Algorithms, Graph Coloring, Timetabling, Multiprocessor scheduling, Assignment Problems, Genetic Operators.

7.1 Introduction

In this chapter we introduce a new evolutionary approach to the Graph Coloring Problem (GCP). The Graph Coloring Problem can be analyzed as a theoretical problem, but we want to show that there a exists very small gap to its practical usage. Therefore, in this chapter GCP is analyzed as a generalization of selected practical problems (such as timetabling problem, job-shop scheduling problem or multiprocessor scheduling task problem) and in this context is considered.

The presented method, so-called GRACOM (*GRAph COloring Method*), is based on the usage of information about conflict in coloring (two graph vertices connected by an edge have the same color). Such information makes linking each conflict with its vertices possible and builds more detailed coloring evaluation function, so-called: *partial fitness function* (*pff*). Values of *pff* function give us opportunity to construct the specialized genetic operators. In

P.B. Myszkowski: *Solving Scheduling Problems by Evolutionary Algorithms for Graph Coloring Problem*, Studies in Computational Intelligence (SCI) **128**, 145–167 (2008)
www.springerlink.com

this chapter we propose two such operators: IBIS (*Iteration Build Solution*) as a mutation and BCX (*Best Color Crossover*) as a specialized crossover. This specific evaluation function *pff* gives us also something more: we can define a phenotypic measure of population diversity.

The reminder of the chapter is organized as follows. Section 7.2 presents Graph Coloring Problem: its definition, notations (used in following sections) and the problem complexity. In Section 7.3 are described the selected scheduling problems which are analyzed as generalization of Graph Coloring Problem. Next, Section 7.4 contains details of the GRACOM construction. There are presented specific information about conflict definition and calculating method (*partial fitness function*), definition of genetic operators: IBIS and BCX, also specific evolutionary algorithm schema modifications. Section 7.5 contents are focused on tests results of the presented method. There are analyzed some evolutionary algorithm parameters, such as selection method, probability of genetic operators usage or population size. Section 7.6 presents the summary of GRACOM efficiency in context of benchmark graphs, there are presented GRACOM efficiency results in comparison to other GCP methods. This section includes some proposed extensions of evolutionary algorithm to apply in the given real problems. Section 7.7 describes summary, conclusions and further work.

7.2 Graph Coloring Problem (GCP): definition and notations

The *Graph Coloring Problem* (GCP) is one of the most studied NP-hard problems and can be defined as follows: given undirected graph $G = (V, E)$, where V is a set of $\mid V \mid= n$ vertices and $E \subseteq V \times V$ is a set of $\mid E \mid= m$ edges, and an integer number of colors k, to find such function $\Psi : V \mapsto 1, 2..k$ which, for every edge of the graph $[u, v] \in E, u, v \in V$ should satisfy a constraint $\Psi(u) \neq \Psi(v)$ [1]. We can define a **proper coloring** as a coloring which is compatible with the above definition. In the GCP we use k colors to color properly a given graph, but if a graph can be properly colored with $k - 1$ colors, then the k-coloring is not optimal. In this way we can consider GCP as an optimization problem in which we try to find a minimal number of used colors (chromatic number χ_G) needed to properly color a given graph [2–4].

Obviously, the GCP landscape S size equals $\mid S \mid= k^n$, and is quite large for graphs with 100 or more vertices [5]. The GCP problem is considered as an NP-hard and we do not know any of effective algorithm, which can give a solution in the polynomial time. Its NP-hard complexity practically means that it is very difficult to find a proper coloring in time acceptable for user even for a graph with few vertices [2]. There are many approximation methods applied to the GCP: approximation algorithms [2, 6], heuristics (e.g. DSATUR [1], RLF [7]), local search algorithms [8, 9], or meta-heuristics, such as the tabu search [4, 10], simulated annealing [11], ant colony algorithms [7, 12, 13] or evolutionary algorithms (EA) [8, 14–19].

Let us focus on the last approach. A graph coloring task for evolutionary algorithm can be defined as: the properly coloring given undirected graph $G = (V, E)$ with k (if $k = \chi_G$ it is an optimal coloring) colors with the aid of number of conflicts reduction, where two vertices $u, v \in V$ cause conflict when both are assigned to the same color and there exists an edge $[u, v] \in E$ [9]. In this approach, the GCP is considered as a Constraint Satisfaction Problem (CSP), where edges of the colored graph correspond to constraints.

7.3 Applications of Graph Coloring Problem

Despite the fact that the GCP seems to be a theoretical problem, there are many practical applications of scheduling or assignment, e.g. in examination timetabling, job shop scheduling, timetabling [16], aircraft scheduling, frequency assignment [20], routing problem and many others [2]. This section describes only selected the GCP applications.

7.3.1 University timetabling: Examination Timetabling and Course Timetabling

The timetabling problem is known as NP-hard, involving combinatorial optimization and it is a representative of the multi-constrained class. In literature the Graph Coloring Problem is considered as generalization of timetabling problems [21, 22].

The typical **course timetabling problem (CTP)** assigns a set of events (e.g. classes) to a set of resources (e.g. rooms) and timeslots, satisfying a set of constraints of various types. Constraints stem from nature of timetabling problems (e.g. two events with the same resources involved cannot be planned at the same time) and the specificity of the institution involved. The problem we consider is a typical university course timetabling problem (UCTP) (e.g. [23]). It consists of a set of events (classes) that have to be scheduled in a certain number of timeslots, and a set of rooms with certain features (and sizes) which events can take place in. The timetabling hard constraints (must be satisfied) can be defined as follows: (1) no student attends more than one event at the same time, (2) the room is big enough for all the attending students and satisfies all the features required by the event and (3) only one event is in each room at any timeslot [23, 24].

The **examination timetabling problem (ETP)** is also highly constrained and, in general, the most common hard constraints are to avoid any student being scheduled for two different exams at the same time [25] and it can be defined as the scheduling a number of exams into a set of periods (timeslots). It this problem we can also find a main hard constraint very similar to CTP: no student attends more than one exam at the same time. Also in this problem we can observe soft constraints (should be satisfied, in opposite to hard constraints which must be satisfied), such as "reasonable study time between exams" [25] or limit of exams in the same timeslot.

Solving examination timetabling or course timetabling is proved to be equivalent to GCP: assigning colors to vertices in graph so that adjacent vertices always have different colors [26]. The exam (or an event) we consider as vertex, each constraint (such as two events/exams cannot be realized in the same time) is represented by the edge connected with two vertices. A set of timeslots can be represented by a set of used colors and coloring is no legal if exists a pair of exams avoided to realization at the same time.

7.3.2 Job-shop Scheduling Problem

The **Job-shop scheduling problem (JSSP)** is a classical manufacturing problem and is known as NP-complete [27]. JSSP can be described as a set of n jobs with different processing time with an assignment to a set of m machines [13, 28]. Each job comprises a set of tasks (so-called operations) with some ordering relations and the given duration time. A legal (feasible) schedule is defined as follows: (1) each task order is preserved, (2) machine cannot process two tasks at the same time and (3) two or more tasks of one job cannot be processed in parallel.

To show a problem complexity let's analyze the example of constraints for the given 6x6 (6 jobs and 6 machines) benchmark sample of JSPP (presented on Table 7.1). Table 7.1 shows that the job 1 must be processed on the machine 3 and needs 1 unit time, next it goes to machine 1 for 3 time units, and so on.

Table 7.1: Benchmark JSSP problem: 6 jobs and 6 machines [29].

	[m,t]	[m,t]	[m,t]	[m,t]	[m,t]	[m,t]
job_1	3,1	1,3	2,6	4,7	6,3	5,6
job_2	2,8	3,5	5,10	6,10	1,10	4,4
job_3	3,5	4,4	6,8	1,9	2,1	5,7
job_4	2,5	1,5	3,5	4,3	5,8	6,9
job_5	3,9	2,3	5,5	6,4	1,3	4,1
job_6	2,3	4,3	6,9	1,10	5,4	3,1

A JSSP can be considered as GCP problem as follows: each task can be represented as vertex and the used colors represent the processing time in the machine schedule. If there exists a pair of tasks that is not coherent to their ordering relation, its representation as a graph coloring is not legal.

7.3.3 Multiprocessor Scheduling Tasks Problem

The **Multiprocessor Scheduling Tasks Problem (MSTP)** is known as NP-complete [30, 31] and is connected with parallel tasks execution on set of processors. Between two tasks can occur a relation "predecessor/successor". Such a relation can be considered as a hard constraint, where if there exists a relation $\langle t_1, t_2 \rangle$ means that the task t_2 cannot be executed before returning of t_1 results (e.g. this value is required in t_2 working). This task correlation is presented as Directed Acyclic Graph (DAG), where the vertex corresponds to a task and an edge corresponding to an existing relation between two tasks (vertices). Another aspect of the problem is the task execution time (we assume that we are working on a set of homogeneous processors). Problem solution is schedule $s = \langle t_1, ..., t_n \rangle$ [30] represents execution rank of each task. The execution time of all scheduled tasks is called makespan and it is minimalised parameter in MSTP.

In [32] construction of DAG with makespan minimalisation it is suggested to solve as a clustering, an ordering or an allocation problem. The evolutionary algorithm can applied to MSTP problem [30], where an individual is a set of elements corresponding to each processor task list, e.g $s = \{\{t_2, t_0\}, \{t_1\}, \{t_3, t_4\}\}$ can be a schedule for 5 tasks and 3 processors. Indeed it is GCP problem, where processor symbolizes a color and each task is considered as a graph vertex.

Each of foregoing problems can be solved as a graph coloring problem with the minor fitness function modifications. In the next section we present a definition of GCP problem as a task to evolutionary algorithm and a proposition of evolutionary method GRACOM for solving GCP.

7.4 GRACOM: Evolutionary Algorithm applied to Graph Coloring Problem

The GCP problem is considered as an NP-hard and we do not know any effective algorithm so we apply other methods such as the evolutionary algorithm. An evolutionary algorithm is a meta-heuristic that uses Darwinian evolution paradigm to produce a solution. EA is based on the probabilistic multipointing searching process of all problem solutions landscape. The balance between the exploration and the exploitation of landscape and usually is implemented by correlation between a mutation (the operation works on one solution) and a crossover (links selected elements of two or more solutions) usage.

EA can be presented in schema (see Fig. 7.1). Firstly, the population of individuals (solutions) is initialized. In the second step it gives a value of each corresponding to its degree of solving problem (fitness function). In the next step stop conditions are tested (if coloring with 0 conflict is found or generation number limit is exceeded). Next two steps select (by a proportional, tournament or random method) individuals and operate individuals by genetic operators to build a new population to close a cycle of EA.

In Fig. 7.1 is presented evolutionary algorithm schema where we provided two modifications: fitness function form and genetic operators collaboration.

The small modification of fitness function form refers to *partial fitness function* and its values. It means that to evaluate the coloring we need to assign a number of conflicts to each vertex of the colored graph. The second EA schema modification concerns with genetic operators collaboration and a usage of one operator excludes the second one. This modification is strongly imposed by the genetic operator requirement of evaluated coloring (reduction of number the *pff* calculations). The second reason is that both genetic operators give coloring with changes which should be evaluated. This is important particularly in GCP as it is a combinatoric problem, where a small modification of a gene value may cause significant change in phenotypic level.

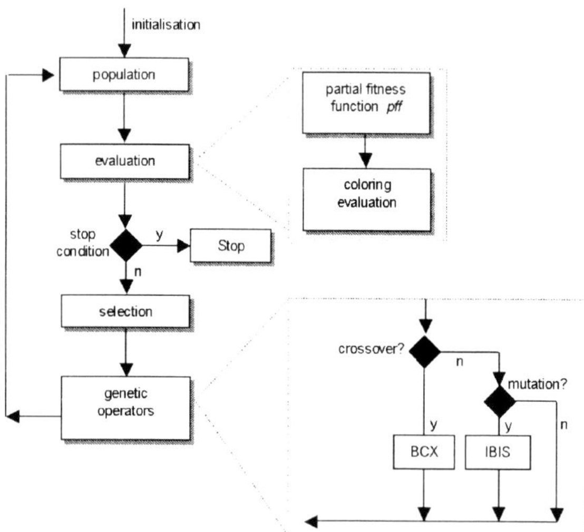

Fig. 7.1: GRACOM schema with provided modifications.

The GRACOM method is based on a number of conflicts assigned to each vertex of the colored graph. Such detailed information can be used to define *partial fitness function* and it has some advantages. In this section are presented details of the provided GRACOM method in the following order: solution coding form, *partial fitness function* calculation method, population diversity measure and proposition of specialized genetic operators: IBIS and BCX.

7.4.1 Coding of solution and a partial fitness function (*pff*)

We decided that an individual in the EA represents a proposition of the graph coloring as a vector $I = \langle \Psi(v_1), \Psi(v_2), ..., \Psi(v_n) \rangle$, where $\Psi(v_i) \in \{1, 2, ..., k\}$ determines color of v_i vertex. Such a representation is direct, intuitive and commonly used, e.g. [5, 8, 18].

The evaluation function is based on a penalty, value of which depends on a number of pairs of neighbor vertices colored with the same color (conflict). Such value gives information how many conflicts exist in a given graph coloring, however nothing is said about conflict localizations. In literature, the number of conflicts correspond only to the whole coloring; even if, considerations are based on a single vertex, it is only a boolean value (if there is a conflict) [15]. Such a evaluation makes it possible to analyze the graph coloring, but also determines which one vertex is in "better" color and which one in "worse" (causes more conflicts).

The proposed *partial fitness function* (*pff*) evaluates each vertex v_i of a colored graph in the context of other vertices, can be defined as follows:

$$pff(v_i) = \sum_{j=1}^{|V|} p(v_i, v_j),$$
$$where \quad p(v_i, v_j) = \begin{cases} 1, & if \Psi(v_i) = \Psi(v_j) \wedge [v_i, v_j] \in E \\ 0, & otherwise \end{cases} \tag{7.1}$$

If $pff(v_i) = 0$, it means that a vertex v_i does not cause conflicts with connected vertices. To determine a total number of conflicts *conf(I,G)* in a given coloring I of graph G we use:

$$conf(I,G) = \frac{1}{2} \sum_{i=1}^{|V|} pff(v_i) \tag{7.2}$$

In Eq. 7.2 the total number of conflicts are calculated twice (for each vertex in a pair causing conflict) so it should be multiplied by $\frac{1}{2}$.

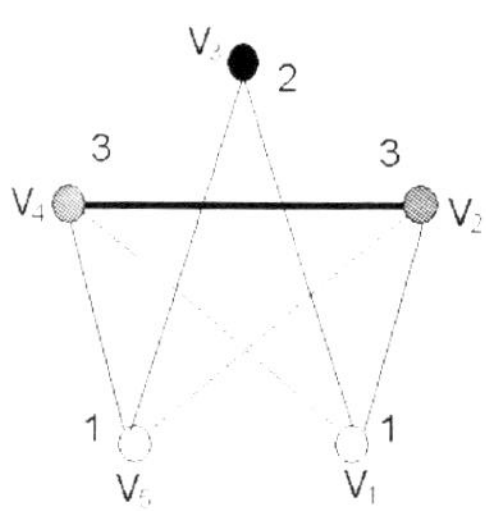

vertex	V1	V2	V3	V4	V5
color	1	3	2	3	1

I=<1,3,2,3,1>

pff value	0	1	0	1	0

Ev=<0,1,0,1,0>
conf(I,G) = 1

Fig. 7.2: Graph coloring with 1 conflict and *pff* values corresponding to each vertex.

In Fig. 7.2 is presented a sample graph with 5 vertices, colored with 3 colors and existing 1 conflict (between v_2 and v_4 vertex). A *pff* function value of each vertex (except v_2 and v_4) equals 0. The total evaluation value of given coloring equals 1.

Thus, the proposed function *pff* extends a scalar evaluation function to a vector $Ev(I,G) = \langle pff(v_1), pff(v_2), ..., pff(v_n)\rangle$, and assigns an evaluation value to every vertex of colored graph. Such approach has many advantages, i.e. we can evaluate each vertex of the colored graph. More about advantages and disadvantages of such form of an evaluation function can be found in [17].

7.4.2 A measure of population diversity

A population diversity in an evolutionary algorithm is one of the main factor of its efficiency. A fast lost of the diversity in population leads to a premature convergence [33]. Therefore, it is important to maintain the high population diversity. There are two types of diversity measures: *genotypic diversity measure* (GDM) and *phenotypic diversity measure* (PDM). The first one involves the genetic material held in a population, while the second concerns the phenotypic level of individuals [34].

The graph coloring is represented as a vector $I = \langle\Psi(v_1), \Psi(v_2), ..., \Psi(v_n)\rangle$ and can measure, in a simple way, distance between two colorings I_K and I_L by applying a *Hamming distance*:

$$D_H(I_K, I_L) = \sum_{i=1}^{|V|} h(\Psi_K(v_i), \Psi_L(v_i)),$$

$$where \quad h(\Psi_K(v_i), \Psi_L(v_i)) = \begin{cases} 1, & if \Psi_K(v_i) \neq \Psi_L(v_j) \\ 0, & otherwise \end{cases} \tag{7.3}$$

The D_H distance has many disadvantages (e.g. does not take into consideration space symmetry [35]), therefore there are also others GDM that can be applied: histograms, statistical dispersion, Euclidean distance or Entropy distance [1,34,35].

A construction of the GDM is rather a simple task, more problematic is the PDM. In [34] it was proposed to control the relation between individuals with the best and the average fitness in population (or average and the worst) to get information about current population diversity. Such measure gives only statistical information for a given population, although with the aid of the *pff* we can have a more precise phenotypic measure of two evaluated graph colorings Ev_K and Ev_L:

$$D_{pff}(Ev_K, Ev_L) = \sum_{i=1}^{|V|} \mid pff_K(v_i) - pff_L(v_i) \mid \tag{7.4}$$

where $pff_K(v_i)$ means: *pff* value of v_i vertex in Kth coloring. A D_{pff} measure gives information how two colorings are different on the phenotypic level. Such measure can be used to e.g. calculate a phenotypic standard deviation in a population.

The PDM can seem to be useless but in [34] it was argued that, the principal feature of this type of measures is that they allow premature convergence to be predicted rather than being detected. This feature gives information about level of population diversity and with the aid of it we can run a preventive strategy of premature convergence.

7.4.3 Genetic operators

It is admitted that classical genetic operators are generally poor for solving optimization problems and have to incorporate the specific domain knowledge [33]. Defined *partial fitness function* gives us information, which we can consider as a CSP domain knowledge, but in fact the graph coloring theory knowledge is not included (e.g. vertex degree or maximal clique in a graph [2]). In this section we present two specialized genetic operators: IBIS mutation and BCX crossover.

Iteration Build Solution (IBIS)

IBIS operator is based on a human heuristic, which can be applied in solving similar problems. Firstly, graph is colored by quasi-random method, and then coloring is evaluated. In the next step, we analyze the number of conflicts caused by each vertex and with reference to its value, it is suggested to recolor or retain a vertex color. Such a heuristic is the base of IBIS operator. However, IBIS implements this heuristics with probabilistic aspects: if vertex causes greater number of conflicts then probability of changing color has a higher value.

In implementation, operator IBIS takes evaluated coloring and analyzes each vertex *pff* value. This function is calculated by applied fuzzy logic driver (Takagi-Sugeno), which determines the values of color remaining probability ($1 - recolor\_prob$, inverse relationship to *pff* value) and the tournament size (*tournament_size*, positive relationship to *pff* value). The first value decides if vertex should be recolored and the second one gives number of concurrent colors (the best one is chosen). If probability of recoloring vertex is too small, vertex color is taken from parent coloring. The IBIS operator pseudocode is presented on Fig. 7.3.

The color tournament selection method takes randomly *tournament_size* for all used colors and checks number of conflicts of a given vertex in each concurrent color. Then it chooses the color with the least number of conflicts caused by given vertex.

The main element of IBIS operator is a Takagi-Sugeno fuzzy logic driver. We decided to apply the Takagi Sugeno driver as we can define simply and intuitive fuzzy rules to compute IBIS parameters (see Fig. 7.4). Also TS driver has small consuming processor time (there is no a defuzzification phase). The

```
function operator_IBIS(coloring in parent, out off)

begin
 calculate pff_min, pff_avg and pff_max values in parent coloring;
 build_TS_driver(pff_min, pff_max, pff_avg);

for each vertex v_i of colored graph
    recolor_prob := TS_recolor(pff_parent(v_i));
    tournament_size := TS_tournament(pff_parent(v_i));

    r := random_number⟨0,0; 1,0⟩;

   if (r<recolor_prob)    color_off(v_i) := color_parent(v_i);
   else   color_off(v_i) := tournament_color(tournament_size);
end.
```

Fig. 7.3: Operator IBIS pseudocode.

If vertex v_i causes **a few** conflicts:
recolor_prob is near 1,0 and *tournament_size* is near 0,0

If vertex v_i causes **average** number of conflicts:
recolor_prob is near 0,75 and *tournament_size* is near 0,75

If vertex v_i causes **many** conflicts:
recolor_prob is near 0,0 and *tournament_size* is near 1,0

Fig. 7.4: Example of Takagi-Sugeno driver rules.

usage of fuzzy logic in evolutionary algorithms is no a novel idea, for example in [34,36] is used to manage exploration and exploitation process in EA. Also in other papers we can find a fuzzy logic usage to adapt EA parameters, individual representation, fitness function or stop conditions. To the best of our knowledge, there is not any usage of fuzzy logic in the mutation operator.

The implementation of TS driver needs fuzzy sets details. We set a number of fuzzy sets, shapes and other parameters experimentally. The result is shown in Fig. 7.5: there are three fuzzy sets (*best*, *avg* i *worst*) based on the average value of number of conflicts (*avg*) in given coloring located on x-axis.

To compute a Takagi-Sugeno driver output values we use a below formula:

$$y = \frac{\sum_{i=1}^{m} \mu_{A_i}(x) f_{A_i}(x)}{\sum_{i=1}^{m} \mu_{A_i}(x)} \tag{7.5}$$

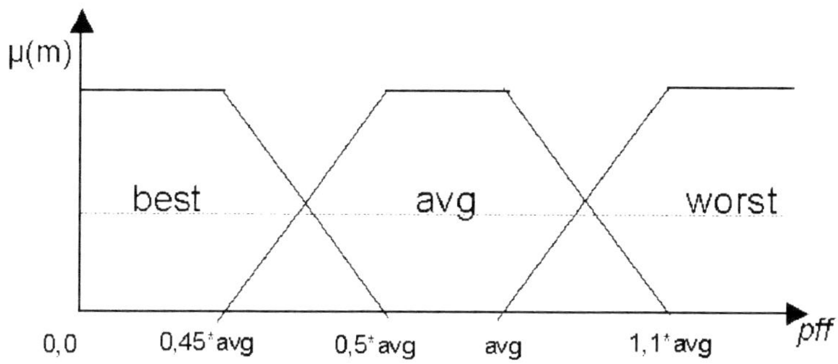

Fig. 7.5: Example of fuzzy sets used in TS driver.

where y symbolizes one of output values: *TS_recolor* or *TS_tournament*. A $\mu_{A_i}(x)$ symbol means given fuzzy sets (μ_{best}, μ_{avg} or μ_{worst}), and $f_{A_i}(x)$ represents output value of used fuzzy rules.

To show how IBIS works let's analyze an example of usage IBIS operator (see Fig. 7.6) on a simple coloring using 3 colors on a graph with 6 vertices. We have to remark that the genetic operator has a probabilistic character and this example should be regarded as an idea not as a deterministic effect.

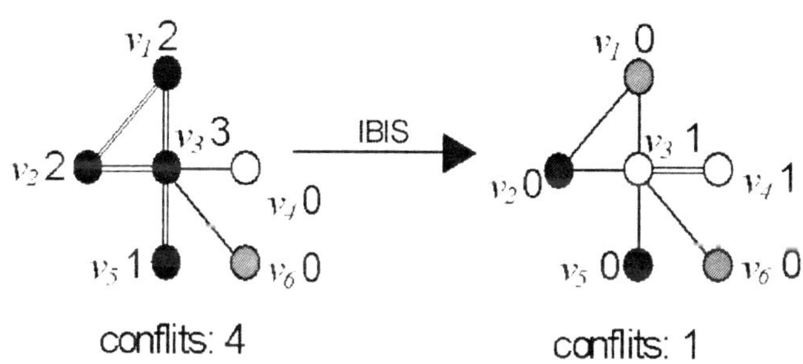

Fig. 7.6: Example of IBIS usage.

We start to analyze v_1 - it causes 2 conflicts and its probability of recoloring is average and IBIS probably will decide to recolor. There is no difference if it will be a gray or white color, the tournament selection can give a gray color. Next vertex v_2 also causes 2 conflicts and its probability of recolor is average too, but let's assume that its color is unchanged. Vertex v_3 causes 3 conflicts and has the biggest *TS_recolor* value and recolor probability. There are 3 analyzed the colors: black (causes 1 conflict), gray (1 conflict) and white (0 conflict), and tournament selection gives white color, as it causes 0 conflict. The other 3 vertices (v_4, v_5 and v_6) does not cause any conflict and its values of recolor probability are so small that we can assume that colors are the same in a new coloring. In this way, a new coloring causes only 1 conflict and it is generated from coloring with 4 conflicts. The IBIS operator is used as a genetic operator in the mutation role. It tries to discover the "best" coloring elements and keeps them in a newly build solution, also elements "average" and "worst" are changed according to a fuzzy logic Takagi-Sugeno driver suggestion.

Best Color Crossover (BCX)

A classical genetic algorithm with standard genetic operator is poor for solving optimization problems [33], so it is suggested to build specialized genetic operators which apply specific domain knowledge. In our approach this role fulfills a conflict localization (implemented as the *pff* value).

The main goal of the presented crossover is to connect two colorings (extended BCX can work on more than two parental colorings) to build a new coloring with a better or comparable quality (number of conflicts). The BCX operator analyzes color of each vertex in both parent colorings and according to number of conflicts selects a color of vertex with the least conflicts. Then a new coloring is analyzed whether it does not make the quality of coloring worse; it checks if number of conflicts are smaller than in the parent coloring and the operator colors of the given vertex. Otherwise, the color of vertex v_i in the new coloring is selected by a tournament method (select the color which causes the less number of conflicts for v_i vertex). A pseudocode of BCX operator is presented on Fig. 7.7.

```
function operator_BCX(coloring in parent_1, in parent_2, out off)

begin
for each vertex v_i of colored graph
    if (pff_parent_1(v_i) >= pff_parent_2(v_i))
        conflict_limit := pff_parent_2(v_i);
        parent_color := color_parent_2(v_i);
    else
        conflict_limit := pff_parent_1(v_i);
        parent_color := color_parent_1(v_i);

    conf := count_conflicts(off, v_i, parent_color);

    if (conf < conflict_limit )    color_off(v_i) := parent_color;
    else   color_off(v_i) := tournament_color(tournament_size);
end.
```

Fig. 7.7: Operator BCX pseudocode.

The BCX and IBIS operators use the heuristic to build a new coloring, and also in this operator *tournament* method is used to a color selection and provide the nondeterministic element to BCX work. Indeed the tournament selection is a smart compromise between the time consuming method of the full search (*tournament_size* parameter equals to a number of all used colors) and the nondeterministic random method (*tournament_size* parameter equals to 1).

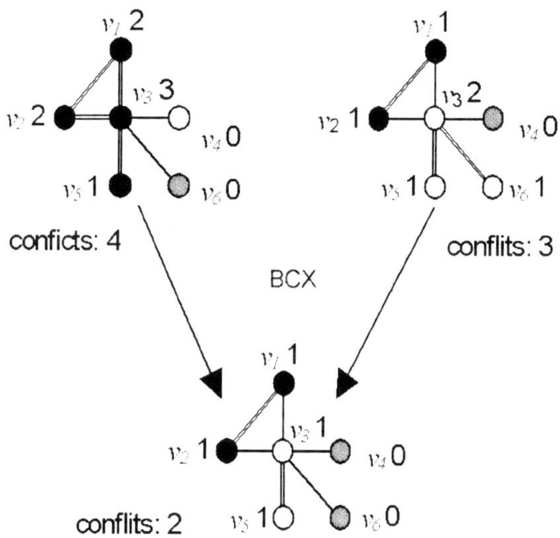

Fig. 7.8: Example of BCX operator usage.

Let's consider a BCX example simple graph with 6 vertex colored by 3 colors (see Fig. 7.8). Two evaluated colorings are analyzed vertex by vertex. Vertex v_1 in both coloring is black, and this color is assigned in the new coloring. Vertex v_2 has an analogical situation and black color is also assigned. For next vertex (v_3) can be assigned black color (causes 3 conflicts in $parent_1$) or white (causes 2 conflict in $parent_2$) - BCX operator chooses white color. In the new coloring it causes 0 conflict (but not all vertices have assigned colors) and this color is assigned. Vertex v_4 does not cause any conflicts in both parental colorings, and is assigned gray color from $parent_2$. Almost the same situation refers to v_5 where it causes 1 conflict in both coloring and also the color from $parent_2$ is assigned (white). The last vertex (v_6) causes less conflicts in color gray from $parent_1$ and this color is assigned.

The BCX crossover replaces a standard semi blind crossover by using specific domain knowledge (conflict localization). This allows to consider "good" the parental coloring elements and build a new coloring with better (or comparable) to the parental coloring quality.

7.5 Experiments and results

The experiments were performed using the DIMACS Challenge[1] set of 63 graphs, which include generated and real-world GCP graphs. All experiments were performed using tournament of 2 individuals selection method without elitism, using IBIS and BCX operators with a modified EA schema: none coloring is generated by two operators only one is applied, with the BCX priority. Such a change in EA schema is a consequence of the fact that both genetic operators may give significant changes in coloring and both need evaluated coloring. The maximum number of generation was set to 30000 (this value is based on experiments).

7.5.1 Genetic operators

The probability of BCX and IBIS has a large influence on the character of evolution process and EA effectiveness. Below we present 3 results selected from all performed experiments: "frequent" BCX with "rare" IBIS ($P_{BCX} = 0,7$ $P_{IBIS} = 0,2$), "frequent" BCX with "frequent" IBIS ($P_{BCX} = 0,7$, $P_{IBIS} = 1$) and "rare" BCX with "frequent" IBIS ($P_{BCX} = 0,3$ $P_{IBIS} = 0,9$). The last configuration is known as optimal. Experiments presented in this section are based on the representative for the DIMACS set graph *queen14_14*, which is colored with 17 colors, in a reasonable time, using population size equal 5. Because of the random tie breaking, EA becomes a stochastic method; results of 100 runs need to be averaged (for successful runs) to obtain useful statistics.

The evolution with configuration ($P_{BCX} = 0,7$ $P_{IBIS} = 0,2$) has a chaotic character. The average genotypic diversity equals 3,1 (standard deviation 4,46), the average phenotypic diversity equals 0,56 (standard deviation 0,68) and in 55% generation there is no diversity in population (premature convergence). In Table 7.2 we present a standard deviation of genotypic and phenotypic diversity in the evolution process. The average time needed to get a solution equals 5,4 seconds (the average number of generations equal 13775), a solution confidence is poor and equals 54%.

EA configuration ($P_{BCX} = 0,7$ $P_{IBIS} = 1,0$) is more efficient: the time needed to obtain a solution equals 5,0 seconds (it corresponds to 5023 generations) and solution confidence equals 95%. The premature convergence is still a large problem and occurs in 9,5% generations. A configuration known as optimal gives solution in 1,1 seconds (the average number of generations 2303) and the total confidence (100%). The evolution process (see Table 7.2)

[1] http://mat.gsia.cmu.edu/COLORING03/

runs regularly with phenotypic and genotypic diversity on a high level, suitably 1,51 (std. deviation 0,39) and 9,8 (std. deviation 5,5). The number of generations with premature convergence were reduced to 0,1%.

The comparison of results of three evolution processes with selected configurations is presented in Table 7.2.

Table 7.2: Genetic operators influence on character of evolution.

genetic		Dp_{avg}	Dg_{avg}	premature	t_{avg}	solution
P_{BCX}	P_{IBIS}	(Dp_{std_dev})	(Dg_{std_dev})	convergence[%]	[s]	[%]
0,7	0,2	0,56 (0,68)	3,1 (4,46)	55	5,4	54
0,7	1,0	1,28 (0,52)	7,69 (4,73)	9,5	5,0	95
0,3	**0,9**	**1,51 (0,39)**	**9,8 (5,5)**	**0,1**	**1,1**	**100**

Computer: Intel Celeron 2.53GHz 512MB RAM
GRACOM config.: $P_{BCX} = 0,3$ $P_{IBIS} = 0,9$, pop_size=5
graph: *queen14_14* 17 colors

The data presented in Table 7.2 shows that configuration has a large influence on the evolution character: the optimal configuration reduces the number of generations with premature convergence, keeps phenotypic and genotypic diversity on a high level, and also increases the EA effectiveness.

7.5.2 Population size

In experiments presented in previous sections, population size was kept on a constant level (5 individuals). However, the population size is also an important factor in EA. In this section, we present EA correlation between effectiveness and population size.

In DIMACS graph set, generally, two subsets of graphs were found: one needs population size 5, the other needs more individuals in population (100, 200 or even 400). Graph *queen14_14* is representative for one subset, the second subset can be represented by the graph *3-fullins-4*. In addition, the other reason why these graphs are considered is that its solving time is acceptable and permits to complete all necessary tests.

In Table 7.3 population size influence on the EA effectiveness in solving two graphs: *queen14_14* and *3-fullins-4* is presented. For coloring graph *queen14_14* the optimal population size is 5 individuals: the average time needed to receive a solution is 1,1 seconds (1868 generations) with low level of standard deviation of time and number of generations. Even though, EA has the best effectiveness for the given graph with such a population size, in 0,1% generations we can observe premature convergence.

The presented graph *3-fullins-4* is a representative of the other subset of tested graphs and EA optimal population size equals 100 or 200. EA with population of 100 individuals needs the average time equals 1,1 seconds (39,1 generations) but 200 individuals in population cause the EA more regular working: standard deviation of time and number of generation is on the lowest level.

Table 7.3: Population size influence on GRACOM effectiveness.

		population size				
		5	10	20	100	200
queen14_14 colors 17	t [s]	**1,1**	5,1	6	46,8	61,9
	t_{std_dev}	**0,4**	6,1	6,1	58	73,2
	gen	**1868**	4298	2568	3884	3162
	gen_{std_dev}	**613**	5048	2602	4832	3775
	pre.conv[%]	**0,1**	0,0	0,0	0,0	0,0
3-fullins_4 colors 8	t [s]	9,7	36,9	57	**1,8**	**2,3**
	t_{std_dev}	16,8	45,8	62	**1,1**	**0,25**
	gen	3879	7623	5939	**39,1**	**25**
	gen_{std_dev}	6717	9503	6558	**23,4**	**2,7**
	pre.conv[%]	0,47	0,0	0,0	**0,0**	**0,0**

Computer: Intel Celeron 2.53GHz 512MB RAM
GRACOM config.: $P_{BCX} = 0,3$ $P_{IBIS} = 0,9$, pop_size = 5

It is interesting that population of 5 individuals leads to premature convergence. It is connected with the genetic drift phenomena, when a small population causes bigger changes in population on phenotypic and genotypic level in the EA process. Each individual, even the worst one, in a small population has more chances to have an offspring and it causes that the evolution process can explore regions with the worse fitness. Furthermore, a genetic drift causes that evolution is similar to random wandering and no longer stays in the local optima and constantly looks for new regions for exploration.

The premature convergence is caused by a fitter individual, which dominates in the population, and gradually displaces individuals with a worse fitness. In addition, tests show that premature convergence usually occurs in one of the local optima where in the population dominates already best known individual. In classical EA premature convergence can be considered as one of EAs stop conditions because EA with the same values of parameters cannot cope with premature convergence. That is why it is a problem in a classical EA, but in the proposed EA with operators BCX and IBIS such a problem does not exist.

Thus, we can draw a conclusion that for some subsets of DIMACS graphs it is worth trying to solve it by EA with population of 5 individuals. Such a population size despite causing premature convergence is optimal to some graphs and gives a solution in the best time. Into considerations were taken many factors that should decide on optimal population size for a given graph, e.g. the number of vertices, the number of edges, the graph density and others graph theory tools. There are no confirmed conclusions, even some of the considered relationships give contradiction, e.g. a comparison of two DIMACS graphs with 3% density[2]: *will199GPIA* (701 vertices, 6772 edges) and *4-fullins-4* (690 vertices, 6650 edges) suggests some similarity in their difficulties. Actually, graph *will199GPIA* needs 5 or 10 individuals in population to solve it and it is one of the simplest DIMACS graph. The Graph *4-fullins-4* needs population size of 200 and is much more difficult to solve.

Although, the determination difficulty of a given graph coloring is an open issue, the intuition gives a hint that there is a relation between the graph diversity and its GCP difficulties. We can draw some suggestions that graphs with the high density probably have a larger number of local optima and are more difficult to solve. Thus, they need a small population size. This conclusion is confirmed partly by DIMACS set of graphs, e.g. two analyzed graphs: *queen14_14* (196 vertices, 837 edges, density 44%) and *3-fullins-4* (405 vertices, 3524 edges, density 4%). The first analyzed graph (*queen14_14*) has 44% density and needs small population size (5 individuals), the second one (density 4%) needs population size of 100 or more individuals.

7.5.3 Individual selection method

We tested two individual selection methods: tournament selection and proportional selection (roulette). The most efficient in GRACOM is the tournament selection, the proportional selection gives the solution in a longer time. We also tested a random selection: as there is no selection pressure, the GRACOM with a random selection is no longer a genetic algorithm. This selection should stop the evolution process but the specialized operators (applied a local search in IBIS and in BCX) allow to continue the evolution process and gives solution.

We analyzed the elite parameter (elitarism as surviving individuals with the best evaluation value) influence to GRACOM efficiency, as well. We noticed that small population (5-10 individuals) is dominated by a "super individual" and stuck the evolution process in one of many local optima. In larger populations (100-200 individuals) this process can be also observed but it has a smaller range. In GRACOM we do not recommend to use the elitarism parameter.

7.6 Tests of GRACOM efficiency

In this section we present the results we were able to obtain using GRACOM method in test instances. We tested GRACOM on the following instances:

- Nine random graphs (*dsjc125.x*, *dsjc250.x* and *dsjc500.x*) with an unknown chromatic number. Those graphs are considered as hard to solve,

[2] Density of graph $G(V,E)$ is calculated as $den(G) = \frac{2*|E|}{|V|*(|V|-1)}$.

- Leighton's graphs *le450_xx* are very difficult to solve in spite of the known chromatic number. More about that graphs in section 7,
- Two queen problem graphs *queen15_15* and *queen16_16*,
- Two timetabling benchmark graphs: *school_xx*,
- Two various graphs: *3-insertions_5* and *4-fullins_4*

We have tested GRACOM method on DIMACS benchmark graphs. The DIMACS set is varied graphs and our method can be tested comprehensively. This set contains e.g. random, scheduling or queen problem graphs. Also tests on a set of DIMACS benchmark graphs allows to compare results with other methods. We assumed that in these tests maximal processing time is less than 5000 seconds (or 10 million generations). GRACOM results presented on Table 7.4 are statistical, the averaged from 10 (100) runs.

Table 7.4: GRACOM test results on DIMACS benchmark graphs.

graph	**best known**	**GRACOM** colors	**runs** success (fails)	**generations** g_{avg} (g_{std_dev})	**time** [s] t_{avg} (t_{std_dev})
*dsjc125.1**	5	6	10 (0)	11162 (1770)	1,2 (1)
		5	10 (0)	15934 (9523)	2,79 (1,6)
dsjc125.5	17	19	10 (0)	2799 (1650)	15,9 (9)
		18	10 (0)	149279 (154885)	26,6 (27)
dsjc125.9	44	45	10 (0)	199254 (110273)	34,9 (19)
		44	10 (0)	1072684 (811322)	195,1 (147)
dsjc250.1	8	10	10(0)	959 (406)	0,4 (0,1)
		9	10(0)	18231 (20505)	7,9 (9,0)
dsjc250.5	28	30	10 (0)	1452290 (1508411)	504,6 (573)
dsjc250.9	72	78	10 (0)	215388 (139133)	74,3 (48)
dsjc500.1	12	13	10 (0)	675424 (429119)	638,9 (408)
dsjc500.5	48	52	10 (0)	4082001 (2645723)	2730 (1772)
*dsjc500.9**	126	140	10 (0)	1015813 (1115271)	2602 (2708)
le450_15a	15	16	10 (0)	154712 (127689)	116,1 (95)
le450_15c	15	21	10 (0)	2226900 (1601067)	1578,4 (1148)
le450_25c	25	26	10 (0)	5730955 (2683631)	3811 (1778)
*3-insertions_5***	?	10	10 (0)	21 (2)	4,8 (0,4)
		6	10 (0)	30 (6)	10,1 (2,2)
queen16_16	18	18	10 (0)	49502 (943)	7,2 (4,1)
queen15_15	16	17	10 (0)	37041 (17910)	121,5 (5,8)
school1_nsh	14	14	100(0)	275211 (613007)	152,5 (339)
*4-fullins_4***	7	8	10 (0)	22 (3)	7,1 (3,7)

Computer: Intel Pentium IV 2.8GHz (dualCore), 512MB RAM
Computational time is limited to 5000 seconds.
GRACOM config.: $P_{BCX} = 0,3$ $P_{IBIS} = 0,9$ pop_size = 5
pop_size = (*)20 (**)200 individuals

In Table 7.4 we present results obtained by GRACOM on the graphs selected from a DIMACS set. The presented results include graphs and configuration that are solved in (a) reasonable assumed time: less than 5000 second, (b) solution is always found. These two conditions make that some graphs cannot be colored optimally - results of extended experiments are presented in Table 7.4.

We compared GRACOM efficiency (minimal number of used colors) with other methods. We compared it with methods the most effective in

DIMACS challenge: HCA (Hybrid of Evolutionary Algorithm and Tabu Search method) [33], CLGA (Genetic Algorithm with heuristics) [14], ACO (Ant Colony Optimization) [12] and ILS (Iterated Local Search) [8]. Methods were tested on DIMACS benchmark graphs and on this base we compare its results. The selected graphs are considered as a difficulty to solve and make possible to do objective method comparisons. Results of comparisons are presented in Table 7.5, where each method is evaluated by a minimal number of colors used to the graph coloring. Results of GRACOM are based on longer time experiments (less than 2 hours of processor[3] working time).

Table 7.5: Comparison of selected methods efficiency.

graph	best known	GRACOM	CLGA	ILS	HCA	ACO
dsjc125.1	5	**5**	-	-	-	5
dsjc125.5	17	**17**	-	-	-	18
dsjc125.9	44	**44**	-	-	-	44
dsjc250.1	8	**8**	-	-	-	9
dsjc250.5	28	29	37	28	28	53
dsjc250.9	72	74	-	-	-	74
dsjc500.1	12	13	-	-	-	15
dsjc500.5	48	51	66	50	48	53
dsjc500.9	126	135	-	-	-	135
le450_15a	15	**15**	18	-	15	-
le450_15c	15	20	27	15	15	15
le450_25c	25	**26**	-	26	26	27
3-insertions_5	?	**6**	6	-	6	-
queen16_16	18	**18**	-	-	18	-
queen15_15	16	**17**	-	-	-	17
school1_nsh	14	**14**	-	-	-	14
4-fullins_4	8	**5**	7	-	-	-

In Table 7.5 we can see GRACOM high efficiency as the GCP solving method. It gives an optimal or suboptimal solution in a reasonable time (less than 2 hours of the processing time). Also, very interesting information we can get from summary of DIMACS'03 challenge. It was said that there is no "the best" GCP method dedicated to all graphs, rather some methods have better results with some sets of graphs.

7.6.1 Scheduling experiments

It was said that scheduling problems can be generalized to GCP. In this section we present some experimental results for Timetabling problems and consideration on Job-Shop Scheduling Problem in the GCP context.

[3] Intel Pentium IV 2.8GHz (dualCore), 512MB RAM.

To prove GRACOM practical usage in the scheduling problem we have tested our method on benchmark graphs constructed to Timetabling Problem. We selected *school11_xx* and *le450_xx* graphs families.

Graphs *school1_xx*
Many high schools ask their students to select a set of courses for the coming year to construct a timetable that allows students to take chosen courses. The problem is to construct such a graph that vertices represents courses and edges represent students. If exists any student that attends two courses there is an edge. The edge also exists if two courses have the same teacher.

Two problems *school1_xx* concern about 500 students and two semesters (the entire year). Each graph includes 385 vertices and is guaranteed 14-colorable. Graph *school1* includes entire timetabling graph, and *school1_nsh* subgraph does not take into consideration study halls [37].

Table 7.6: Timetabling benchmark graphs.

graph	color	$\mid V \mid$	$\mid E \mid$	$density(G)$	**GRACOM** runs(fails) t_{avg}
school1	14	385	19095	26%	100(0) 4,0[s]
school1_nsh	14	352	14612	24%	100(0) 152,5[s]

Test results are presented on Table 7.6. We can see that GRACOM method solves problems *school1_xx* in the acceptable time. Also GRACOM efficiency is high: the solution it is always found and it is an optimal solution.

Graphs *le450_xx*
These graphs were introduced by F.T.Leighton (Leighton F.T., "A graph coloring problem for large scheduling problems", Journal of Research of the National Bureau of Standards 84, pp.489-505, 1979) and its main feature is χ_G guaranteed (5, 15 or 25 colors, it is given in a graph name). Each of these graphs have 450 vertices and the density equals 25%. In Leighton's opinion, such a density is characteristic to exams scheduling problems. GRACOM results for selected *le450_xx* graphs are presented on Table 7.4 and Table 7.5.

The Graph Coloring Problem is defined as a generalization of scheduling problems and can be applied to solve this problem directly, e.g. in **TimeTabling** or **Examination TimeTabling Problem** we can include a classroom assignment method [38]. This method has a set of non-conflicting exams and for each timeslot it should be taken by students and a set of classrooms (according to its capacity).

The other method we can apply solving **Job-Shop Scheduling Problem**. We can provide some "local" modification of GCP definition to *Mixed Graph Coloring* [13] [39]. In this model we define a graph $G =< V, A, E >$ where V, E are adequately a set of vertices and a set of edges, A is a set of *arcs*. Each *arc* connects two vertices and represents a relation "predecessor/-successor", it means that if $(v_i, v_j) \in A$ means $\Psi(v_i) < \Psi(v_j)$ and the edge inclusion $[v_i, v_j] \in E$ implies $\Psi(v_i) \neq \Psi(v_j)$ [39].

We have also tested an early form of GRACOM on University TimeTabling Problem benchmark instances. Results of this test were presented in [40]. Also others researches tested such approach (e.g. [16]).

7.7 Conclusions and future work

In this chapter we introduced a novel evolutionary approach (GRACOM) to GCP and its application in scheduling problems. The basic scheduling definitions, problems and GCP notations are presented. The GRACOM method is described in details: *partial fitness function*, phenotypic and genotypic and specialized genetic operators (IBIS, BCX). The chapter shows that there exist a very small gap between GCP and scheduling real world problems. We hope that its would arouse interest in GCP scheduling application and evolutionary solving method.

The provided GRACOM method based on evolutionary algorithm prove its high efficiency but there are some future work directions. We plan to realize parallel GRACOM implementation to speed up the search process. We hope that GRACOM with multiplied population causes the evolution process more stable. The second reason of parallelism realization is testing GRACOM scalability.

References

1. Hamirez J-P , Hao J-K., An analysis of solution properties of the graph coloring problem, Applied Optimization, Meta-heuristics: Computer decision-making, pp.325-345, ISBN:1-4020-7653-3, 2004.
2. Kubale M. (ed.), Optymalizacja dyskretna: modele i metody kolorowania grafów (Discrete Optimisation: models and methods of graph coloring) Wydawnictwo Naukowo-Techniczne, Warszawa 2002. (in polish)
3. Kubale M. (ed.) Graph Colorings, Contemporary Mathematics 352, American Mathematical Society 2004.
4. Paquete L., Stuetzle T., An experimental Investigation of Iterated Local Search for Coloring Graphs, Applications of Evolutionary Computing, Proceedings of EvoWorkshops 2002, volume 2279 of Lecture Notes in Computer Science, pages 122-131. Springer-Verlag, 2002. Best Paper Award EvoCOP, 2002.
5. Dorne R., Hao J-K, A New Genetic Local Search Algorithm for Graph Coloring, Proceedings of the 5th International Conference on Parallel Problem Solving from Nature, pp.745-754, September 27-30, 1998.
6. Culberson J.C., Iterated Greedy Graph Coloring and the Difficulty Landscape, Technical Report 92-07, University of Alberta, Canada, 1992.
7. Vesel A., Zerovnik J., How good can ants color graphs?, Journal of Computing and Information Technology CIT (8), pp.131-136, 2000.

8. Chiarandini M., Stuetzle T., An application of Iterated Local Search to Graph Coloring Problem, In D.S. Johnson, A. Mehrotra, and M. Trick, (eds.), Proceedings of the Computational Symposium on Graph Coloring and its Generalizations, pp.112-125, Ithaca, New York (USA), 2002.
9. Galinier P., Hertz A., A survey of local search methods for graph coloring, Computers & Operations Research 33, pp.254256, 2006.
10. Eiben A.E., van der Hauw J.K., van Hemert J.I., Graph Coloring with Adaptive Evolutionary Algorithms, Journal of Heuristics 4(1), pp.25-46, 1998.
11. Fotakis D.A., Likothanassis S.D., Stefanakos S.K., An evolutionary annealing approach to graph coloring, E.J.W. Boers and al. (Eds.) EvoWorkshop 2001, LNCS 2037, pp.120-129, Springer-Verlag 2001.
12. Salari E., Eshghi K., An ACO algorithm for Graph Coloring Problem, Congress on Computational Intelligence Methods and Applications (ICSC) 2005.
13. Heinonen J., Pettersson F., Hybrid ant colony optimisation and visibly studies applied to a job-shop scheduling problem, Applied Mathmatics and Computation 187, pp.989-998, 2007.
14. Croitoru C., Luchian H., Gheorghies O., Apetrei A., A New Genetic Graph Colouring Heuristic, COLOR02, Ithaca, N.Y., 2002.
15. Kokosinski Z., Kolodziej M., Kwiaciarny K., Parallel Genetic Algorithm for Graph Coloring Problem, Proc. 4th Int. Computational Science Conference ICCS'2004, Krakow, Poland [in:] Lecture Notes in Computer Science, Vol. 3036, pp. 217-224, 2004.
16. Myszkowski P.B., Kwasnicka H., IBIS: A new evolutionary algorithm for the timetable problem, Proccedings Intelligent Information Processing and Web Minning, IIS:IIPWM04, [in] Advances in Soft Computing, pp. 454-458, Springer 2004.
17. Myszkowski P.B., A Partial Fitness Function in Evolutionary Algorithms applied to Graph Coloring Problem, AE 1064, pp.180-189, Proceedings of 13th Conference Knowledge Acquisitions and Management (KAM'05), Karpacz, May 12-14 2005.
18. Myszkowski P.B., New evolutionary approach to the GCP: premature convergence and an evolution process character, Proceedings of 5th International Conference on Intelligent Systems Design and Applications: ISDA'05 (Wroclaw, 8-10 Sep 2005) [Eds: Kwasnicka H., Paprzycki M.], Los Alamitos [in.]: IEEE Computer Society [Press], pp. 338-343, 2005.
19. Juhos I., Toth A., Tezuka M., Tann P., van Hemert J.I., A new permutation model for solving the graph k-coloring problem, proceedings Kalmàr Workshop on Logic and Computer Science, pp.189-199, 2003.
20. Marx D., Graph Coloring Problems and their applications in scheduling Periodica Polytechnica Ser. El. Eng. 48(1-2):5-10, 2004.
21. Astarian A.S., de Werra D., A generalized class-teacher model for some timetabling problems, European Journal of Operational Research 143 (2002), pp.531-542, 2002.
22. Terashima-Martin H., Ross P., Valenzuela-Rendón M., Clique-Based Crossover for solving the Timetabling Problem with GA, [in] M.Schoenauer et al (ed.), Proceedings of CEC 99 Conference, Washington, IEEE Press, 1999.
23. Paechter B., Rankin R.C., Cumming A., Fogarty T.C., Timetabling the Classes of an Entire Univeristy with an Evolutionary Algorithm, [In] A. E. Eiben, M. Schoenauer, and H.-P. Schwefel (editors), Parallel Problem Solving From Nature – PPSN V, Amsterdam, Holland, 1998. Springer-Verlag, 1998.

24. Rossi-Doria O., Paechter B., An hyperheuristic approach to course timetabling problem using evolutionary algorithm, Technical Report CC-00970503, Napier University, Edinburgh, Scotland 2003.
25. Burke E.K., Newall J.P., Solving Examination Timetabling Problems through Adaption of Heuristic Orderings, Annals of operations Research No. 129, pp. 107-134, 2004.
26. Azimi N.Z., Hybrid heuristics for Examination Timetabling problem, Applied Mathematics and Computation 163, pp.705-733, 2005.
27. Fang H-L, Ross P., Corne D., A promising genetic algorithm approach to Job-Shop Scheduling, Rescheduling, and Open-Shop Scheduling Problems, SAI Research Paper no. 623, 1993.
28. Xu X-d., Li C-x., Research on immune genetic algorithm for solving the job-shop scheduling problem, The International Journal of Advanced Manufacturing Techology, Springer London 2006.
29. Hart E., Ross P., Corne D., Evolutionary scheduling: a review, Genetic Programming and Evolable Machines 6, pp.191-220, 2005.
30. Correa R.C., Ferreira A., Rebreyend P. Scheduling Multiprocessor Tasks with Genetic Algorithms, IEEE Transactions on Parallel and Distributed Systems, Vol.10, No.8, 1999.
31. Montazeri F., Salmani-Jelodar M., Najmeh Fakhraie S. and Mehdi Fakhraie S., Evolutionary Multiprocessor Task Scheduling, Proceedings of the International Symposium on Parallel Computing in Electrical Engineering (PARELEC'06), IEEE Computer Society 2006.
32. Rzadca K., Seredynski F., Heterogeneous multiprocessor scheduling with differential evolution, Evolutionary Computation: The 2005 IEEE Congres, Volume: 3, pp: 2840- 2847 Vol. 3, 2005.
33. Galinier P., Hao J-K, Hybrid evolutionary algorithms for graph coloring, Journal of Combinational Optimization 3(4), pp.379-397, 1999.
34. Herrera F., Lozano M., Fuzzy Genetic Algorithms: Issues and Models, Technical Report DECSAI-98116, Dept. of Computer Science and A.I., University of Granada, 1998.
35. Hamirez J-P., Hao J-K., Scatter Search for Graph Coloring, Selected Papers from the 5th European Conference on Artificial Evolution, pp.168-179, 2001.
36. Herrera F., Lozano M., Adaptation of Genetic Algorithm Parameters Based on Fuzzy Logic Controllers, [in:] Herrera F. and J.L. (Eds.), Genetic Algorithms and Soft Computing, pp.95-125, Physica-Verlag, 1996.
37. Lewandowski G., Condon A., Experiments with parallel graph coloring, heuristics and applications of graph coloring, Second DIMACS Implementation Challenge, volume 26 of 26 DIMACS Series in Discrete Mathematics and Theoretical Computer Science, pp. 309-324, American Mathematical Society, 1996.
38. Dammak A., Elloumi A., Kamoun H., Classroom assignment for exam timetabling, Advanced in Engineering Software 37, pp.659-666, 2006.
39. Al-Anzi F.S., Sotskov Y.N., Allahverdi A., Andreev G.V., Using mixed graph coloring to minimize total completion time in job shop scheduling, Applied Mathematics and Computation 182, pp.1137-1148, 2006.
40. Myszkowski P.B., A hybrid genetic algorithm for timetable problem, Mendel'03 proceedings, pp.102-107, 9th International Conference on Soft Computing - MENDEL, Brno (Czech Republic), 2003.

8

Heuristics and meta-heuristics for lot sizing and scheduling in the soft drinks industry: a comparison study

D. Ferreira[1], P.M. França[2], A. Kimms[3], R. Morabito[1], S. Rangel[4], and C.F.M. Toledo[5]

[1] Departamento de Engenharia de Produção, Universidade Federal de Sao Carlos, C.P. 676, 13565-905, Sao Carlos, SP, Brazil deise@dep.ufscar.br; morabito@power.ufscar.br

[2] Departamento de Matemática, Estatística e Computação, Universidade Estadual Paulista, C.P. 1234, 19060-400, Presidente Prudente, SP, Brazil paulo.morelato@fct.unesp.br

[3] Dept. of Technology and Operations Management, University of Duisburg-Essen, 47048, Duisburg, Germany alf.kimms@uni-duisburg-essen.de

[4] Departamento de Ciência da Computação e Estatística, Universidade Estadual Paulista, Rua Cristóvão Colombo, 2265, 15054-000, S. J. do Rio Preto, SP, Brazil socorro@ibilce.unesp.br

[5] Departamento de Ciência da Computação, Universidade Federal de Lavras, C.P. 3037, 37200-000, Lavras, MG, Brazil claudio@dcc.ufla.br

Summary. This chapter studies a two-level production planning problem where, on each level, a lot sizing and scheduling problem with parallel machines, capacity constraints and sequence-dependent setup costs and times must be solved. The problem can be found in soft drink companies where the production process involves two interdependent levels with decisions concerning raw material storage and soft drink bottling. Models and solution approaches proposed so far are surveyed and conceptually compared. Two different approaches have been selected to perform a series of computational comparisons: an evolutionary technique comprising a genetic algorithm and its memetic version, and a decomposition and relaxation approach.

Key words: Two-level Production Planning, Lot Sizing, Scheduling, Soft Drinks Industry, Genetic Algorithm, Memetic Algorithm.

8.1 Introduction

The motivation behind writing this contribution is to offer the academic and practitioner industrial engineering community dealing with planning and scheduling tasks in the soft drinks industry a text with the most recent

D. Ferreira et al.: *Heuristics and meta-heuristics for lot sizing and scheduling in the soft drinks industry: a comparison study*, Studies in Computational Intelligence (SCI) **128**, 169–210 (2008)
www.springerlink.com

contributions to the field and also a comparative study with some selected approaches. A major concern that inspired the chapter was to review modern techniques especially designed for building production schedules applied to real world settings. The technical literature devoted to planning and scheduling is vast and there are plenty of sophisticated methods. However, the specificities of the soft drinks industry require dedicated models and specific solution methodologies that justify a text like this one. Thus, the objective of this chapter is first to discuss the planning features of a soft drinks plant and then to assess the main suitable mathematical models, as well as to explore and evaluate the quality and computing time of some selected solution methods.

8.1.1 Soft Drinks Plant

The consumption of soft drinks has grown considerably worldwide. In Brazil, where part of the present research has been carried out, there are more than 800 plants supplying a 13-billion liter annual consumer market, which is the third in the world. This figure represents an amount which is twice as large as ten years ago. The diversity of products offered to consumers, the scale of plants and the complexity of modern filling lines require the adoption of optimization-based programs to produce efficient production plans. Indeed, a plenty of specialized commercial packages have been launched over the last years as an effort to overcome the difficulties human schedulers have faced. However, in most cases the complexity of the planning task imposes hard manual adjustments for the production schedules produced by those packages. The biggest contribution of the approaches studied in this chapter is to propose integrated optimization-based models able to encompass both the two interdependent production levels, namely the tank level and the bottling level. Due to its inherent complexity, the needed synchronization between these two levels is disregarded by commercial packages thus often leading to ineffective production schedules.

The production process found in medium to large plants consists of an upper level with capacitated mixing tanks used to prepare and store liquids which are pumped to the lower level constituted by bottling or canning lines disposed in parallel (Fig. 8.1). At the tank level, decisions concerning the amount and the time the raw materials have to be stored in every available tank must be made. Analogously, at the production line level the lot size of each demanded item and its corresponding schedule in each line must also be determined. However, a line is able to meet the weekly demand only if the necessary amount of raw material can be stored in a connected tank. Indeed, a solution which integrates these two lot sizing and scheduling problems has to be determined. Moreover, once the necessary amount of raw material is stored, it can not stay for a long time waiting to be pumped to lines. There is a synchronization problem here because the production in lines and the storage in tanks must be compatible with each other throughout the time horizon. Hence, a lot sizing and scheduling problem has to be solved at each one of these two-levels taking into account that the corresponding decisions must be synchronized.

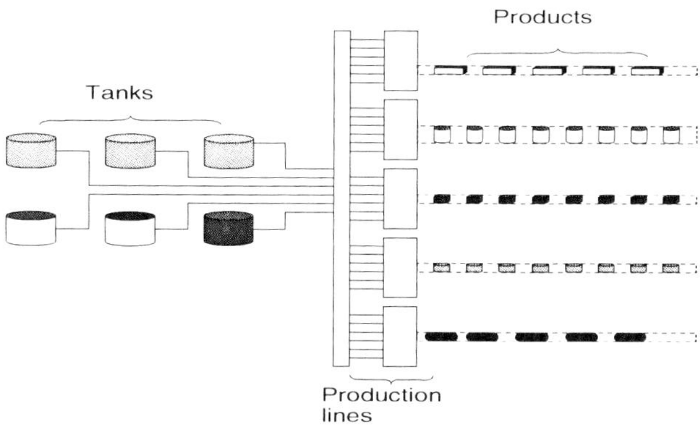

Fig. 8.1: The two-level production process.

Possibly the two-level synchronization is the most challenging aspect of this problem. Due to this fact, this problem is called the Synchronized and Integrated Two-level Lot sizing and Scheduling Problem (SITLSP) [1].

The raw materials are the flavors of the liquids which are bottled in the production lines. For technical reasons, a tank is only filled up when empty and two different raw materials cannot be stored at the same time in the same tank. A sequence-dependent changeover time (setup) of up to several hours occurs to clean and fill up a tank, even if the same soft drink is replaced. A sequence-dependent setup time means that the time required to prepare and fill a tank for the next liquid depends on the liquid previously stored. Indeed, the setup time when a diet drink follows a plain flavor drink is much longer when the sequence is inverted. Nothing can be pumped to a production line from the tank during the setup time. One tank can be connected to several production lines which will share the same raw material. Moreover, the production lines can be connected to any tank. However, it can receive raw material from only one tank at a time. The final product (item) is defined by the flavor of the soft drink and the type of container (glass bottles, plastic bottles or cans) of different sizes. In large plants it is common to find situations where various products can share a common production line and various lines can produce the same product in parallel. The production schedule also has to take into consideration the impact of product changeovers on the effective capacity of the production lines. As in the mixing tanks, these changeover times (setups) are also sequence-dependent and occur whenever a line has two different products switched.

The weekly demands have to be met within a time horizon of a certain number of weeks. Since the forecasts of customer orders are error-prone, there is little interest in seeking solutions in large horizons. Instead, it is more realistic to work in a rolling-horizon basis with a 3 or 4-week time horizon. The excessive number of final products leads to inventory costs. There are also inventory costs for the storage of raw materials in tanks in various time periods. The sequence-dependent setup costs for products and raw materials are proportional to the sequence-dependent setup times in lines and tanks, respectively.

As synchronization is a key feature to be taken into consideration, a deeper explanation in this respect is now in order. As said before, the lines must wait until the liquids are ready to be pumped to them. On the other hand, the liquids stored in tanks can not be sent to the lines unless they are ready to initiate the bottling process. Fig. 8.2 illustrates the commitment between the two levels.

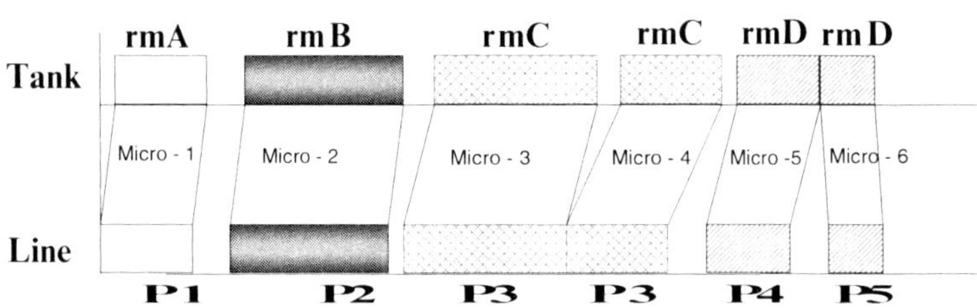

Fig. 8.2: Batches sequenced but not synchronized.

Observe that batches of liquids and items are properly sequenced in the tank and in the line, respectively, but they are not synchronized. The gaps between two batches of liquids (rmA, rmB, rmC, rmD) and items ($P1$, $P2$, $P3$, $P4$, $P5$) represent given changeover times. The planning horizon is divided into 6 micro-periods. Notice that item $P3$ is produced in both micro-periods 3 and 4; it uses the same liquid rmC but needs tank replenishment. Also observe that the same liquid rmD is used in distinct products $P4$ and $P5$ because they make use of, say, different bottle sizes. Due to the discrepancies between tank and line setup times, the product batches have to be delayed by inserting idle times (black rectangles) in the production line in micro-periods 1-4 while the liquid batches must be delayed as well by inserting idle times (empty rectangles) in micro-periods 5 and 6, as shown in Fig. 8.3.

Given that the matter of synchronization has to be treated by postponing batches, it causes impact on the capacities of tanks and production lines and may result in infeasible schedules. Therefore, these actions must be considered together with lot sizing and scheduling decisions. Thinking in terms of mathematical models, they must incorporate decision variables especially designed to deal with the crucial issue of synchronization thus adding substantial complexity to model building and implementation of solution techniques.

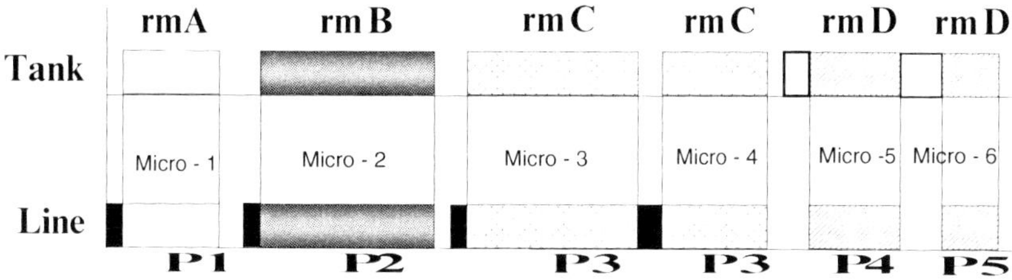

Fig. 8.3: Sequenced and synchronized batches.

8.1.2 Literature Review

The SITLSP is a two-level lot sizing and scheduling problem with some particular considerations - such as the two-level synchronization - which render it much more complex. In this literature review the main published articles addressing the SITLSP itself or similar problems dealing with the planning and scheduling tasks in the soft drinks industry will be summarized and commented on. As a result of this review, the two most promising methods are selected and then described in detail in the next sections, followed by a computational study which compares their performances and application fields.

The SITLSP addresses various issues of classical lot sizing and scheduling problems that have been dealt with in the literature before. We refer to [2] and the references therein for overviews in capacitated lot sizing and scheduling problems. The capacitated lot sizing and scheduling problem is a NP-hard optimization problem [3], but finding a feasible solution is easy (e.g., a lot-for-lot like policy) if no setup times are to be taken into account. If setup times are present, the problem of finding a feasible solution is NP-complete already. A discussion of lot sizing and scheduling with sequence-dependent setup costs or sequence-dependent setup times can be found in e.g. [4]- [8]. Publications addressing multi-level lot sizing (scheduling) problems can be found in e.g. [9]- [13]. Studies regarding these problems with parallel machines are found in e.g. [4], [14]- [18].

However, to the best of our knowledge, the only work that comes close to the SITLSP is the one described in [16], which is a multi-level extension of [8]. In this approach, however, it might be necessary to split up a lot into smaller ones in order not to lose generality. This would not be a practical idea in solving the SITLSP because every new lot for the tanks requires a new setup, which is not desired.

Focusing the attention only on articles which specifically deal with the SITLSP, one can cite [19] where an extensive mixed-integer mathematical model describing the problem is presented. Unfortunately, due to its complexity and size, the model had to be omitted in this chapter. Instead, a brief explanation regarding its formulation is given next. The underlying idea to create a model for the SITLSP combines issues from the General Lot sizing and Scheduling Problem (GLSP) and the Continuous Setup Lot Sizing Problem (CSLP). A comparison of the CSLP and GLSP and more details about these problems can be found in [2], [20]- [22].

As shown in [19], the SITLSP model supposes that a planning horizon is divided into T (macro-) periods of the same length. A maximum number of slots (S for each line and $\overline{S}$ for each tank) is fixed for each macro-period $t = t_1, t_2, \ldots, t_T$. The limited number of slot assignments is an idea taken from the GLSP. This enables us to determine in the SITLSP for which liquid (raw material) a particular slot in a particular line (tank) is reserved, and which lot size (a lot of size zero is possible) should be scheduled. Fig. 8.4 illustrates the idea. Consider $T = 2$ macro-periods, 5 raw materials (rmA, rmB, rmC, rmD and rmE), 6 products ($P1$, $P2, \ldots, P6$), 3 tanks ($Tk1$, $Tk2$ and $Tk3$) and 3 lines ($L1$, $L2$ and $L3$). Suppose that the raw material rmA produces the product $P1$, rmB produces $P2$ and $P3$, rmC produces $P4$, rmD produces $P5$ and rmE produces $P6$. The total number of slots is $S = \overline{S} = 2$ and it can not be exceeded in each macro-period.

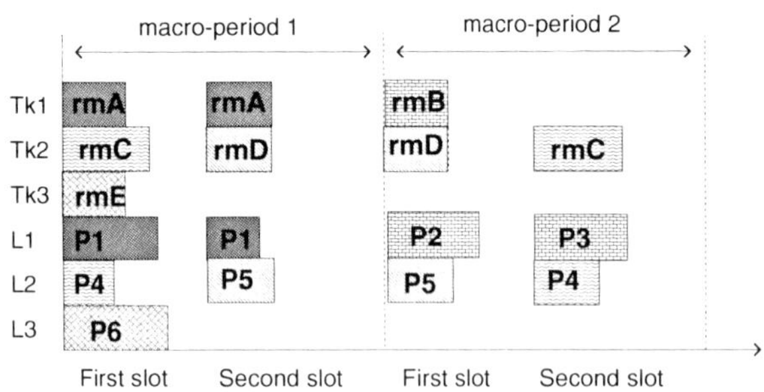

Fig. 8.4: Sequence of slots for tanks and lines.

Fig. 8.4 shows that only one slot is occupied by lots of raw materials and products in tank $Tk3$ and line $L3$, respectively, during the first macro-period. In the second macro-period, there is no slot occupied in $Tk3$ and $L3$. On the other hand, two slots need to be occupied by lots of rmA in the first macro-period of $Tk1$. This raw material is used to produce $P1$ which is assigned to the two possible slots in the first macro-period of $L1$. In this case, the necessary amount of rmA to produce $P1$ filled up $Tk1$ completely in the first slot assignment. Another slot assignment of rmA is necessary to conclude the production of $P1$. Refilling $Tk1$ with rmA leads to an interruption in $L1$, so a second slot assignment of $P1$ in $L1$ is also made. Variables indexed by slots in the SITLSP mathematical model made it possible to write constraints that integrate the line and tank occupation [1, 19].

As well as the assignment of liquids to tanks and products for lines in each macro-period, it is also necessary to synchronize the slots scheduled in a two-level problem like this. This is done using the micro-period idea found in the CSLP, where each macro-period t is divided into T^m micro-periods with the same length. Furthermore, according to the CSLP assumptions, the capacity of each micro-period can be total or partially occupied and only one product type (item) can be produced per micro-period. These assumptions are used in the SITLSP. Fig. 8.5 illustrates this idea using the assignment shown in Fig. 8.4.

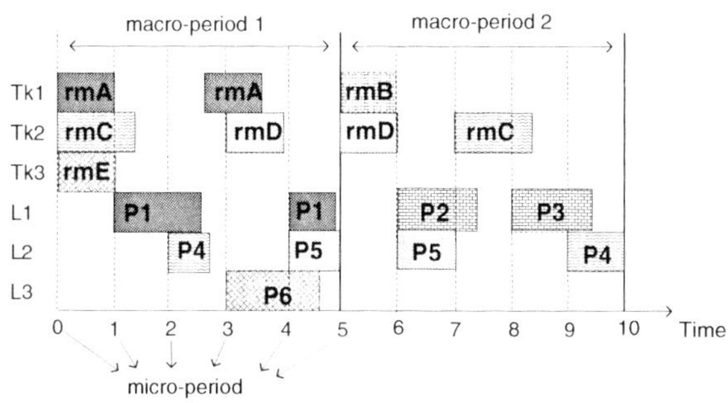

Fig. 8.5: Synchronization between lines and tank slots.

The macro-periods are divided into 5 micro-periods of the same length. The times necessary to prepare each tank (setup time) are represented now by the slots with acronyms rmA, rmB, rmC, rmD and rmE. In the same way, the processing time of each product is represented by the slots with symbols $P1$, $P2$, $P3$, $P4$ and $P5$. For instance, the two slots of $Tk1$ occupied by rmA indicate that the tank is ready to be used at the end of the first micro-period and it is refilled with the same raw material between the third and fourth micro-period. The micro-periods enable us to synchronize the beginning of $P1$ production in the second micro-period of $L1$ after the end of rmA setup time in $Tk1$. Moreover, we can see when the refilling of $Tk1$ occurs in the second slot of $Tk1$. This requires the interruption of $P1$ and the beginning of the second slot occupation of $P1$ in $L1$, after the end of rmA setup time in $Tk1$. The constraints and variables indexed by the micro-period make it possible to describe these situations in the mathematical model of the SITLSP [1,19].

The objective function is to minimize the total sum of setup costs, inventory holding costs, and production costs for tanks and production lines. Observe that for a given demand a feasible solution may not exist. In order to guarantee a feasible solution the model allows for shortages. This is modelled by allowing every item a certain quantity of units to be "produced" in the first period without using capacity. Naturally, a very high penalty M is attached to such shortages so that, whenever there is a feasible solution that fulfills all demands, one would prefer this one.

In short, one can say that the main contribution of the approach used to model the SITLSP is to integrate and synchronize two lot sizing and scheduling problems. This is done using variables and constraints indexed by slots and micro-periods. The variables and constraints indexed by slots help to integrate a feasible sequence of occupation in the two levels, while variables and constraints indexed by micro-periods help to determine a feasible synchronization in the two levels within the time horizon. The entire set of variables and constraints is able to mathematically describe most of the decisions and constraints present in the industrial problem studied here.

The SITLSP model is coded and solved using the GAMS/CPLEX package that uses a branch-and-cut solution approach to find an optimal solution. For small instances the method is fast and reliable but as problem size grows, the number of distinct integer solutions increases exponentially, causing the search to take too long even for finding the first integer feasible solution. The computational results in a series of instances with $T = 1, 2$, 3 and 4 macro-periods and $T^m = 5$ micro-periods revealed that for $T > 2$ the method failed in finding optimal solutions within 1 hour of execution time.

A second paper [23] dealing with the SITLSP introduces an evolutionary approach capable of overcoming the limitations faced by the previous method [19] which is not useful to solve real world instances. Being an approximate approach, optimal solutions are not guaranteed but performance comparisons carried out in the paper using a set of small instances with known optimal solutions have shown good behavior in reasonable computing time.

Evolutionary algorithms (EAs) belong to a class of computational methods called bio-inspired systems that simulate biological processes such as crossover, mutation and natural selection. The simulation of these processes follows the method when searching for a solution (individual) of a specific problem. The method begins by analyzing a set of problem solutions (population) and determines new solutions applying genetic operators (selection, crossover and mutation) to the previous set of solutions. Some solutions (new or old ones) are selected and form the new set of solutions for the next iteration. This procedure is repeated until some stop criterion is satisfied. The first work concerning EA was presented by [24] and details on their implementation can be found in [25] and [26]. A memetic algorithm (MA) is a hybrid population-based approach [27] which combines the recognized strength of population methods such as genetic algorithms (GA) with the intensification capability of a local search. In a MA, all agents or individuals evolve solutions until they are local minima of a certain neighborhood, i.e. after steps of recombination and mutation, a local search is applied to the resulting solutions.

The GA presented in [23] proposes a multi-population approach that can be understood as a variant of island models. The method was developed using the NP-Opt that is an object-oriented framework written in JAVA code [28]. This framework has optimization procedures based on evolutionary computation techniques to address NP-hard problems.

The computational experiments carried out with the GA approach on a set of large instances with up to 12 macro-periods, 10 micro-periods and 15 final products show that the method is able to solve real-world instances. Consequently it was one of the approaches selected to take part in the comparisons using real data.

Recently another interesting method capable of handling the SITLSP and also applied to the planning and scheduling of soft drinks has been proposed. Unlike the approach presented in [1] where a highly general (and complex) model is introduced and solved by a commercial MIP package, the MIP model formulated in [29] is more restricted and amenable to be solved by approximate approaches. Amongst other differences that will be pointed out in detail in Section 8.3, the mathematical formulation is less general than the one presented in [1] as it forces the quantity of tanks to be the same as for the lines. On the other hand it opens up the possibility that a tank can be filled with all liquids needed by its line.

Alternatively from [1], which uses a MIP package for finding optimal solutions, the solution approach adopted in [29] relies on different heuristic methods to obtain approximate production schedules. The details of these heuristics will be explained fully in the next section.

In [30] a mathematical programming model to deal with the planning of a canning line at a drinks manufacturer is introduced. Compared to the models presented in [1] and [29], this one is much more restricted because it disregards any sequencing considerations when designing production plans. Instead, it focuses only on planning issues, i.e., lot sizing problems. The MIP model objective function minimizes inventories and backorder penalties. The model allows for different setup times depending on whether the changeover between canned products involves a change of liquid or not. However, the model does not consider sequence-dependent setup times. Solution strategies consist of a "lazy" default - simply let an industrial strength branch-and-bound MIP solver try to find a good solution within a pre-specified amount of time - to more sophisticated heuristic approaches. One of these that is worth mentioning is a local search-based meta-heuristic which the author called diminishing-neighborhood search (DNS). In this approach one starts the method with the largest possible neighborhood to avoid bad local optima, and then narrow the neighborhood in a continuously diminishing way as the search proceeds towards a good solution, even if it is still a non-global optimum. The methods were tested in real data instances with up to 41 distinct products to be filled from 14 different liquids in a planning horizon made up of 13 consecutive weeks (periods). The final conclusion is that the methods present a classical trade off between quality and CPU time with the best results being obtained by DNS at the expense of 2 hours of computing time. In spite of the interesting solution methods and results reported in [30], the approach will not be included in the comparisons mainly due to the more restrictive character of the production system for which it was developed.

Closing this introduction, one can conclude that the methods for dealing with the SITLSP embedded in a soft drinks production planning scenario, and most importantly with potential to be applied in real world situations in a broad sense, are the EA approaches as the one proposed in [23] and the heuristic model-based method presented in [29]. Therefore, a computational comparison involving these two approaches are conducted in a series of practical instances obtained from a large soft drinks manufacturer. One of the purposes of the comparison is to suggest practical guidance on which method best suits the different possible situations found in the industry. The other sections are organized as follows. In Section 8.2 an MA approach based on the GA developed in [23] is proposed while the methods introduced in [29] are presented in Section 8.3. Section 8.4 shows the instances used in the computational comparisons and discussions on the test results. Finally, Section 8.5 concludes the chapter and discusses some topics for further research.

8.2 Evolutionary Approaches

In this section a memetic version of the GA presented in [23] is proposed to solve the SITLSP . The term "Memetic Algorithms" [27,31] (MAs) was introduced in the late 80s as a class of meta-heuristics that have the hybridization of different algorithmic approaches as a crucial aspect in their conceptions. The majority of the MA applications are population-based approaches in which a set of cooperating and competing agents are engaged in periods of individual improvement of the solutions while they sporadically interact. The adjective 'memetic' comes from the term 'meme', coined by Dawkins [32] as an analogy to the term 'gene' in the context of cultural evolution. It stands for the unit of information which reproduces itself as people exchange ideas. MAs are also referred in the technical literature as hybrid genetic algorithms and usually they take a less sophisticated conception as a combination of GAs with a local search procedure applied to some of the individuals in the population.

8.2.1 The MA structure

EAs are global search procedures inspired by biological evolution processes [24, 25]. Amongst the EAs, the GAs are the most popular. A GA differs from local search or constructive heuristics because it has an initial set of solutions (individuals) which have been randomly established. Also called chromosomes, these individuals are solution representations for the problem to be solved. At each GA generation, the individual's fitness is measured and genetic operators are executed in the population. These operators are based on genetic behavior such as crossover, mutation and selection. The individuals with better fitness values remain in the population from one generation to another. MAs and GAs have been applied to solve complex and real-world problems (see [33]- [35]).

The evolutionary methods presented in this section are conceived as a multi-population approach with a hierarchical ternary tree structure. The multi-population approach was chosen because populations that evolve separately usually have different characteristics according to the genetic drift idea [36]. This can lead to a more effective exploration in the solution space of the problem. A better performance of hierarchically structured populations over non-structured population schemes in EAs has been attested in previous experiments concerning different problems (e.g. machine scheduling, asymmetric travelling salesman, capacitor placement). Computational results solving these optimization problems were reported by [18], [28] and [37] using a multi-population GA with a hierarchical ternary tree structure. Furthermore, the authors in [38] reported that the results obtained with the total tardiness single machine scheduling problem using GA with hierarchically structured populations are better than the ones with non-structured populations. These findings have been confirmed by the multi-population MA developed to solve the SITLSP. Moreover, the adoption of the multi-population approach has enhanced the convergence features of the GA, postponing premature convergence and improving its whole effectiveness [23].

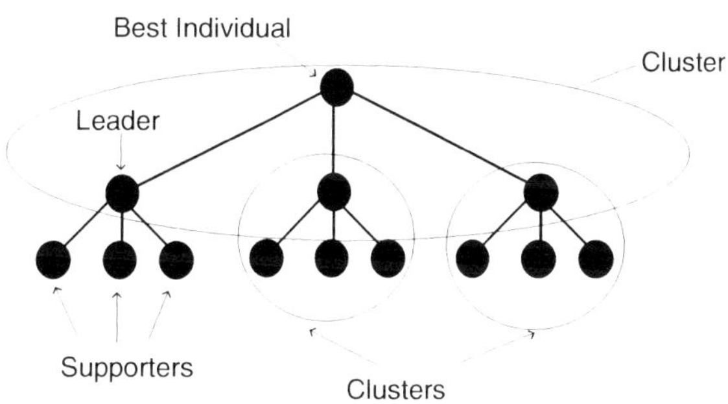

Fig. 8.6: MA clustered population.

The MA population structure consists of several clusters, each one having a leader solution and three supporter solutions, as shown in Fig. 8.6. The leader of a cluster is always fitter than its supporters. As a consequence, top clusters tend on average to have fitter individuals than bottom clusters. As new individuals are constantly generated, replacing old ones, periodic adjustments are necessary to keep this structure well ordered. The number of individuals in the population is restricted to the numbers of nodes in a complete ternary tree, i.e. $(3k - 1)/2$ where k is the number of levels of the tree. That is, 13 individuals are necessary to construct a ternary tree with 3 levels, 40 to make one with 4 levels and so on. Previous experiments with distinct tree structures (binary, ternary, etc.) and a variable number of levels (two, three, etc.) attested that the best results were obtained by a 3-level ternary tree [6]. Observe that the population is constituted by four distinct clusters, three in the bottom level and one in the upper level, with four leaders, respectively. The upper level leader (best individual) is always the fitter individual in the population.

The MA procedure applied to the structured population is summarized in Algorithm 8.1. The initial population *Pop*() is generated and submitted to a generation loop. Operators *recombinePop*() and *mutatePop*() produce a new individual (or more than one) which is improved by a local search algorithm in *optmizePop*(). In *structurePop*() the population is re-structured to maintain the hierarchy between agents.

Algorithm 8.1: The MA procedure.

---

initializePop();
repeat
    recombinePop();
    mutatePop();
    optimizePop();
    structurePop();
until *Termination condition*;

The operator *recombinePop*() performs a crossover over a cluster (selected at random) and always involves a supporter node (selected at random) and its corresponding leader node. The new individual (Child) is submitted to a mutation procedure *mutatePop*() depending on the mutation probability test result. The operator *optimizePop*() applies a local search to Child and if it is now better than some parents, the Child will replace the parent with the worst fitness value (Fig. 8.7). Otherwise, the new individual is not inserted into this population. After every crossover/mutation/optimization/replacement operation, adjustments carried out by *structurePop*() are necessary to keep the cluster structure well ordered where the best is always the leader (Fig. 8.8).

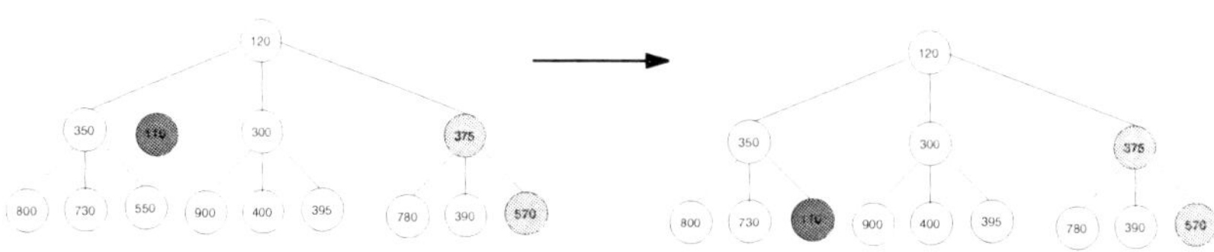

Fig. 8.7: Population before and after Child insertion.

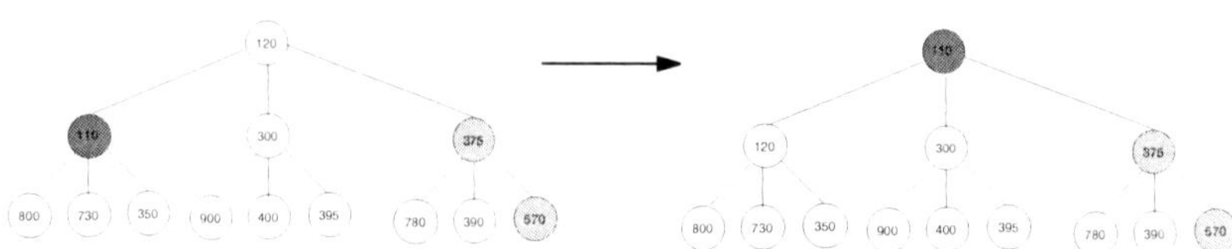

Fig. 8.8: Adjustments in the population.

8.2.2 The multi-population structure

Multi-population MAs are common in implementations in which computational tests are executed on parallel computers. Usually in such case, each processor is responsible for one population. Results obtained on parallel computers are in general much better than the ones obtained on sequential machines. Even though the computational implementation and tests have been carried out on a single-processor computer, the multi-population scheme has been implemented to take advantage of the hierarchical population structure. This decision is also supported by the fact that the SITLSP is a complex combinatorial problem for which simplistic evolutionary solution approaches tend to fail. Fig. 8.9 shows how four populations interact. After a certain number of executions of the genetic operators crossover and mutation performed in each population, followed by a local search step applied to the best individual, one can check if the population convergence occurred, i.e. if no new individuals are inserted in it. If not, the process is repeated until convergence of all the populations. In this case the migration step takes place with a copy of each best individual being inserted into the next population and by replacing some individual randomly selected - except the best one.

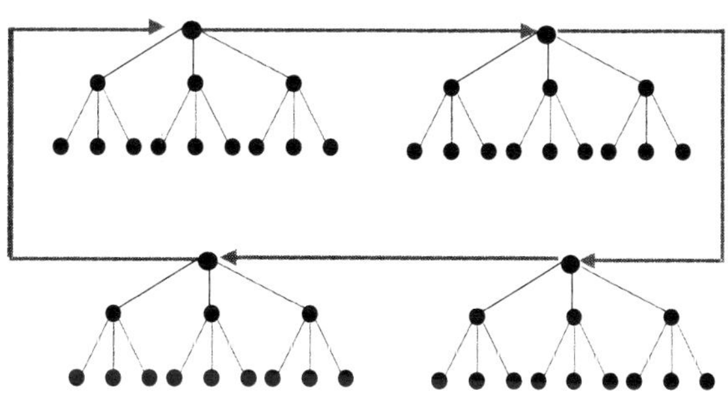

Fig. 8.9: Migration policy.

The pseudo-code shown in Algorithm 8.2 summarizes the whole multi-population MA algorithm. The algorithm executes a fixed number of generations in each population while there is no convergence. This process involves a parent selection (selectParents), a new individual creation by crossover execution (crossover(individualA, individualB)), a possible mutation execution to the new individual (mutation(newInd)), its fitness evaluation (evaluateFitnessIndividual(newInd)) and its insertion or not into the population (insertPopulation(newInd, pop(i))). The population convergence occurs when there are no new individuals inserted after a fixed number of generations given by the parameter $\gamma * PopSize$, where γ is the crossover rate and $PopSize$ is the population size. A migration between populations (executeMigration) is executed when all populations have converged and the stop criterion has not been satisfied yet. A new initialization of the populations (initializePopulation(pop(i))) will occur, but the best individual and the migrated individuals are kept. The stop criterion is usually a pre-specified computing time.

The MA described in this section was implemented using the NP-Opt [28, 37], an object-oriented framework written in JAVA code which contains procedures based on evolutionary computation techniques to address NP-hard problems.

Algorithm 8.2: Pseudo-code for the multi-population memetic algorithm.

```
repeat
    for i=1 to numberOfPopulations do
        initializePopulation(pop(i));
        evaluatePopulationFitness(pop(i));
        structurePopulation(pop(i));
        repeat
            for j=1 to numberOfGenerations do
                selectParents(individualA,individualB);
                newInd=crossover(individualA,individualB);
                if execute mutation newInd then
                    newInd=mutation(newInd);
                    evaluateFitnessIndividual(newInd);
                    insertPopulation(newInd,pop(i));
                end
                structurePopulation(pop(i));
                localSearch(pop(i));
            end
        until populationConvergence pop(i);
    end
    for i=1 to numberOfPopulations do
        executeMigration(pop(i));
    end
until stop criterion;
```

8.2.3 Individual representation

A MA approach was developed to solve a lot sizing and scheduling problem with sequence-dependent setup times in [39], where an individual is represented by a string of paired values (type of product and lot size) for each scheduling period. A similar representation of solution as an individual is proposed by [40] for a GA used to solve the capacitated lot sizing and loading problem with setup times, parallel facilities and overtime. In this approach, the individual is also a string of paired values, where the first value is the lot size and the second is the facility. A GA with more elaborated solution representation is proposed by [41], where a binary matrix PxT (P products and T periods) represents an individual for the multi-level lot sizing problem.

Each binary entry $y_{i,t} = 1$, if a setup for product i occurs in t; otherwise $y_{i,t} = 0$. All the GA publications mentioned have specific genetic operators. Their crossover, mutation and selection procedures were designed to deal with those individual representations for lot size and scheduling problems.

A new solution representation is proposed in this work. It is close to one presented in [42] that uses assignment rules in a multi-level proportional lot sizing and scheduling problem with multiple machines. Fig. 8.10 introduces the individual representation proposed for the SITLSP.

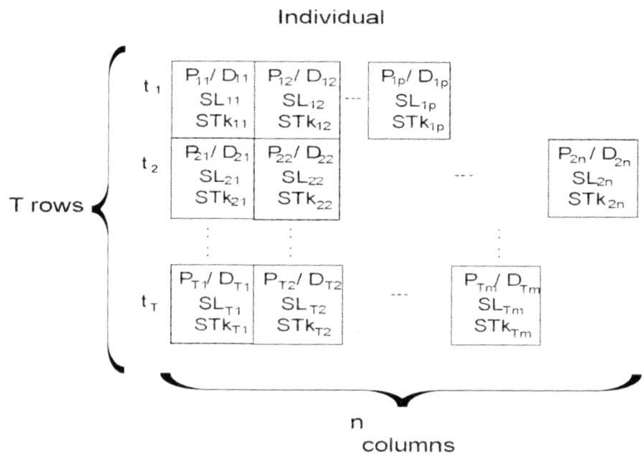

Fig. 8.10: Individual for the SITLSP.

An individual is a two-dimensional matrix with T rows and N columns. The number of rows represents the number of macro-periods $t_1, t_2, \ldots, t_T$. The number of columns represents the number of genes and there can be a different number of genes per macro-period. Each gene corresponds to a cell (m, n), $m \in T$, $n \in N$, in the individual matrix and contains the following data:

- P_{mn}: product in gene n to be produced in macro-period m.
- D_{mn}: lot size of product P_{mn}.
- SL_{mn}: sequence of lines where D_{mn} can be produced.
- STk_{mn}: sequence of tanks where the raw material of D_{mn} can be stored.

The demand d_{it} of product i in micro-period t is divided into several lots (D_{mn}) and randomly distributed among the genes in micro-periods t, $t-1$, $t-2, \ldots, 1$. The sequences SL_{mn} and STk_{mn} are randomly generated with length k. Sequence $SL_{mn}{=}(\alpha_1, \ldots, \alpha_k)$ with $\alpha_i \in \{1, \ldots, L\}$, where α_i is a possible line number and L is the number of lines. The value α_i is taken from L possible values. Sequence $STk_{mn}{=}(\beta_1, \ldots, \beta_k)$ with $\beta_i \in \{1, 2, \ldots, 2\overline{L}\}$, where β_i defines where and how the raw material will be stored. Parameter $\overline{L}$ is the number of tanks. The β_i is taken from $2\overline{L}$ possible values, but the real tank number j is obtained from β_i using:

$$j = \begin{cases} \beta_i, & 1 \leq \beta_i \leq \overline{L}; \\ \beta_i - \overline{L}, & \overline{L} < \beta_i \leq 2\overline{L}. \end{cases} \tag{8.1}$$

If $1 \leq \beta_i \leq \overline{L}$, the tank $j = i$ will be occupied after the raw material previously stored has been used. This forces the method to find solutions where there is a partial use of the tank capacity. If $\overline{L} < \beta_i \leq 2\overline{L}$, the tank $j = \beta_i - \overline{L}$ will be immediately occupied. This forces the method to find solutions where the tank capacity is completely used. These conditions have some exceptions. If tank j, selected by one of the previous criterions, stores a raw material different from the raw material of the product P_{mn}, it will be necessarily occupied after the raw material which was previously stored has been used. The same will happen if tank j is completely full. On the other hand, if tank j is empty, it will be immediately occupied. If j is not empty, but the raw material stored is the same as P_{mn} and the minimum tank capacity has not been satisfied, this tank will be also occupied immediately. The individuals in each initial population are generated following the pseudo-code illustrated in Algorithm 8.3. There are T macro-periods, J products, L lines and $\overline{L}$ tanks. The variable Dem receives the total demand $d_{Pi,t}$ of product P_i in the macro-period t. This demand is randomly divided and distributed among the genes. At this point, the sequences of lines (SL_{mn}) and tanks (STk_{mn}) are also randomly generated.

Algorithm 8.3: Individual initialization algorithm.

for $t = t_1, t_2, \ldots, t_T$ **do**
  repeat
    Select a product $P_i \in \{P_1, P_2, \ldots, P_J\}$ randomly with $d_{P_i,t} > 0$;
    Set Dem=$d_{P_i,t}$;
    while $Dem > 0$ **do**
      Select the matrix row $m \in \{t_1, t_2, \ldots, t\}$ of the individual randomly;
      Determine n as the first gene available in line m of the individual;
      Determine $D_{m,n} \neq 0$ with $D_{m,n} \in [0, Dem]$ randomly generated;
      Insert $D_{m,n}$ and P_i in the gene position (m, n) of the individual;
      Generate $SL_{m,n} = (\alpha_1, \ldots, \alpha_k)$ with $a_i \in \{1, \ldots, L\}$ randomly selected;
      Generate $STk_{m,n} = (\beta_1, \ldots, \beta_k)$ with $\beta_i \in \{1, \ldots, 2\overline{L}\}$ randomly selected;
      Set $Dem = Dem - D_{m,n}$;
    end
  until *All demands have been distributed among the genes*;
end

The following example clarifies the solution representation of an individual. Suppose two products ($P1$ and $P2$) where each product has a demand of 100 units to be filled in macro-period t_1 and another 200 units to be filled in macro-period t_2. The products use different raw materials ($Rm1$ and $Rm2$, respectively). Moreover, there are two lines available to produce both products and two tanks available to store both raw materials. Fig. 8.11 shows two possible representations, both based on the individual initialization algorithm. The demands are distributed in their respective macro-periods in individual 1. However, the demand of $P1$ in t_1 is split between two genes. The same happens with the demand of $P2$ in t_2. In individual 2, part of the $P1$ demand in t_2 is split between two genes in t_1. Notice that the sequence of lines (SL_{mn}) and tanks (STk_{mn}) can repeat values of $\alpha_i \in \{1, 2\}$ and $\beta_i \in \{1, 2, 3, 4\}$ for $L = \overline{L} = 2$ and $k = 4$ (length).

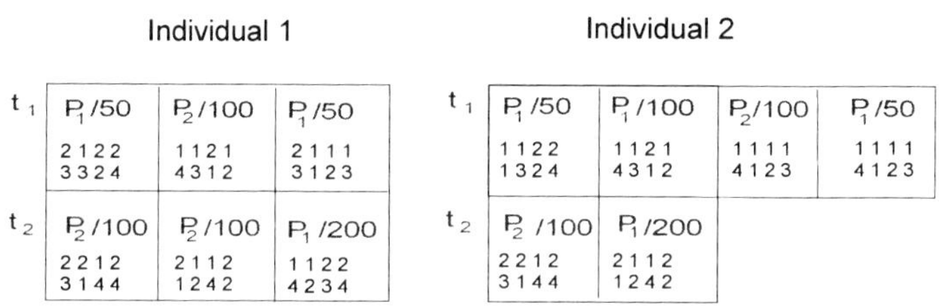

Fig. 8.11: Two possible individuals.

8.2.4 Decoding and evaluation

The decoding procedure is responsible for determining a problem solution from the data encoded in an individual. The procedure starts from the first gene in the last macro-period up to the last gene in the first macro-period. This backward procedure enable us to postpone setups and processing time of products and raw materials in lines and tanks. However, there is no guarantee that all demands will be produced at the end and a penalty in the fitness is taken into account for that.

An example illustrates the decoding procedure. Consider the same data used in the example of the last section and individual 1 shown there. Let's also suppose that for each macro-period there are 5 micro-periods with the same length. Therefore, the time horizon is divided into 10 micro-periods. The process begins by the first gene in the last macro-period (Fig. 8.12). A lot of product $P2$ ($D21 = 100$) has to be produced using the first pair $(\alpha_1, \beta_1) = (2, 3)$ from sequences SL_{21} and STk_{21}. Product $P2$ is produced in line 2 because $\alpha_1 = 2$ and let's assume that its processing time takes two micro-periods. The other processing times used in this example are suppositions as well. According to equation (1), raw material $Rm2$ of $P2$ has to be stored in tank j=3-2=1 because β_1=3 and $2 < \beta_1 < 4$ with $\overline{L} = 2$. Given the criteria defined in Section 8.2.2, tank j=1 is empty and it must be occupied immediately. A tank should be ready at least one micro-period before the production starts on the lines. Therefore, the setup time of $Rm2$ occurs one micro-period before the $P2$ production starts in $L2$ (see Fig. 8.12).

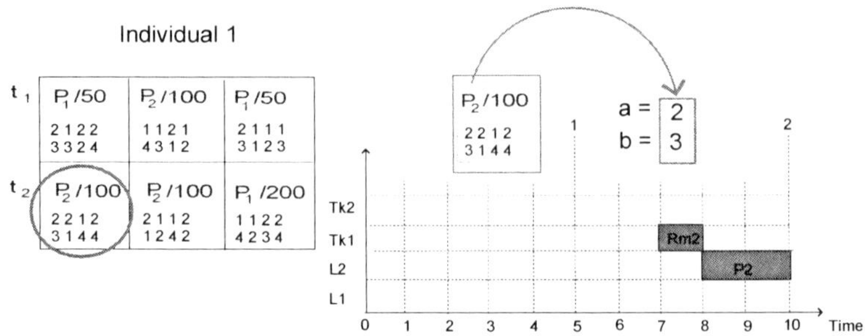

Fig. 8.12: Decoding of the first gene in t_2.

Let us suppose that the setup time of raw materials will take one micro-period in any tank in this example. The demand of this gene has been completely scheduled, so the next gene in t_2 is decoded now (Fig. 8.13). The first pair of rules $(\alpha_1, \beta_1) = (2, 1)$ is selected to schedule 100 units of $P2$ which are produced in $L2$ ($\alpha_1 = 2$). Value β_1=1 means $j = 1$ ($1 \leq \beta_1 \leq 2$) by equation (1) and this tank must be occupied after the raw material which was previously stored has been used. At this point, it is worth noticing that this criterion allows for establishing schedules with a tank partially filled. Fig. 8.14 shows the decoding of the third gene in t_2. A lot from product $P1$ ($D21 = 200$) has to be scheduled in line $L1$ ($\alpha_1 = 1$) and its raw material has to be stored in the empty tank j = 4-2 = 2 ($2 < \beta_1 \leq 4$). However, let's suppose now that tank $j = 2$ has a capacity of storing raw material sufficient to produce only 100 units of $P1$. In this case, the next pair (α_2, β_2)=(1,2) is selected to schedule the remaining lot $D12$ = 200-100 = 100 (Fig. 8.15). The remaining lot is also produced in $L1$ ($\alpha_2 = 1$) and a new setup time of $Rm1$ occurs in tank $j = 2$.

The decoding process continues in the first gene of t_1 (Fig. 8.16). Product $P2$ is produced in $L2$ ($\alpha_1 = 2$). A setup time occurs because $P1$ is produced next in $L1$. Let's suppose that the setup time from $P1$ to $P2$ takes one micro-period. The tank $j = 1$ (β_1=3 with $2 < \beta_1 \leq 4$) should be immediately occupied, but it already stores $Rm1$. In this case, a new lot assignment to this tank is necessary.

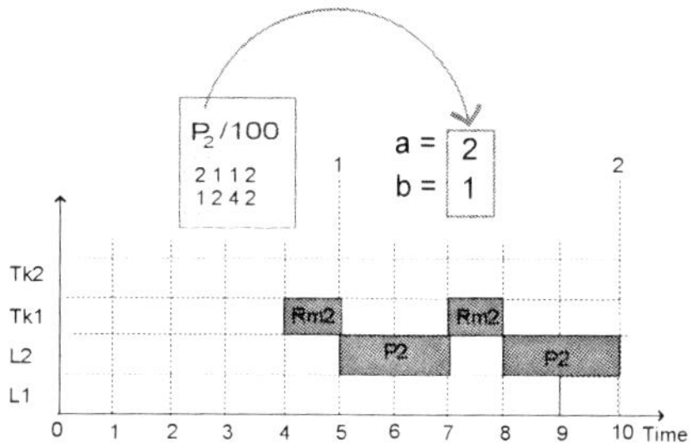

Fig. 8.13: Decoding of the second gene in t_2.

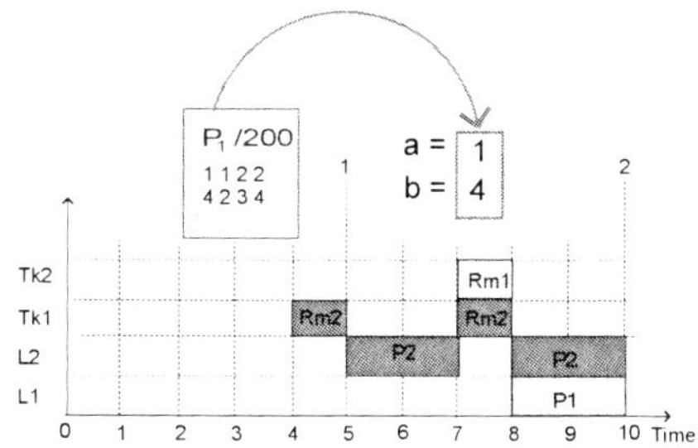

Fig. 8.14: Decoding of the third gene in t_2.

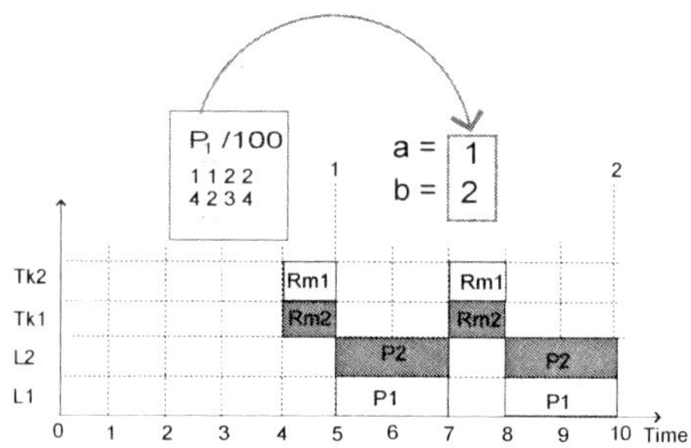

Fig. 8.15: Decoding of the remaining demand of the third gene in t_2.

The next gene decoding is shown in Fig. 8.17. Line $L1$ ($\alpha_1 = 1$) is selected to produce $D12 = 100$ units of $P2$. Let's suppose that the setup time from $P2$ to $P1$ takes two micro-periods. Raw material $Rm2$ is assigned to $j = 4$-$2 = 2$ ($\beta_1 = 4$) and it must be ready one micro-period before $L1$ produces $P2$. Fig. 8.18 has the last gene decoding. Product $P1$ is scheduled in $L2$ ($\alpha_1 = 2$) and there is no setup time because $P1$ is also produced next. $Rm1$ will integrate the lot previously stored in tank j = 3-2 = 1 because $\beta_1 = 3$. For this reason, the setup time of tank $j = 1$ is anticipated to the second micro-period.

8.2.5 Crossover and mutation

Previous computational experiments with various crossover operators revealed that the best behavior was attained by the uniform crossover. In this recombination operator, genes from two different parents that occupy the same position in the individuals have some probability of being inherited by the child. Individuals 1 and 2 (Fig. 8.11) are used to show how the uniform crossover

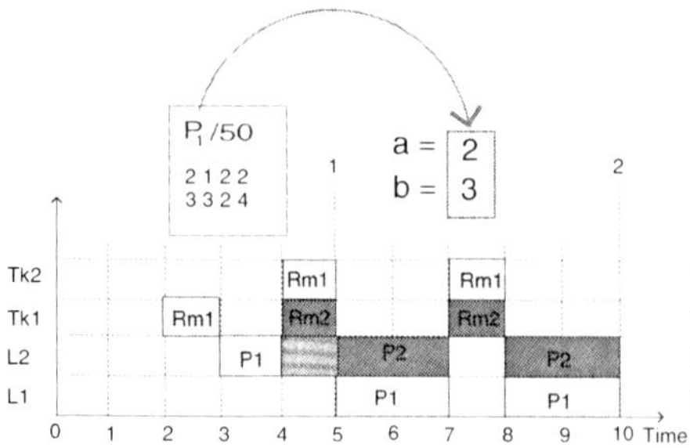

Fig. 8.16: Decoding of the second gene in t_2.

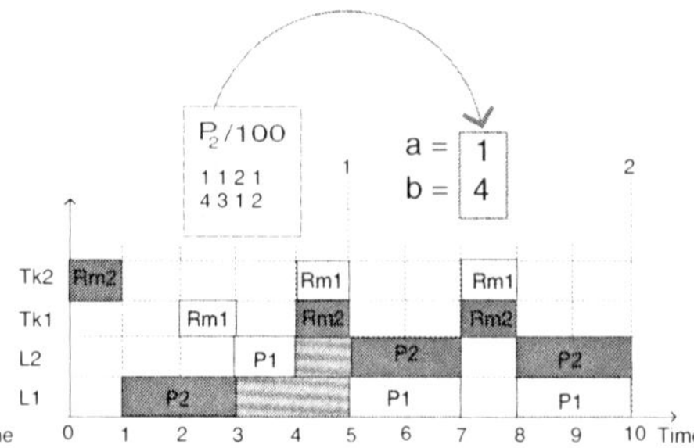

Fig. 8.17: Decoding of the third gene in t_2.

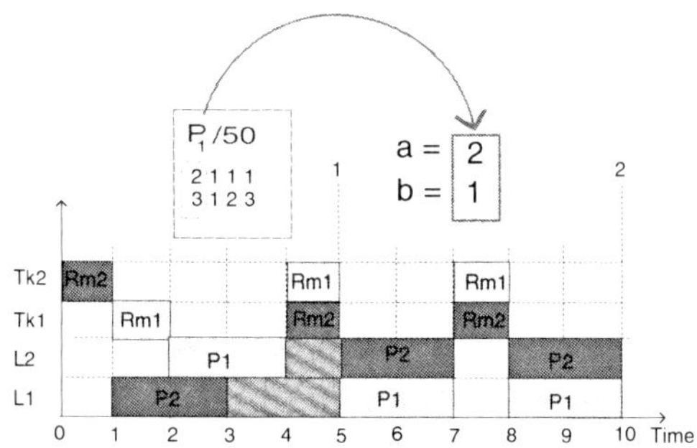

Fig. 8.18: Decoding of the remaining demand of the third gene in t_2.

operator works. Sequences SL_{mn} and STk_{mn} are not relevant because the new individual (Child) will inherit these sequences without changes. Fig. 8.19 illustrates the crossover of Ind1 and Ind2.

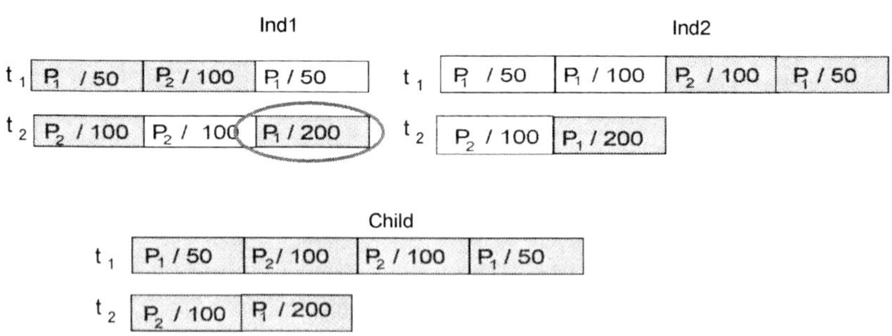

Fig. 8.19: Uniform crossover example.

For each gene in the same position in Ind1 and Ind2, a random value $\lambda \in [0, 1]$ is generated. If $\lambda < 0.5$, the Child inherits the gene from Ind1; otherwise, the Child inherits the gene from Ind2. The genes selected following this procedure are shaded in Fig. 8.19. Notice that there are more genes in macro-periods t_1 of Ind2 than in the same macro-period of Ind1. In this case, the procedure continues in Ind2 selecting those genes where $\lambda \geq 0.5$. We do not allow excessive demands in a new individual. For example, the gene from Ind1 marked by a circle in Fig. 8.19 is not inherited because it would exceed the total demand of $P1$ in the Child. If there is a gene in the same position in Ind2, this gene must be inherited by the Child if the same problem does not occur. On the other hand, a lack of demand can take place at the end of the crossover. For this reason, a repair procedure is necessary and the demand deficits in some macro-period are inserted.

Mutation aims to keep diversity in a population avoiding premature convergence. A mutation rate determines the number of individuals that are changed. The mutations adopted basically swap gene positions (Fig. 8.20).

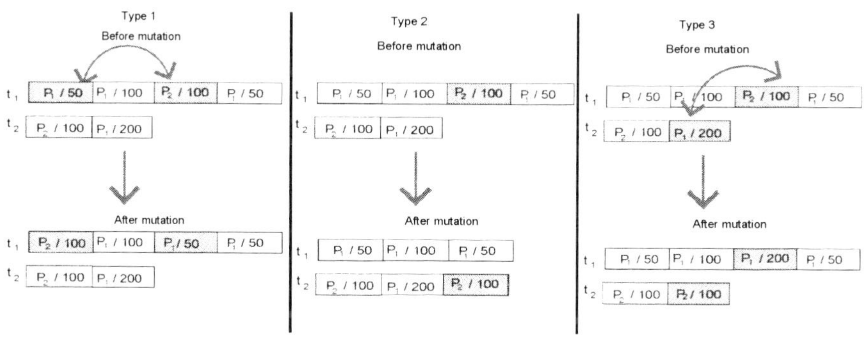

Fig. 8.20: Mutation types.

The first type swaps the positions of two selected genes in the same macro-period. In the second type, the selected gene is removed and inserted into another position which is also randomly selected. The third type swaps the positions of two chosen genes that are in different macro-periods. The new gene positions have to respect the macro-period demand of each product. A product swap will not take place if it can violate the demand satisfaction. The mutation procedure randomly chooses which mutation type will be applied to the individual.

8.2.6 Local search algorithm

A substantial part of the computational effort related to MA implementations is due to the local search. Bearing this in mind, and also that SITLSP is a complex combinatorial problem to be solved in a real-world context, a simple local search method is a natural choice. A threshold accepting (TA) procedure was elected as the local search built-in method to be used in this MA implementation [43]. The pseudo-code of the TA is shown in Algorithm 8.4.

The neighborhood movements can lead to worse individuals and the TA accepts them, since their fitness value remain inside a threshold. Doing this the method can skip from local minima. Threshold reductions lead the method to convergence. Two neighborhood movements were carried out to modify the best individual making changes in the gene positions and changes in the lot size. The changes in the gene positions follow the same behavior of the three types of mutations described in the previous section. Regarding the mutation types, the allowed movements are:

- Swap positions of two genes in the same macro-period.
- Remove one gene and insert it into another position.
- Swap positions of two genes in different macro-periods.

The changes in the lot sizes consider two cases:

- Split lots of one gene in two pieces. One piece stays in the gene and the other piece is inserted into another position of the individual.
- Merge lots of the same product. The lot of one gene is removed from its original position and inserted into another gene with the same product.

Problems with demand satisfaction are taking into account, so neighborhood movements that lead to infeasibility are forbidden. After the fixed number of generations has been completed, the MA takes the best individual of the population and executes the TA local search. First, one of the three possible changes in the gene positions are randomly chosen and executed. After a maximum number of iterations, the TA stops and restarts executing now changes in the lot sizes. Moreover, the two possible changes in the lot sizes are randomly chosen and executed over randomly selected genes. In the end, if a better individual is obtained, it is inserted as the new best individual of this population.

Algorithm 8.4: Local search procedure.

```
individual = bestIndividual;
repeat
    newIndividual = moveExecution(individual);
    Δf = fitness(Individual)-fitness(newIndividual);
    if Δf > −Th*fitness(individual) then
        individual = newIndividual;
    else
        reduce(Th);
    end
until maxNumberOfIteration;
```

The MA as conceived here transforms itself into a GA simply by the deleting the local search procedure. Although our experience with previous research has demonstrated that in most cases the MA versions outperform their GA versions, a comparison within a fixed amount of time will be carried out in Section 8.4. The underlying issue under consideration is whether the time spent by the local search executions in the MA is favorably used to reach better solutions or if it could be used better by the GA version which will spend this time executing a higher number of generations.

8.3 The Decomposition and Relaxation Approach

In this section the mathematical model to represent the SITLSP and the solution approaches proposed in [29] are described. The model considers the synchrony between the production levels and integrates the lot sizing and scheduling decisions as well as the model in [19]. However, as pointed out in Section 8.1.2, it differs from the latter in many aspects. A simplification of the problem is considered supposing that each filling line, thereafter called machine, has a dedicated tank. Each tank can be filled, in turn, with all the liquids needed by the associated machine. The planning horizon is divided into T macro-periods. It is a big bucket model and to obtain the order at which the items will be produced each macro-period is divided into a number of micro-periods. The total number of micro-periods is defined by the user, but should be set as the maximum number of setups in each macro-period. The micro-period size is flexible and is defined by the model since it depends on the item's lot size. The total number of micro-periods in both levels is the same and only one item (liquid flavor) can be produced in each micro-period.

8.3.1 Model development

To describe the Two-Level Multi-Machine Lot Scheduling Model (P2LMM) given in [29, 44], let the following parameters define the problem size:

J = number of soft-drinks (items);
M = number of machines (and tanks);
F =number of liquid flavors;
T = number of macro-periods;
N = total number of micro-periods (i.e. total number of setups);

and let (i, j, m, k, l, t, s) be the index set defined as:

$$i, j \in \{1 \dots J\}; t \in \{1 \dots T\}; k, l \in \{1 \dots F\}; s \in \{1 \dots N\}; m \in \{1 \dots M\}.$$

Consider also, that the following sets and data are known:

Sets:

S_t = set of micro-periods in each macro-period t;
P_t = first micro-period of period t;
ρ_j = set of machines that can produce item j;
δ_m = set of items that can be produced on machine m;
θ_m = set of liquid flavors that can be produced on tank m;
ω_{ml} = set of items that can be produced on machine m and need flavor l.

The data and variables described below with superscript I relate to Level I (tank) and with superscript II relate to Level II (bottling):

Data:

d_{jt} = demand for item j in macro-period t;
h_j = non-negative inventory cost for item j;
g_j = non-negative backorder cost for item j;
s^I_{kl} = changeover cost from liquid flavor k to l;
s^{II}_{ij} = changeover cost from item i to j;
b^I_{kl} = changeover time from liquid flavor k to l;
b^{II}_{ij} = changeover time from item i to j;
a^{II}_{mj} = production time in machine m of item j;
K^I_m = total capacity of tank m, in liters of liquid;
K^{II}_{mt} = total time capacity in machine m in period t;
r_{jl} = quantity of liquid flavor l necessary for the production of one lot of item j;
q^I_{ls} = minimum quantity to produce liquid flavor l in micro-period s;
I^+_{j0} = Initial inventory for item j.

Variables:

I^+_{jt} = inventory for item j at the end of macro-period t;
I^-_{jt} = backorder for item j at the end of macro-period t;
x^{II}_{mjs} = production quantity in machine m of item j in micro-period s;

$$y^I_{mls} = \begin{cases} 1, \text{ if the tank } m \text{ is setup for syrup } l \text{ in micro-period } s; \\ 0, \text{ otherwise} \end{cases}$$

$$y^{II}_{mjs} = \begin{cases} 1, \text{ if the machine } m \text{ is setup for item } j \text{ in micro-period } s \\ 0, \text{ otherwise} \end{cases}$$

$$z^I_{mkls} = \begin{cases} 1, \text{ if there is changeover in tank } m \text{ from syrup } k \text{ to } l \text{ in } s; \\ 0, \text{ otherwise} \end{cases}$$

$$z^{II}_{mijs} = \begin{cases} 1, \text{ if there is changeover in machine } m \text{ from item } i \text{ to } j \text{ in } s; \\ 0, \text{ otherwise} \end{cases}$$

To include the synchrony between the two production levels in the P2LMM model another set of variables is necessary. As discussed in Section 8.1, a machine must wait until the liquid is ready in the tank. The set of continuous variables, $v_{ms}^{II} \geq 0$, computes this waiting time for each machine m, in each micro-period s. The waiting time is equal to the difference between the tank changeover time and the machine changeover time. That is:

$$v_{ms}^{II} \geq \sum_{k\in\delta_m} \sum_{l\in\theta_m} b_{kl}^{I} z_{mkls}^{I} - \sum_{i\in\delta_m} \sum_{j\in\delta_m} b_{ij}^{II} z_{mijs}^{II} \qquad m = 1,\ldots,M, s = 1,\ldots,N$$

If the machine changeover time from item i to j is greater than the tank changeover time from liquid flavor k to l, the waiting variable is zero and only the machine changeover time is considered in the associated capacity constraint. Otherwise, the total waiting time of the macro-period is taken into account.

The P2LMM model is then:

$$Min\ \ Z = \sum_{j=1}^{J} \sum_{t=1}^{T} (h_j\, I_{jt}^{+} + g_j\, I_{jt}^{-}) + \sum_{m=1}^{M} \sum_{s=1}^{N} \sum_{k\in\theta_m} \sum_{l\in\theta_m} s_{kl}^{I}\, z_{mkls}^{I} + \sum_{m=1}^{M} \sum_{s=1}^{N} \sum_{i\in\delta_m} \sum_{j\in\delta_m} s_{ij}^{II}\, z_{mijs}^{II} \qquad (8.2)$$

Subject To

Level I (Tank)

$$\textstyle\sum_{j\in\omega_{ml}} r_{lj} x_{mjs}^{II} \leq K_m^{I} y_{mls}^{I}, \qquad m = 1,\ldots,M, l \in \theta_m, s = 1,\ldots,N; \ (8.3)$$

$$\textstyle\sum_{j\in\omega_{ml}} r_{lj} x_{mjs}^{II} \geq q_{ls}^{I} y_{mls}^{I}, \qquad m = 1,\ldots,M, l \in \theta_m, s = 1,\ldots,N; \ (8.4)$$

$$\sum_{l\in\theta_m} y_{ml(s-1)}^{I} \geq \sum_{l\in\theta_m} y_{mls}^{I}, \qquad m = 1,\ldots,M, t = 1,\ldots,T, s \in S_t - \{P_t\}; \ (8.5)$$

$$z_{mkls}^{I} \geq y_{mk(s-1)}^{I} + y_{mls}^{I} - 1, \qquad m = 1,\ldots,M, k,l \in \theta_m, s = 2,\ldots,N; (8.6)$$

$$z_{mkls}^{I} \geq \sum_{j\in\omega_{ml}} y_{mj(s-1)}^{II} + y_{mls}^{I} - 1, \qquad m = 1,\ldots,M,\ k,l \in \theta_m, \ t = 2,\ldots,(T-1), s = P_t; \qquad (8.7)$$

$$\sum_{k \in \theta_m} \sum_{l \in \theta_m} z^I_{mkls} \leq 1, \qquad m = 1, , \ldots, M, s = 1, \ldots, N; \tag{8.8}$$

$$\sum_{k \in \theta_m} z^I_{mkl1} \geq y^I_{ml1}, \qquad m = 1, \ldots, M, l \in \theta_m \ ; \tag{8.9}$$

Level II (bottling)

$$I^+_{j(t-1)} + I^-_{jt} + \sum_{m \in \rho_j} \sum_{s \in S_t} x^{II}_{mjs} = I^+_{jt} + I^-_{j(t-1)} + d_{jt}, \qquad j = 1, \ldots, J, \; t = 1, \ldots, T \tag{8.10}$$

$$\sum_{j \in \delta_m} \sum_{s \in S_t} a^{II}_j x^{II}_{mjs} + \sum_{i \in \delta_m} \sum_{j \in \delta_m} \sum_{s \in S_t} b^{II}_{ij} z^{II}_{mijs} + \sum_{s \in S_t} v^{II}_{ms} \leq K^{II}_{mt}, \quad m = 1, \ldots, M, t = 1, \ldots, T \tag{8.11}$$

$$v^{II}_{ms} \geq \sum_{k \in \theta_m} \sum_{l \in \theta_m} b^I_{kl} z^I_{mkls} - \sum_{i \in \delta_m} \sum_{j \in \delta_m} b^{II}_{ij} z^{II}_{mijs}, \quad m = 1, \ldots, M, s = 1, \ldots, N; \tag{8.12}$$

$$x^{II}_{mjs} \leq \frac{K^{II}_{mt}}{a^{II}_{mj}} y^{II}_{mjs}, \qquad m = 1, \ldots, M, \; j \in \delta_m, t = 1, \ldots, T, s \in S_t; \tag{8.13}$$

$$\sum_{j \in \delta_m} y^{II}_{mjs} = 1, \qquad m = 1, \ldots, M, s = 1, \ldots, N; \tag{8.14}$$

$$z^{II}_{mijs} \geq y^{II}_{mi(s-1)} + y^{II}_{mjs} - 1, \qquad m = 1, \ldots, M, i, j \in \delta_m, \; s = 2, \ldots, N \tag{8.15}$$

$$\sum_{i \in \delta_m} \sum_{j \in \delta_m} z^{II}_{mijs} \leq 1, \qquad m = 1, \ldots, M, s = 1, \ldots, N; \tag{8.16}$$

$$\sum_{i \in \delta_m} z^{II}_{mij1} \geq y^{II}_{mj1}, \qquad m = 1, \ldots, M, j \in \delta_m; \tag{8.17}$$

$$\begin{aligned} & I^+_{jt}, I^-_{jt} \geq 0, \qquad j = 1, \ldots, J, t = 1, \ldots, T, \\ & z^I_{mkls}, v_{ms}, x^{II}_{mjs}, z^{II}_{mijs} \geq 0, \quad y^I_{mls}, y^{II}_{mjs} \in \{0, 1\} \\ & m = 1, \ldots, M, \quad i, j \in \delta_m, \quad k, l \in \theta_m, \quad s = 1, \ldots, N. \end{aligned} \tag{8.18}$$

The objective function 8.2 is to minimize the total sum of inventory costs, backorder costs, machine and tank changeover costs. In Level I, the demand for liquid flavor l is computed in terms of the production variables. That is, the demand for liquid l in each tank m in each micro-period s is given by and $\sum_{j\in\omega_{ml}} r_{lj}x^{II}_{mjs}$. The constraints (8.3), similar to constraints (8.13) in Level II, together with constraints (8.4) guarantees that if tank m is setup for syrup l in micro-period s ($y^{I}_{mls} = 1$) there will be production of liquid flavor l (between the minimum quantity necessary for liquid homogeny and the tank maximum capacity). The constraints (8.5) force the idle micro-periods to happen at the end of the associated macro-period. Constraints (8.6), similar to constraints (8.15) in Level II, control the liquid flavor changeover. Note that the tank setup does not hold from one macro period to another if the setup variable in the last micro-period is zero. Therefore, constraints (8.7) are needed to count the changeover between macro-periods. Note also that the setup variables in level II indicate which liquid flavor was prepared in the last non-idle micro-period of each macro-period. Constraints (8.8), similar to constraints (8.16) in Level II, count the first changeover of each tank. Constraints (8.9), similar to constraints (8.17) in Level II, guarantee that there is at most one changeover in each tank m in each micro-period s.

In Level II, constraints (8.10) represent the flow conservation constraints for each item in each macro-period. Since the production variable is defined for each micro-period, to obtain the total production of item j in a given macro-period t it is necessary to sum the associated production variables over all machines where it can be produced ($m \in \delta_j$) and micro-periods ($s \in S_t$) of macro-period t. Constraints (8.11) represent the machine capacity in each macro-period. Note here the inclusion of the waiting variable, v^{II}_{ms}, to ensure that the lot schedule will be feasible. The waiting time in machine m in each micro-period s is computed by constraints (8.12) as explained above. Constraints (8.13) guarantee that there is a production of item j only if the associated setup variable is set to one, and constraints (8.14) and (8.15) count the changeover in each machine m in each micro-period s. Constraints (8.14) refer to a single mode production in each micro-period s. Note that production may not occur although the machine is always ready to produce an item.

Finally, constraints (8.18) define the non-negativity and integrality restrictions. Note that the changeover variables z^{I}_{mkls} and z^{II}_{mijs} are continuous. Constraints (8.5), (8.6), (8.14), (8.15), and the optimization sense (minimization) ensure that these variables will take only 0 or 1 values.

As happened to the model presented in [19], the solution of practical instances of model P2LMM using the exact methods included in standard software such as CPLEX [45] was not satisfactory. This indicated the need to develop specific solution strategies.

8.3.2 Relax and fix strategies

The relax and fix heuristic has been largely used as a method to obtain good primal bounds (feasible solutions) for hard mixed-integer programs either on its own or in hybrid algorithms e.g. [47]- [50]. In this approach, first the integer variable set is partitioned into P disjunctive sets Q_i, $i = 1, \ldots, P$. At iteration n, the variables of Q_n are defined as integers while all others are relaxed. The resulting problem is then solved. If it is infeasible, the procedure finishes since it is not possible to find a feasible solution with the variables in Q_i, $i = 1, \ldots, n-1$ fixed at their actual values. Otherwise the variables of Q_n are fixed at their current values and the process is repeated for all the P sets. Besides the variable set partition, criteria to fix the variables in set Q_n must also be defined before applying the procedure. The main feature of this heuristic is the solution of submodels that are smaller, and possibly easier, than the original one. The partition of variables and the criteria used to fix the variables have a strong connection with the degree of the submodel difficulty.

In the usual relax and fix strategy the variables are grouped by periods (macro-periods) and only the integer variables are fixed at each iteration. In [47] these criteria were used in the solution of a multi machine multi items lot sizing model. The heuristics iterations number is thus the number of periods. In [48] this heuristic is applied to a class of project scheduling problems and explores various strategies to partition the set of binary variables. The relax and fix heuristic has also been used in combination with meta-heuristics such as Tabu Search. In [49] a hybrid tabu search procedure in which the relax and fix heuristic is used either to initialize a solution or to complete partial solutions is presented. The hybrid approach is applied to solve a big bucket lot sizing problem with setup and backlog costs. At each iteration of the relax and fix heuristic only the variables of a given period that concern a single product is fixed. The strategy is called relax-and-fix-one-product. The main advantage of this strategy is to solve smaller submodels since some mono-period, mono-machine multi-items lot sizing problems are hard to solve. In [1] a relax and fix heuristic to solve the SITLSP model formulated in [19] is proposed. The criterion used is to first fix the binary variables in Level I, then the ones in Level II, in a backward fashion. That is from the last period to the first. Other relax and fix strategies also fix continuous variables. In [50] the relax and fix heuristic is classified as a particular case of a progressive interval heuristics. They also mention that fixing continuous variables reduce the flexibility of the heuristic and propose various strategies varying the number of continuous variables fixed.

The P2LMM model presents various possibilities to build sets Q_i, $i = 1, \ldots, P$ [29]. The setup and changeover variables are indexed by levels, machines, items and periods. These indexes are explored when defining various variable partition strategies. Different criteria were proposed to fix variables. For example, after solving a submodel in a given iteration, it is possible to only fix the binary variables associated to non-zero production variables. Table 8.1 shows 12 relax and fix strategies, divided into two groups. The Group 1 has five strategies (G1.1 - G1.5) and the Group 2 has seven (G2.1 - G2.7). The first column shows the strategy name (Strat.), the second and third columns show the criteria used for the partition (Part.) and fixing (Fix) the variables, respectively. The variables presented in Table 8.1 are the same ones used to describe the model P2LMM, however some of its indexes were omitted.

Table 8.1: Relax and Fix Strategies.

Strat.	Part.	Fix
G1.1	Period	y^I , y^{II}
G1.2	Period	y^I, z^I, y^{II} ,z^{II}
G1.3	Period	y^I, z^I, y^{II}, z^{II}, x^{II}
G1.4	Period	y^I, z^I, y^{II}, z^{II} i.th.p.*
G1.5	Period	y^I, z^I, y^{II}, z^{II} i.th.p. with reevaluation
G2.1	Machine/Period	y^I, y^{II}
G2.2	Level I then Level II	y^I, y^{II}
G2.3	Level II then Level I	y^I, y^{II}
G2.4	Period/ Level I then Level II	y^I, y^{II}
G2.5	Period/ Level II then Level I	y^I, y^{II}
G2.6	Machine/Period/ Level II and Machine/Period/ Level I	y^I, y^{II}
G2.7	Machine/Period/ Level II and Machine/Period/ Level I	y^I, z^I, y^{II}, z^{II} i.th.p. with reevaluation

The first five strategies (G1.1 - G1.5) use the usual criteria of partitioning the variables according to periods. They differ from each other by the criteria used to fix the variables in a given iteration. These criteria are based on the idea that the submodels should have a dimension that favors the decision process. The objective was to evaluate the influence of the variables in the submodels solution.

The Group 2 strategies explore the multi-machine, two level structure of the model to partition the set of variables. The objective was to evaluate the influence of each machine (level) in the solution of the submodels. Note that the criterion used to fix the variables in this group is the same one used in the strategy G1.1, except for the strategy G2.7. In this criterion, when the variable y^{II} is fixed to one, it only assures that the machine will be prepared, it does not say if the item will be produced or not. The variable y^I besides defining that the tank will be prepared, also states that there will be production of an item, between the tank capacities (constraints (2) and (3) in the P2LMM model). In the strategy G2.7, besides the binary variables, the continuous variables (z^I, z^{II}) are also considered to be fixed when there is production ($x^{II} > 0$). In this strategy the idle micro-periods in previous iterations which did not have any variables fixed are also reevaluated for further variable fixing.

At each iteration of the relax and fix heuristic an instance of a mixed integer optimization submodel has to be solved. In general, they are solved by exact methods included in standard software (e.g. the branch and cut method in CPLEX). Although the submodels in each iteration are smaller than the original model, they are still difficult. If the optimal solution of the submodel solved at each iteration is not achieved in a pre-defined amount of time, the branch and cut execution is halted and the best solution is used to fix the variables.

8.3.3 The relaxation approach

In some industries the liquid preparation in Level I does not represent a bottleneck for the production process. That is, the tank capacities are large enough to ensure that whenever a machine needs a liquid of a given flavor it will be ready to be released. Therefore, there is no need to control either the changeover in the tanks or the synchrony between the two production levels, only the minimum tank capacity constraints to ensure the liquid homogeny is necessary in Level I. This situation was explored as a solution approach to model P2LMM. The Relaxation Approach (RA) is based on the idea that once the production decision is taken in Level II, the decision for Level I is easily taken.

In [44] a one level model to the production planning of a small soft drink manufacturer that has only one machine and several tanks is presented. To extend their model to the multi-machines case, only constraints (8.3), (8.4), (8.10), (8.12)-(8.18) has to be considered, dropping the changeover variables in Level I, z^I_{mkls}, and adding the constraint:

$$\sum_{j\in\delta_m}\sum_{s\in S_t} a_j^{II} x_{mjs}^{II} + \sum_{i\in\delta_m}\sum_{j\in\delta_m}\sum_{s\in S_t} b_{ij}^{II} z_{mijs}^{II} \le K_{mt}^{II}, \quad m = 1,\ldots,M, t = 1,\ldots,T \tag{8.19}$$

The objective function is:

$$Min \sum_{j=1}^{J}\sum_{t=1}^{T}(h_j I_{jt}^{+} + g_j I_{jt}^{-}) + \sum_{m=1}^{M}\sum_{s=1}^{N}\sum_{i\in\delta_m}\sum_{j\in\delta_m} s_{ij}^{II} z_{mijs}^{II} \tag{8.20}$$

These sets of constraints and the objective function 8.20 define a one level multi-machines lot scheduling (P1LMM) model. It can be used in the first phase of the RA algorithm (described below) to define the lot sizing and scheduling of items in Level II. An adjustment of the solution of this model might be necessary to take into account the synchrony between the two levels. This can be obtained by model P2LMM with the setup variables fixed according to the solution of model P1LMM. That is, if item j is produced then the tank and the machine must be setup. This procedure is outlined in Algorithm 8.5, where σ_j is the liquid flavor necessary to produce item j.

Algorithm 8.5: The RA Algorithm.

```
Step 1   Solve the P1LMM model;
Step 2   if P1LMM is feasible then
             for m=1 to M; j ∈ δ_m, s = 1,...,N do
                 if x^II_mjs > 0 then
                     Fix the setup variables of P2LMM model according
                     to:
                     for j ∈ δ_m, l ∈ σ_j do
                         y^II_mjs = 1 and y^I_mls = 1
                     end
                 end
             end
         end
Step 3   Solve the P2LMM model obtained in Step 2.
```

The relax and fix strategies presented in Section 8.3.2 can also be used to solve the models in Steps 1 and 3 of Algorithm RA. Note however that in model P1LMM in Level I only the capacities constraints are considered and there are no changeover variables associated to this level. Therefore the relax and fix strategies described in Table 8.1 have to be modified accordingly. If the models in Algorithm RA are solved by standard software and their optimal solution are not achieved in a pre-defined amount of time, the branch and cut execution is also halted and the best solution is considered. Other solution approaches for SITLSP are presented and tested in [44] and [51].

8.4 Computational Tests

In this section we present and analyze the computational results of the evolutionary algorithms (MA and GA) described in Section 8.2, and the decomposition and relaxation approaches described in Section 8.3, when applied to solve industrial instances of SITLSP. Among all approaches reported in Section 8.3, we present the results of only P2LMM_G2.7 (relax-and fix strategy G2.7 applied to model P2LMM) and RA_G2.1 (relax-and-fix strategy G2.1 applied to the model P1LMM of Algorithm RA), since they were the ones that produced the best solutions for this set of instances tested. Other computational tests with randomly generated examples of small-to-moderate and moderate-to-large sizes and other industrial instances were performed with the MA and the GA approaches presented in Section 8.2. The P2EMM_G2.7, RA_G2.1 and the other approaches of Section 8.3 are as well tested in other examples. Their results are reported in [23] and [29], respectively.

8.4.1 Generation of instances

The solution methods presented in Sections 8.2 and 8.3 can be easily adapted to represent the particularities of different soft drinks companies. To follow, we describe next the necessary adjustments to represent the situation encountered at a Brazilian soft drink manufacturer, Plant A, as well as the data used to generate the instances used in the computational tests described in Section 8.4.2.

Plant A produces many types of soft drinks (items) characterized by the liquid flavor and bottle type. It has various tanks and filling lines (machines) with different capacities. Some tanks (machines) are dedicated to produce only a given subset of flavors (items), whereas others can produce any one. There is a single liquid flavor ($l = 4$) whose demand is by far superior to the others. While most of the liquid flavors have demands around 20,000 units per period, flavor 4 has demand around 150,000 units. It also has a high setup time and cost. Therefore, in Plant A there is a tank which is fully dedicated to continuously produce flavor 4. That is, whenever a machine is ready to produce an item that uses this flavor, the tank is also ready to release it.

To represent this situation in model P2LMM of Section 8.3, the changeover time in Level I from any flavor k to flavor 4 and from flavor 4 to any other is set to zero ($b^I_{k4} = b^I_{4k} = 0$). Similar adjustments are made in the evolutionary algorithms of Section 8.2. Although there is no need to setup the tanks for flavor 4, there is still a need to setup the machines when items that need this flavor are produced. However, the tank setup variable, y^I_{m4s} cannot be set to zero, since when this is done, constraints (8.3) together with (8.4) impose that the production of any item that uses this flavor is zero (if $y^I_{m4s} = 0$ then $x^{II}_{mjs} = 0$, for $j \in \omega_{m4}$). Therefore the tank capacity (constraints (8.3) and (8.4)) for $l = 4$ is dropped.

When defining the lot size and schedule, Plant A also considers that the inventory in a given period must be enough to cover the demand in the next period. To have a fair comparison between the model and the company solutions, a new set of constraints should be included in the model:

$$I^+_{jt} = d_{j(t+1)} \quad j = 1, \ldots, J; t = 2, \ldots, T+1; \tag{8.21}$$

Two minor modifications were made in the MA and GA to ensure that their solutions are comparable to the solutions of P2LMM_G2.7 and RA_G2.1, as well as the solutions used by Plant A. The first change refers to the assumption that each filling line has a dedicated tank, that is, there is only one tank allocated to only one line. Therefore, the criterion of the MA/GA for selecting a tank, instead of determined by (Section 8.2):

$$j = \begin{cases} \beta_i, & 1 \le \beta_i \le \overline{L}; \\ \beta_i - \overline{L}, & \overline{L} < \beta_i \le 2\overline{L}. \end{cases} \tag{8.22}$$

It was re-defined as: $j = \beta_i$, where $\beta_i = \alpha_i$. In addition, the fitness function of the MA/GA was modified to include only the product inventory, product changeover and demand shortage costs. As the company primarily favors production schedules with no demand shortages, the shortage unit cost was considered as a sufficiently large penalization so as to avoid shortages in the solutions. In cases whereby the MA and GA were unable to find a schedule without shortages, their solution was simply considered infeasible. The same was considered with the solutions of approaches P2LMM_G2.7 and RA_G2.1. We refer to these modified versions of MA and GA as MA1 and GA1, respectively.

The second change in the MA/GA is the consideration of penalization costs over the liquid flavor inventories maintained in the tanks from one period to another. In other words, besides the costs taken into account in the fitness function of the MA1 and GA1, we also considered penalization costs to avoid holding liquid flavor inventories between consecutive periods. This is because the company prefers production schedules that do not hold liquid inventories from one period to other, because they are perishable. Recall from Section 8.3 that approaches P2LMM_G2.7 and RA_G2.1 do not hold liquid inventories. We refer to these modified versions of the MA and GA as MA2 and GA2, respectively.

Several visits to Plant A were made in order to understand its production processes and to collect the data necessary to simulate their SITLSP. Data associated to demands, changeover times in both levels, tank and machine capacities, etc., were obtained during these visits. The details of the collected data are presented in [29, 51]. The data was used to generate 15 instances of the SITLSP.

The first instance (P1) was generated based on data related to two machines that can produce items in common. The first one (machine 1) can produce 23 items and the second one (machine 2) only 10 out of these. That is, there are 13 items that can be produced on any one of these two machines. Eighteen different flavors are necessary to produce this set of items.

Three weeks were considered in the planning horizon. Machine 1 was available for four working days per week (total of 5,760 minutes per week) and machine 2 six working days (total of 8,640 minutes per week). It was estimated that the tank could have up to five changeovers per day. Taking the average of the number machine working days (5 days), it is possible to have up to 25 changeovers per week. Therefore, the P1 instance has three macro-periods (3 weeks) with a total of 75 micro-periods (25 per macro-period). The production scheduling for this instance was provided by Plant A, making the comparison with the strategies proposed in Sections 8.2 and 8.3 possible.

To simulate different scenarios, four other instances (P2-P5) were generated by modifying part of the data used in instance P1. The inventory costs were doubled (P2), the changeover costs halved (P3), the total demand was redistributed among the periods (P4), and the machines capacities were reduced (P5). These modifications are detailed in Table 8.2.

Demand data related to a period of 30 weeks was also available. This data allowed the generation of another group of 10 instances (P6-P15). Each one of these instances is associated to three consecutive weeks. Instances P6, P7, P14 and P15 are associated to periods of higher demands when compared to P8-P13. Except for the demands, all the other parameters used to generate these instances were the same ones used to generate instance P1. More details of the procedure used to generate the instances can be found in [29,51].

Table 8.2: Modifications in instance P1 to generate P2-P5.

Instances	Modification
P1	Plant A data
P2	Inventory costs of P1 doubled
P3	Inventory costs of P1 doubled and the changeover costs of P1 halved.
P4	The total demand of P1 was randomly redistributed among the periods.
P5	The machines capacities were reduced

8.4.2 Computational results

The experiments described in this section ran on a microcomputer Pentium IV with 1 GB Ram and 2.8 GHz. The approaches P2LMM_G2.7 and RA_G2.1 were implemented using the modelling language AMPL [52] and the optimization solver CPLEX 10.0 [45] with default parameters, while the MA and the GA were implemented using the NP-Opt [28,37], an object-oriented framework written in JAVA code which contains procedures based on evolutionary computation techniques to address NP-hard problems. An execution time limit of 4 hours was established to solve each example by each method. The MA and the GA were applied three times to each example and the best solution obtained was chosen. In order to satisfy the total time limit of 4 hours, a limit of 4800 seconds was also imposed to each MA or GA run. It should be mentioned that the 4-hour limit is acceptable to support the decisions involved in the production scheduling of the company.

The MA and GA have been adjusted with 3 populations structured in ternary trees of 13 individuals each, which means a total of 39 individuals. The crossover rate ρ was set to 1.5 leading the methods to execute 19 crossovers over each population. The mutation rate was fixed at 0.7 where a γ value is randomly chosen in $[0, 1]$ with uniform distribution. The mutation operator is executed on the new individual, if $\gamma < 0.7$. All these values are based on previous tests which are reported in [1].

Table 8.3 presents the total cost values (in thousands of monetary units) of the solutions obtained by the MA1, GA1, MA2 and GA2, in comparison with the solutions of P2LMM_G2.7 and RA_G2.1, for the instances P1-P15. As mentioned, for each example, the values of the memetic and genetic algorithms correspond to the lowest cost solution found among the three executions of the algorithm. Note that the values of the table correspond only to inventory and changeover costs, since there are no demand shortages. The symbol "inf." in the table indicates that the method was unable to find a feasible solution (i.e., a solution without demand shortages) within the time limit. The best solution value of each example is highlighted in bold.

Table 8.3: Solution values of methods P2LMM_G2.7, RA_G2.1, MA1, GA1, MA2 and GA2 for instances P1-P15.

Ex.	P2LMM_G2.7	RA_G2.1	MA1	GA1	MA2	GA2
P1	399.3	322.3	256.8	285.6	**244.6**	271.4
P2	509.6	325.9	306.7	280.3	287.5	**268.4**
P3	270.3	212.9	146.6	148	**141.8**	158.8
P4	480.3	330.5	217.6	221.5	214.4	**208.1**
P5	inf.	inf.	inf.	inf.	inf.	inf.
P6	inf.	inf.	inf.	inf.	inf.	inf.
P7	inf.	inf.	inf.	inf.	inf.	inf.
P8	inf.	**529**	inf.	inf.	inf.	inf.
P9	560.9	**261.7**	inf.	372.5	348.7	391.6
P10	744.4	**266.2**	303.9	317.1	358.3	375.4
P11	461.6	**294**	inf.	inf.	inf.	inf.
P12	inf.	**344.8**	inf.	inf.	inf.	inf.
P13	inf.	**358.5**	405.7	422.4	393	483.6
P14	inf.	inf.	inf.	inf.	inf.	inf.
P15	inf.	inf.	inf.	inf.	inf.	inf.

"inf." - No feasible solution found within the time limit

As discussed in Section 8.4.1, instance P1 is the only one of the problem set for which we are aware of the corresponding production schedule used by Plant A. This company solution meets all product demands with no delays, yielding a total cost of 422.7 in thousands of monetary units. Comparing this solution to the ones presented in Table 8.3, we notice that all methods P2LMM_G.7, RA_G2.1, MA1, GA1, MA2 and GA2 were able to find better solutions than the company, with relative cost reductions of 5.5%, 23.8%, 39.2%, 32.4%, 42.1% and 35.8%, respectively. This indicates that these approaches have a potential to generate competitive solutions - note that the cost reductions can be significant. Moreover, it can be observed that the memetic versions outperformed the genetic ones.

In instances P1 and P2-P4 (which are based on instance P1), the solutions obtained by MA1, GA1, MA2 and GA2 are better than the ones obtained by P2LMM_G.7 and RA_G2.1 (Table 8.3). For these instances, the P2LMM model involved in approach P2LMM_G.7 has 86,359 variables (4,575 binary variables) and 86,140 constraints, while the P1LMM model in approach RA_G2.1 has 54,559 variables (4,575 binary variables) and 49,544 constraints. Some performance variations between the MA1 and MA2 (and between the GA1 and GA2) are due to the randomly generated initial populations of the algorithms. In particular, we do not know if instance P5 is feasible (from the point of view of demand shortages), since none of the methods found a solution without shortages. This instance was generated using the same data as P1, except for the reduction in the machine capacities (Table 8.2).

The remaining instances P6-P15 refer to collected data of the product demands for different months, provided by Plant A. Since we do not have the production schedules used by the company for these examples, we are not aware if they are feasible from the point of view of the demand shortages. In these examples, approach RA_G2.1 produced better solutions than MA1, GA1, MA2 and GA2. Moreover, for all instances P1-P15, the RA_G2.1 solutions dominate the ones of P2LMM_G.7 (Table 8.3). As well as for instance P5, we do not know if instances P6, P7, P14 and P15 are feasible, since none of the methods found a solution without shortages. The minimum capacity necessary to produce all the items in these instances is higher than in the others. However, it is worth mentioning that by using realistic backlogging or lost sales unit costs (instead of large penalties) in these methods, they can be employed to generate effective schedules balancing the trade-offs between the inventory, changeover and shortage costs.

Table 8.4 resumes the relative deviations of the solution values of MA1, GA1, MA2 and GA2 regarding approach RA_G2.1. These deviations were calculated using the expression: $Dev(\%) = 100(z - \overline{z})/\overline{z}$, where z is the solution value of MA1 (or GA1, MA2, GA2) and $\overline{z}$ is the value found by approach RA_G2.1. Note that the deviations of the evolutionary algorithms with respect to RA_G2.1 are relatively large, varying from -37.0% to 49.6%. The superiority of the memetic approach over its genetic version is also corroborated by these results.

Given that the methods were unable to find a solution with no demand shortages for instances P5, P6, P7, P14 and P15, we roughly approximate the backlogging unit costs as the profit contributions of the products and we applied the methods again to solve the examples using these parameters as the shortage unit costs. Table 8.5 presents the modified total cost values (in thousands of monetary units) of these experiments - note that, unlike Table 8.3, these values correspond to inventory, changeover and shortage costs. The best solution found for each example alternates between the methods RA_G2.1, MA1, GA1 and MA2, and these methods do not dominate each other. Table 8.6 depicts the relative deviations of the evolutionary approaches with respect to method RA_G2.1 for the examples with demand shortages.

Considering the results in Tables 8.3 and 8.5, we note that MA1 (or MA2) outperforms GA1 (or GA2) in 8 out of the 12 examples for which

Table 8.4: Relative deviations of the solution values of MA1, GA1, MA2 and GA2 with respect to approach RA_G2.1.

Ex.	MA1	GA1	MA2	GA2
P1	-20.3	-11.4	-24.1	-15.8
P2	-5.9	-14	-11.8	-17.6
P3	-31.1	-30.5	-33.4	-25.4
P4	-34.1	-33	-35.1	-37
P8	***	***	***	***
P9	***	42.3	33.2	49.6
P10	14.2	19.1	34.6	41
P11	***	***	***	***
P12	***	***	***	***
P13	13.1	17.8	9.6	34.9

*** The deviation was not computed because the approaches MA1, GA1, MA2 or GA2 did not find a feasible solution within the time limit

Table 8.5: Modified solution values of methods P2LMM_G2.7, RA_G2.1, MA1, GA1, MA2 and GA2 with demand shortages.

Ex.	P2LMM_G2.7	RA_G2.1	MA1	GA1	MA2	GA2
P5	603.7	379.5	**371.7**	422.2	485.7	393.7
P6	663.9	526.5	565.4	**492.8**	527	567.1
P7	591.5	509.5	**370.4**	398.1	372.8	413.9
P14	588.5	449.5	357.6	343.3	**310.4**	378.5
P15	671.3	**446.2**	476.6	541.1	555.4	491.2

Table 8.6: Relative deviations of the solution values of MA1, GA1, MA2 and GA2 with respect to approach RA_G2.1 (infeasible examples).

Ex.	MA1	GA1	MA2	GA2
P5	**-2**	11.2	27.9	3.7
P6	7.3	**-6.4**	0.1	7.7
P7	**-27.3**	-21.9	-26.8	-18.8
P14	-20.4	-23.6	**-30.9**	-15.8
P15	6.8	21.3	24.5	10.1

these algorithms found a feasible solution, indicating a better performance of MA over GA in these instances. A similar result was observed in the other instances randomly generated and tested in [1]. Moreover, comparing MA2 and RA_G2.1, we note that MA2 outperforms RA_G2.1 in 6 examples and RA_G2.1 outperforms MA2 in 9 examples, showing that these methods are competitive. The GA2 and MA2 approaches provide a better solution in scenarios where the total capacity is loose and the RA_G2.1 in scenarios where the total capacity is tight.

It is worth remarking that, for all instances P1-P15, the gap between the best feasible solutions of model P2LMM (found within the time limit) and its linear relaxation solution is over 90%, which does not provide much information about the optimality gap of the solutions in Tables 8.3 and 8.5.

8.5 Final Remarks and Conclusions

In this chapter we study a two-level production planning problem involving, on each level, a lot sizing and scheduling problem with parallel machines, capacity constraints and sequence-dependent setup costs and times. The problem is referred to as the Synchronized and Integrated Two-Level Lot sizing and Scheduling Problem (SITLSP) and it can be found in some industrial settings, as, for example, in soft drink companies where the production process involves two interdependent levels with decisions concerning raw material storage and soft drink bottling.

A mixed integer programming model for the SITLSP was introduced in [19], integrating and synchronizing the two lot sizing and scheduling problems involved. This model can be useful when dealing with small-to-moderate size instances, however, as the problem size grows, the number of distinct integer solutions increases exponentially, causing the branch-and-bound search to take too long, even for finding the first integer feasible solution.

In order to overcome these limitations, and deal with larger and more realistic problem instances, a genetic algorithm with a particular representation of solutions for individuals and a hierarchically structured multi-population is proposed in [23]. In Section 8.2 the genetic approach is extended by a memetic algorithm version. A tailor-made decoding procedure is used to evaluate the solution encoded in the gene of each individual. Moreover, a tailor-made recombination over population clusters takes place and migrations among different populations are allowed. The memetic approach is capable of generating competitive solutions if compared with the ones utilized in practice, as shown in Section 8.4.

As an alternative for these meta-heuristic approaches, another solution method capable of dealing with realistic problem instances of the SITLSP is presented (Section 8.3). Unlike the highly general and complex MIP model in [19], a simplified MIP formulation, in the sense that it forces the number of tanks to be the same as the filling lines so that each line has a dedicated tank, is described. The solution approach of the simplified model relies on different relax-and-fix heuristics and model decomposition strategies.

The performances of the memetic algorithm-MA (as well as its genetic version- GA) and the decomposition/relaxation approaches (P2LMM_G2.7 and RA_G2.1) are evaluated solving a set of instances based on actual data provided by a Brazilian soft drink company. All approaches are capable of producing competitive solutions when compared to the production schedule used by the company, requiring affordable computational runtime. In some cases the cost savings of the solutions are substantial (Section 8.4). Comparing the relative performances of the approaches when solving this problem set, it is observed that the MA outperforms the GA and that the MA and the RA_G2.1 do not dominate each other. In particular, the MA provides a better solution in scenarios where the total capacity is loose and the RA_G2.1 in scenarios where the total capacity is tight.

Topics for future research include new experiments with the methods studied in this chapter involving real instances to be obtained from other soft drinks plants. The evolutionary approaches can be improved by introducing new genetic operators, especially other crossovers not tested yet and other more powerful local search-based algorithms such as Simulated Annealing or Tabu Search. The introduction of valid inequalities is a research topic to be attempted to improve the decomposition/relaxation approaches.

As point out in the Introduction, the main purpose of this chapter was bring to the attention of the academic community and also to the industrial engineering practitioners the state-of-the art of computer-aided tools used to generate effective production schedules in the soft drink industry. Focusing the SITLSP, a hard combinatorial optimization problem, we believe that the most important conclusions that can be extracted from the studied methods as well as from the computational assessment performed in this chapter are two-fold: first, mathematical modelling provides deeper insight into a very complex industrial planning problem, and, second, tailor-made optimization methods can yield less costly and more effective production schedules in reasonable computing times when compared with the ones provided by general purpose commercial packages commonly used by the industry.

Acknowledgements

We are indebted to Flávio Veronese and Aldo Fernandes Júnior from Companhia de Bebidas Ipiranga, Brazil for their kind collaboration. This research was partially supported by grants from FAPESP and CNPq.

References

1. Toledo C. F. M. (2005) The integrated two-stage lot sizing and scheduling problem, Doctoral Thesis (in Portuguese), State University of Campinas, Brazil
2. Drexl A., Kimms A. (1997) Lot sizing and scheduling - survey and extensions, European Journal of Operational Research 99: 221-235.
3. Bitran G. R., Yanasse H.H. (1982) Computational complexity of the capacited lot size problem, Management Science 28(10): 1174-1186.

4. Clark A. R., Clark S. J. (2000) Rolling-horizon lot sizing when set-up times are sequence-dependent, International Journal of Production Research 38(10): 2287-2307.
5. Fleischmann B. (1994) The discrete lot sizing and scheduling problem with sequence-dependent setup costs, European Journal of Operational Research, 75: 395-404.
6. Gupta D., Magnusson T. (2005) The capacitated lot sizing and scheduling problem with sequence-dependent setup costs and setup times, Computers & Operations Research 32: 727-747.
7. Haase K., Kimms A. (2000) Lot sizing and scheduling with sequence dependent setup costs and times and efficient rescheduling opportunities, International Journal of Production Economics 66: 159-169.
8. Meyr H. (2000) Simultaneous lot sizing and scheduling by combining local search with dual reoptimization, European Journal of Operational Research120: 311-326.
9. Berreta R. E., França P. M., Armentano V. (2005) Meta-heuristic approaches for the multilevel resource-constrained lot sizing problem with setup and lead times, Asia-Pacific Journal of Operational Research 22(2): 261-286.
10. França P. M., Armentano V., Berretta R. E., Clark, A. R. (1997) A heuristic method for lot sizing in multi-stage systems, Computers & Operations Research 24 (9): 861-874.
11. Kimms A. (1997) Demand shuffle - A method for multi-level proportional lot sizing and scheduling, Naval Research Logistics 44: 319-340.
12. Kimms A. (1997) Multi-level lot sizing and scheduling - Methods for capacitated, dynamic, and deterministic models, Physica, Heidelberg.
13. Özdamar L., Barbarosoglu G. (2000) An integrated lagragean relaxation-simulated annealing approach to the multi-level multi-item capacitated lot sizing problem, International Journal of Production Economics 68: 319-331.
14. Kang S., Malik K., Thomas L. J. (1999) Lot sizing and scheduling on parallel machines with sequence-dependent setup costs, Management Science 45: 273-289.
15. Kuhn H., Quadt, D. (2002) Lot sizing and scheduling in semiconductor assembly - A hierarchical planning approach. In: Mackulak G. T., Fowler J. W., Schömig A. (eds), Proceedings of the International Conference on Modeling and Analysis of Semiconductor Manufacturing, Tempe, USA: 211-216.
16. Meyr H. (2002) Simultaneous lot sizing and scheduling on parallel machines, European Journal of Operational Research 139: 277-292.
17. Quadt D., Kuhn H. (2003) Production planning in semiconductor assembly, Working Paper, Catholic University of Eichstätt-Ingolstadt.
18. Stadtler H. (2003) Multilevel lot sizing with setup times and multiple constrained resources: internally rolling schedules with lot sizing windows 51(3): 487-502.
19. Toledo C.F.M., Kimms A., França P.M, Morabito R.(2006) A mathematical model for the synchronized and integrated two-level lot sizing and scheduling problem, Journal of Operational Research Society: under review.
20. Bitran G. R., Matsuo H. (1986) Approximation formulations for the single product capacitated lot size problem, Operations Research 34: 63-74.
21. Fleischmann B., Meyr, H. (1997) The general lot sizing and scheduling problem, OR Spektrum 19: 11-21.

22. Drexl A., Haase K. (1997) Proportional lotsizing and scheduling, International Journal of Production Economics 40: 73-87.
23. Toledo C. F. M., França P. M., Morabito R., Kimms A. (2007) A multi-population genetic algorithm to solve the synchronized and integrated two-level lot sizing and scheduling problem, International Journal of Production Research: in press.
24. Holland J. H. (1975) Adaptation in natural and artificial systems, The University of Michigan Press.
25. Goldberg D. E. (1989) Genetic algorithms in search, optimization, and machine learning, Addison Wesley.
26. Michalewicz Z. (1996) Genetic Algorithms + data structure = evolution programs, Springer-Verlag.
27. Moscato P. (1989) On evolution, search, optimization, genetic algorithms, and martial arts: towards memetic algorithms, Technical Report, Caltech Concurrent Computation Program, C3P Report 826.
28. Mendes A. S. (2003) The framework NP-Opt and its applications to optimization problems, Doctoral Thesis (in Portuguese), State University of Campinas - Brazil.
29. Ferreira D., Rangel S., Morabito R. (2007) Solution approaches for the soft drink integrated lot sizing and scheduling problem, European Journal of Operational Research (under review).
30. Clark A. R. (2003) Hybrid heuristics for planning lot setups and sizes, Computers & Industrial Engineering 45: 545-562.
31. Moscato P. (1989) On genetic crossover operators for relative order preservation. C3P Report 778, California Institute of Technology, Pasadena, CA 91125.
32. Dawkins R. (1976) The selfish gene. Oxford University Press Oxford.
33. Gen M., Cheng R. (1997) Genetic algorithms & engineering design. John Wiley & Sons New York NY.
34. Goldberg D.E. (2002) The design of innovation: lessons from and for competent genetic algorithms. Addison-Wesley Reading, MA.
35. Hart W. E., Krasnogor N., Smith J.E. (Eds.) (2005) Recent advances in memetic algorithms series: studies in fuzziness and soft computing 166.
36. Weiner J. (1995) The beak of the finch. Vintage Books New York.
37. Mendes A. S., França P. M., Moscato P. (2001) NP-Opt: an optimization framework for NP problems. Proceedings of POM2001: 82-89 Guarujá, Brazil.
38. França P. M., Mendes A. S., Moscato P. (2001) A memetic algorithm for the total tardiness single machine scheduling problem. European Journal of Operational Research 1(132): 224-242.
39. Sikora R. (1996) A genetic algorithm for integrating lot sizing and sequencing in scheduling a capacitated flow line. Computers & Industrial Engineering 30(4): 969-981.
40. Özdamar L., Birbil S. I. (1998) Hybrid heuristic for the capacitated lot sizing and loading problem with setup times and overtime decisions. European Journal of Operational Research 110: 525-547.
41. Dellaert N., Jeunet J., Jonard N. (2000) A genetic algorithm to solve the general multi-level lot sizing problem with time-varying costs. International Journal of Production Economics 68: 241-257.
42. Kimms A. (1999) A genetic algorithm for multi-level, multi-machine lot sizing and scheduling. Computers & Operations Research 26: 829-848.

43. Dueck G., Scheuer T. (1990) Threshold accepting: A general purpose optimization algorithm appearing superior to simulated annealing. Journal of Computational Physics 90: 161-175.
44. Ferreira D., Morabito R., Rangel S. (2007) A MIP model and relax and fix heuristics for production planning and scheduling in a small soft drink plant (in Portuguese), Produção (São Paulo): in press.
45. ILOG (2001) Using the CPLEX Callable Library, Copyright, ILOG.
46. Wolsey L. A. (1998) Integer Programming, John Wiley & Sons.
47. Dillemberger C., Escudero L. F., Wu Zhang A. W. (1994) On practical resource allocation for production planning and scheduling with period overlapping setups, European Journal of Operational Research 75: 275-286.
48. Escudero L. F., Salmeron J. (2005) On a fix-and-relax framework for a class of project scheduling problems, Annals of Operations Research 140: 163-188.
49. Pedroso J. P., Kubo M. (2005) Hybrid tabu search for lot sizing problems, in Hybrid Meta-heuristics, Lecture Notes in Computer Science 3636: 66-77, Springer, Berlin.
50. Federgruen A., Meissner J., Tzur M. (2007) Progressive interval heuristics for multi-item capacitated lot sizing problems, Operations Research, 55 (3): 490-502.
51. Ferreira D. (2006) Approaches to the integrated problem of the lot sizing and scheduling of the soft drink production, Doctoral Thesis (in Portuguese), Federal University of São Carlos, Brazil.
52. Fourer R., Gay M. D., Kernighan B. W. (1993) AMPL - A modeling language for mathematical programming, The Scientific Press, Danvers, Massachusetts.

9

Hybrid Heuristic Approaches for Scheduling in Reconfigurable Manufacturing Systems

Ricardo Galán

Agua y Estructuras, S.A. (Ayesa) Marie Curie 2, 41092 Seville, Spain
rgalan@ayesa.es

Summary. Reconfigurable Manufacturing Systems (RMS) are the next step in manufacturing, allowing the production of any quantity of highly customized products together with the benefits of mass production. In RMS, products are grouped into families, each of which requires one configuration of the system. The system is configured for producing the first family of products. Once it is finished, the system is reconfigured in order to produce the second family, and so forth. Then, the effectiveness of a RMS depends on the selection of the best set of product families and their scheduling. A methodology has been developed to group products into families, which takes into account the requirements of products in RMS. Once the families have been selected, a model to optimize the production scheduling is presented. This model is based on minimizing the costs required to reconfigure the system while both the capacity and functionality of machines are maximized. The complexity of the problem suggests the use of heuristics methodologies. Several heuristics are candidates to be used. With the aim of covering different approaches, both a specific heuristic for this problem and general heuristics or meta-heuristics have been developed. Tabu search is a traditional meta heuristic that has demonstrated to offer satisfactory results to a broad range of combinatorial problems, and it has been considered to be implemented. In order to use another meta-heuristic to compare results, ant colony optimization techniques have been implemented because they have demonstrated to offer good behaviors to similar problems.

Key words: Reconfigurable Manufacturing Systems, Meta-heuristics, Scheduling, Tabu Search, Ant Colony Optimization.

9.1 Introduction

Markets are driven by globalization, product customization and a continuous improvement of the manufacturing technology [1]. Time reduction to introduce new products to the market with high quality maintaining low cost has become essential to survive in this new scenario [2]. Therefore, the manufacturing

R. Galán: *Hybrid Heuristic Approaches for Scheduling in Reconfigurable Manufacturing Systems*, Studies in Computational Intelligence (SCI) **128**, 211–253 (2008)
www.springerlink.com

system must be able to produce different batch sizes from a portfolio of product types with the capacity and functionality required in each case.

Traditional manufacturing systems, such as dedicated manufacturing systems, produce large volumes from a small range of products under known demand by using fixed resources. Other systems, such as flexible and cellular manufacturing systems, emerged due to the need of satisfying new markets by offering small volumes from a wide range of products, with predictable demand and high inversion costs. Agile manufacturing systems may react to the continuous changes of the current competitive market, but they are focused on organizational aspects more than on manufacturing system operations [3].

Nowadays, the manufacturing industry faces the challenge of satisfying quickly the dynamic requirements of the customers. The key factor is the ability to launch new products to the market with high quality, low cost and fast delivery [4]. This may be reached through *mass customization*, a new paradigm for industries to provide products fulfilling customer needs with mass production efficiency [5]. Its implementation requires managing a variety of customized processes, economies of scale and high flexibility in the production processes. The manufacturing system must be able to produce a wide variety of products, with a quick launch of products to the market, with low cost and high quality. Reconfigurable manufacturing systems (RMS) arise to face this challenge.

RMS can produce a wide variety of products by rearranging the manufacturing system itself. It is composed of combinations of hardware and software modules which can be rearranged in a rapid and reliable way, avoid becoming obsolete. Besides, it is open-ended, so it may be improved regularly through the implementation of new methodologies. A RMS is designed at the outset for rapid adjustment of its production capacity and functionality, with the aim of satisfying the requirements of the market through changing or rearranging its components [1].

Several studies about the future of manufacturing for the 2015–2020 horizon include the RMS as a priority line of research. The Delphi study entitled "Visionary manufacturing challenges for 2020" has identified and catalogued the RMS as one of the "six grand challenges" about the future of manufacturing [6]. The FutMan project "The future of manufacturing in Europe 2015–2520" demands the development of flexible adaptive and reconfigurable manufacturing equipment [7]. Finally, the IMTI project "The integrated manufacturing technology initiative–2015 vision" includes the need of rapidly reconfigurable and self-configuring facilities [8].

The European Union knows the importance of RMS and promotes its development by supporting the I*PROMS (innovative production machines and systems). Network of Excellence within the 6th framework programme of research. Its objective is to develop concepts, tools and techniques to make flexible and reconfigurable production systems [9]. Likewise, in 1996 the National Science Foundation (NSF) of the US created the Engineering Research Center for Reconfigurable Manufacturing Systems with the aim of helping the nation to face the challenges of an ever-growing and ever-changing global economy.

9.2 Scope of the Research

The process of designing and operating a RMS can be split into three main issues: system-level, machine and controls and ramp-up time reduction [10]. System-level issues are focused on the design of the production system. They are guided by tools to link product features with modules of machines, obtaining as a result a layout of the production system and a process plan. Machine and control issues are considered to enable rapid and efficient reuse of modular components, including the design of modular components and open-architecture controllers. Finally, ramp-up time reduction tries to reduce the time required to manufacture products by diagnosing component failure or deviation of product quality.

This research is focused on system-level issues. Methodologies, mathematical models and heuristics will be developed with the aim of selecting and sequencing product families that minimize the cost of reconfiguring the production system.

The first step deals with the formation of product families. Literature presents plenty of methods to obtain families and diverse formation criteria, especially regarding cellular manufacturing. Nevertheless, these methods and criteria cannot be used directly in reconfigurable manufacturing because it has its own singularities that must be taken into consideration, which differ from other manufacturing paradigms. Identification of the key attributes of products that must be considered for their clustering is a key issue. After that, a methodology to group products into families and a model to select the best set of families and their scheduling are required. The selection of different sequences implies different costs and therefore a detailed selection is necessary. The development of heuristics when exact methodologies cannot offer a solution must be considered.

9.3 Formation of Product Families for RMS

9.3.1 Identification of Product Attributes

The formation of product families in RMS has to be based in some grouping criteria. These criteria are the key attributes of products that are considered important for their manufacturing in the production system.

Modularity is considered a key attribute of RMS because it is essential for implementing customized and complex products. Modularity can be defined as the degree to which a product is composed of independent modules, without interactions among them [11]. For customized products, modularity allows the assembly of simple and functionally independent parts. Although the number of parts in the modular design is larger than in the integral design, the benefits of modularity arise from the fact that the total time of machining operations and manufacturing costs are low in the modular design because of the simplicity of those parts [12]. The integral designs result in complex parts that require more complicated and costly resources. Modularity can also be regarded as a strategy for effectively organizing complex products and processes [4].

The object of developing a RMS is to ensure a variety of products that reflects market requirements. Product variety is defined as the diversity of products provided to the marketplace by a production system [13]. Apart from modularity, another important concept related to product variety is *commonality* [14] which can be defined as a measure of how well the product uses standardized parts [15], with the view of reducing the total number of different parts. Commonality ensures that a component of a product is shared by two or more products of the same family [16].

Another key attribute is the *compatibility* between products from the same family. Products that, for example, require the same technical operations or target the same market must be grouped in the same family.

The elements in a RMS must be reusable. Although it is referred to machines, controllers, etc. more than to products, product *reusability* could be maximized through arranging products and assigning them to families of similar products. Reusability measures the use of existing design configurations while reconfiguring manufacturing elements for a new product type [17].

Within the framework of RMS, manufacturers receive orders of products for all the families, so producing a high volume of a certain family may delay the delivery time of the rest of the families [2]. Thus, product *demand* is another key attribute to consider in RMS.

9.3.2 Review and Selection of Grouping Methods

The proliferation of methods for grouping products arose with the development of cellular manufacturing systems (CMS); the first manufacturing paradigm focused on making the cost-effective manufacture of several part types possible. Traditionally, grouping products into families and cell formation in CMS have been closely linked.

Instead of developing new methods for grouping products in RMS, a study of the methods used in CMS will be carried out with the aim of identifying one of them that can be conveniently modified in order to stand the requirements for products in RMS.

Several methods have been developed for families' formation [18], such as descriptive procedures, mathematical programming approaches, array-based clustering methods, and hierarchical clustering, among others.

Most of the descriptive procedures are not highly sophisticated or accurate, although they are cost-effective [19]. They may be appropriate to little and repeated problems, but they do not provide good solutions in general.

Mathematical programming approaches are incompletely formulated [18] and therefore their usefulness is limited in industrial environments. All these models are computationally complex and it is unlikely that they can provide good solutions to large-scale problems.

The array-based clustering methods achieve acceptable solutions with low computational cost. Although ordered matrices can be developed using array-based clustering, disjoint part families or machine groups are not identify. Besides, they have the disadvantage of depending on the initial incidence matrix configuration [19].

The hierarchical clustering agglomerative methods group together similar elements (products) in clusters based on their similarities of attributes. The coefficients that measure similarity between two parts are calculated from the incidence matrix. After that, a dendogram (inverted tree structure) shows the similarity degree for grouping parts. These methods are the most broadly implemented. They use similarity or dissimilarity coefficients among parts to obtain the groups. The most important similarity coefficient for part-family formation is the Jaccard similarity coefficient [20], which measures the similarity between a pair of products (m, n), and it is defined in terms of the machines that each product has to visit.

In the context of cell (and families) formation, only agglomerative clustering techniques have been used [18]. These techniques have a string effect known as *chaining*, which creates a few large clusters while leaving several parts unmerged [21]. Among those techniques, Average Linkage Clustering (ALC) algorithm has the least tendency to chain [22] and it is considered in this chapter as the most appropriate to apply.

9.3.3 Development of the Methodology

Formation of Matrices

The application of ALC starts with the part-machine incidence matrix formation, which is a matrix that indicates whether a part is processed by a machine.

The process of designing matrices in RMS has to take into account the product attributes previously identified. Matrices that indicate, for each pair of products to group, how well those products are similar in terms of modularity, commonality, compatibility, reusability, and demand are necessary.

Modularity Matrix

Modularity can be obtained from the bill of materials (BOM), which represents the product structure, including the components and subassemblies that make up the product.

The implementation of the modularity matrix is a three-stage procedure. In the first stage, the product-part matrix is created according to the component tree. This matrix is composed of n products $i = (A, B, \ldots n)$ and m parts $j = (1, 2, \ldots m)$ in rows and columns, respectively. Values in the matrix (a_{ij}) are 1 if product i requires part j, and 0 otherwise. Those values are obtained from the BOM. For example, Fig. 9.1 represents the BOM of four different products (A, B, C, D).

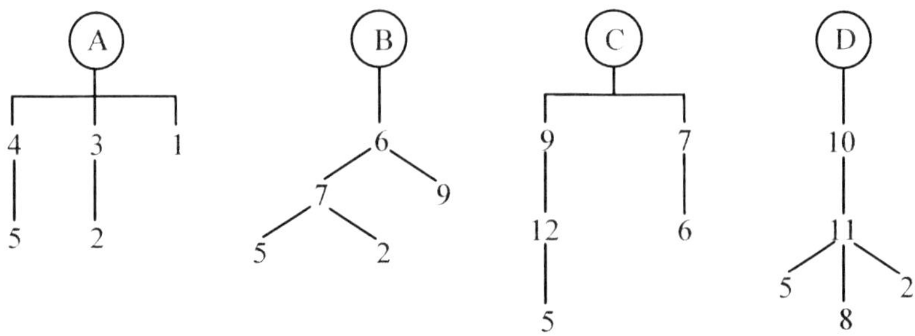

Fig. 9.1: Example of BOM.

From this BOM, the product-part matrix can be obtained, as it is depicted in Table 9.1.

Table 9.1: Product-part matrix.

	1	2	3	4	5	6	7	8	9	10	11	12
A	1	1	1	1	1	0	0	0	0	0	0	0
B	0	1	0	0	1	1	1	0	1	0	0	0
C	0	0	0	0	1	1	1	0	1	0	0	1
D	0	1	0	0	1	0	0	1	0	1	1	0

The level of product modularity is calculated in the second stage. That is achieved using expression 9.1, which determines the number of modular components of the product compared to the total number of components that form the product:

$$M_p = \frac{\Psi_p}{\Phi_p} \qquad 0 \leq M_p \leq 1 \tag{9.1}$$

The modularity level of product p (M_p) depends on the number of components of product p that are shared by more products (Ψ_p), and on the total number of components of product p (Φ_p). Following the example in Table 9.1, the modularity level of products A, B, C, and D is calculated using expression 9.1:

$$M_A = \frac{2}{5} = 0.4 \qquad M_B = \frac{5}{5} = 1 \qquad M_C = \frac{4}{5} = 0.8 \qquad M_D = \frac{2}{5} = 0.4$$

In the third stage, the modularity matrix is formed through the similarities between pairs of products. The similarities coefficients are calculated with expression 9.2:

$$S_{pq} = 1 - |M_p - M_q| \qquad 0 \leq S_{pq} \leq 1 \tag{9.2}$$

S_{pq} is the similarity between products p and q. M_p and M_q are the modularity levels of products p and q, respectively. Referring to the described example, the similarity coefficients between pairs of products are calculated using expression 9.2. Note that $S_{pq} = S_{qp}$.

$$\begin{array}{ll} S_{AB} = 1 - |0.4 - 1| = 0.4 & S_{AC} = 1 - |0.4 - 0.8| = 0.6 \\ S_{AD} = 1 - |0.4 - 0.4| = 1 & S_{BC} = 1 - |1 - 0.8| = 0.8 \\ S_{BD} = 1 - |1 - 0.4| = 0.4 & S_{CD} = 1 - |0.8 - 0.4| = 0.6 \end{array}$$

The modularity matrix is presented in Table 9.2, composed of the four products in rows and columns. The formation process of the modularity matrix can be described with the algorithm presented in Fig. 9.2.

Table 9.2: Modularity matrix.

	B	C	D
A	0.4	0.6	1
B		0.8	0.4
C			0.6

Commonality Matrix

The commonality matrix can be used to identify products that share some parts. The implementation of the commonality matrix is a two-stage procedure. First, a product-part matrix is developed. For the above example, this matrix was shown in Table 9.1. In the second stage, the similarity between pairs of products (p, q) is measured with the *Jaccard's similarity coefficient*, which may be expressed as:

$$J_{pq} = \frac{a}{a+b+c} \qquad 0 \leq J_{pq} \leq 1 \tag{9.3}$$

Where a indicates the number of parts that form both products p and q, b stands for the number of parts that form product p but not product q, and c the number of parts that form product q but not product p. Therefore, if $J_{pq} = 1$ then both products are composed of the same parts, and if $J_{pq} = 0$ then products are composed of different parts. Products with higher similarity coefficients are grouped together.

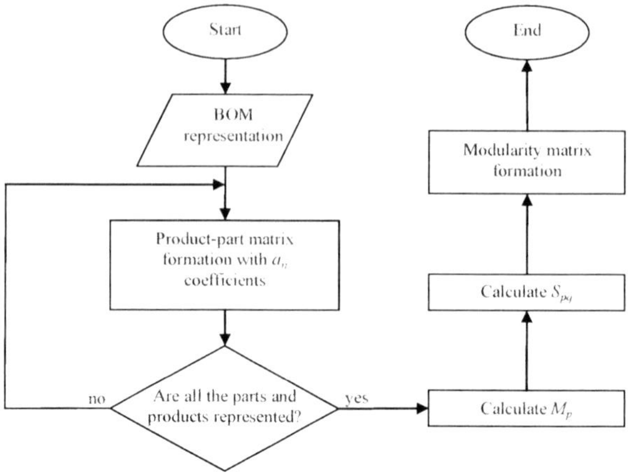

Fig. 9.2: Algorithm for modularity matrix.

Applying expression 9.3 to the example, the Jaccard's coefficients are:

$$J_{AB} = J_{AD} = J_{BD} = \frac{1+1}{1+1+1+1+1+1+1+1} = 0.25 \quad J_{BC} = \frac{1+1+1+1}{1+1+1+1+1+1} = 0.67$$

$$J_{AC} = J_{CD} = \frac{1}{1+1+1+1+1+1+1+1+1} = 0.11$$

The commonality matrix finishes when transferring these results to a matrix, as shown in Table 9.3. The matrix formation process is presented with the algorithm shown in Fig. 9.3.

Table 9.3: Commonality matrix.

	B	C	D
A	0.25	0.11	0.25
B		0.67	0.25
C			0.11

Compatibility Matrix

Compatibility measures the degree to which different products can be joined to form a family of similar products. It can be calculated from the compatibility matrix, which maps the compatibility of each product against all others.

Compatibility is classified into *technological* and *marketing*. Technological compatibility refers to technical similarities among products, which share some operations as, for instance, manufacturing and assembly operations. Marketing compatibility refers to the combination of products into families that together are desirable for a certain market [23].

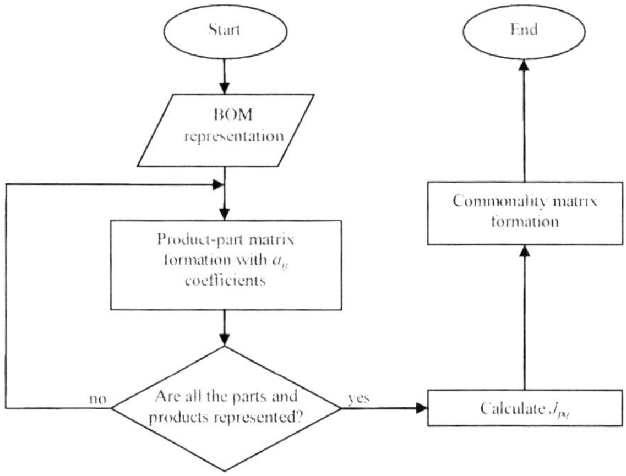

Fig. 9.3: Algorithm for commonality matrix.

Compatibility can be calculated using two matrices, one for measuring the technological compatibility and the other for the marketing one. The matrices are based on the suggestions of a team composed of experts and consumers, and they can be developed sequentially or simultaneously. The technological-compatibility matrix should be designed by a team consisting of experts from all the production stages (such as manufacturing and assembly). The marketing-compatibility matrix should be developed by a team consisting of marketing experts (such as sales personnel, distributors, and retailers), and consumers.

If a new product is introduced into a family by replacing another, both compatibility matrices must be reformulated. This is due to the new product details, which may define new sets of interactions with existing products. In the case of innovative products, expert judgement is the basis for proper marketing compatibility evaluation.

The matrices are composed by products in rows and columns. Referring to the values in the matrices, 0 indicates full product incompatibility, and 1 full product compatibility. Experts may not be able to classify all pairs of products as compatible or incompatible; therefore measures between 0 and 1 are required. Being i and j two different products, their compatibility coefficients in the matrix (a_{ij}) are $0 \leq a_{ij} \leq 1$.

For example, if two products are compatible in manufacturing operations but not in assembly operations or they are compatible in one market segment but not in others, then a ratio measurement between 0 and 1 is required. Possible measurements are shown in Table 9.4.

Another option is to develop two groups of matrices. One group should be composed of technological compatibility matrices, such as three single matrices regarding manufacturing, assembly and inspection operations. The other

Table 9.4: Compatibility values among products.

Compatibility	Values
Not compatible	0
Slightly compatible	0.3
Compatible	0.5
Very compatible	0.8
Highly compatible	1

group should include marketing compatibility matrices, one for each market segment.

An example of a compatibility matrix for four products is shown in Table 9.5. As the design stage has taken into consideration the modularity nature of products, it is expected that a high degree of compatibility among products exists.

Table 9.5: Compatibility matrix.

	B	C	D
A	1	0	1
B		1	1
C			0

The formation process of compatibility matrices is presented with the algorithm presented in Fig. 9.4.

Reusability Matrix

Reusability deals, at product level, with the use of existing product components to manufacture a new product type. Thus, reusability is maximized when all components of a product are used to manufacture the following product.

The implementation of the reusability matrix is a three-stage procedure. In the first stage, the product-component matrix is formed in the same way as in modularity and commonality matrices: matching products and components. This matrix was shown in Table 9.1.

In the second stage, a matrix composed by products in rows and columns is required. The coefficients of this matrix refer to the reusability between products p and q, when product q is manufactured just after being produced product p, and are calculated with expression 9.4:

$$R_{pq} = \frac{\gamma_{pq}}{\Phi_p} \qquad 0 \leq R_{pq} \leq 1 \tag{9.4}$$

Reusability between products p and q (R_{pq}) depends on the number of components of product p shared with product q (γ_{pq}), and on the total number

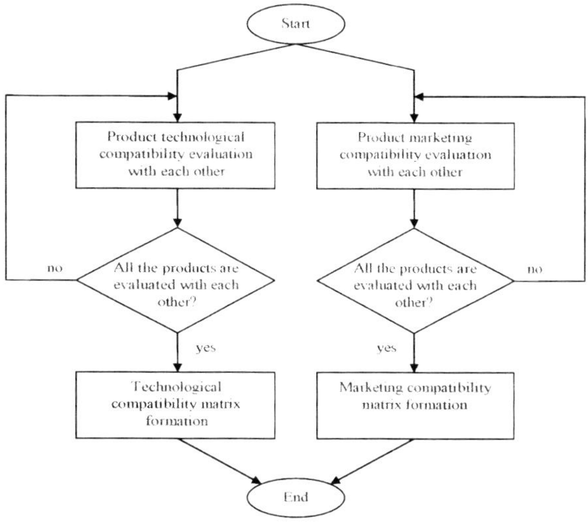

Fig. 9.4: Algorithm for compatibility matrix.

of components of product p (Φ_p). Note that, $R_{pq} \neq R_{qp}$ unless products have the same number of components.

The values for the reusability in the example of Table 9.1 are obtained using expression 9.4:

$$R_{AB} = R_{BA} = \tfrac{2}{5} = 0.4 \qquad R_{AC} = R_{CA} = \tfrac{1}{5} = 0.2 \qquad R_{AD} = R_{DA} = \tfrac{2}{5} = 0.4$$

$$R_{BC} = R_{CB} = \tfrac{4}{5} = 0.8 \qquad R_{BD} = R_{DB} = \tfrac{2}{5} = 0.4 \qquad R_{CD} = R_{DC} = \tfrac{1}{5} = 0.2$$

These values are transferred to the previous matrix shown in Table 9.6.

Table 9.6: Reusability matrix.

	B	C	D
A	0.4	0.2	0.4
B		0.8	0.4
C			0.2

In the reusability matrix, and in order to keep consistency among the developed matrices, R_{pq} and R_{qp} should be equal. Consequently, the third stage consists of calculating the coefficients of the reusability matrix (A_{pq}), through expression 9.5, as the arithmetic mean of the pairs of products, whether they are not the same:

$$A_{pq} = \frac{R_{pq} + R_{qp}}{2} \qquad 0 \leq A_{pq} \leq 1 \tag{9.5}$$

In this example, as $R_{pq} = R_{qp}$ both previous and reusability matrices are the same (Table 9.6). The formation process of the reusability matrix is presented with the algorithm shown in Fig. 9.5.

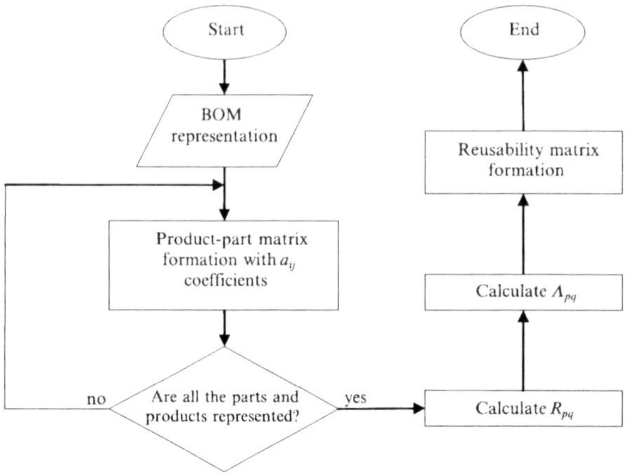

Fig. 9.5: Algorithm for reusability matrix.

Demand Matrix

The elements of a homogeneous system configuration should have similar capacity. In order to obtain a cost effective system configuration, the system capacity should have the highest possible utilization rate. Therefore, a requirement is to group products with similar demands to select a machining system with similar capacity.

Therefore, the interaction value between products p and q is calculated using expression 9.6:

$$D_{pq} = 1 - \frac{|d_p - d_q|}{d_{max} - d_{min}} \qquad 0 \leq D_{pq} \leq 1 \tag{9.6}$$

Where D_{pq} is the interaction value between products p and q, d_i the demand of product $i = \{p, q\}$, d_{max} the maximum value of d_i, and d_{min} the minimum value of d_i. Besides, $D_{pq} = D_{qp}$.

For example, if the demand for four products A, B, C, and D is $D_A = 5$, $D_B = 7$, $D_C = 3$, and $D_D = 5$ units, the product demand matrix coefficients are calculated using expression 9.6.

$$D_{AB} = 1 - \frac{|5-7|}{7-3} = 0.5 \qquad D_{AC} = 1 - \frac{|5-3|}{7-3} = 0.5 \qquad D_{AD} = 1 - \frac{|5-5|}{7-3} = 1$$

$$D_{BC} = 1 - \frac{|7-3|}{7-3} = 0 \qquad D_{BD} = 1 - \frac{|7-5|}{7-3} = 0.5 \qquad D_{CD} = 1 - \frac{|3-5|}{7-3} = 0.5$$

The resulting matrix is shown in Table 9.7.

The matrix formation process is presented with the algorithm shown in Fig. 9.6.

Table 9.7: Demand matrix.

	B	C	D
A	0.5	0.5	1
B		0	0.5
C			0.5

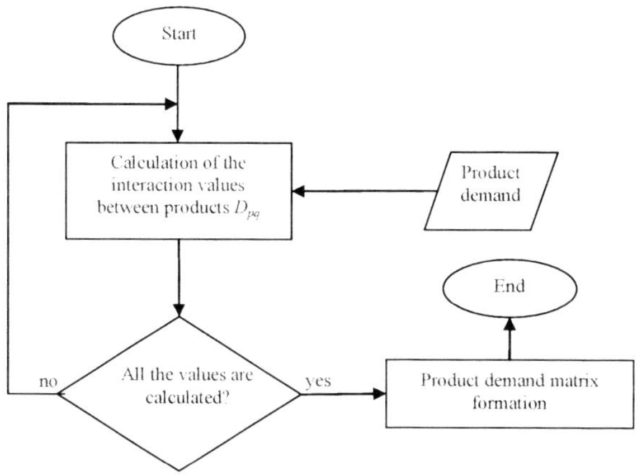

Fig. 9.6: Algorithm for product demand matrix.

Clustering Methodology for RMS

A unique matrix comprising the values of interaction among products in ALC algorithm is required. Thus, values included in the matrices are used to calculate a weighted matrix that balances the requirements. This is a multi-criteria decision making problem, solved using weighting techniques.

The purpose of weighting methods is to assign values to a set of objectives/criteria to indicate their relative importance. The most popular method is the *Analytic Hierarchy Process (AHP)* proposed by Saaty [24]. The importance of the criteria is rated on a nine-point scale, ranging from equal importance (1) to absolutely more important (9). Then, eigenvalues are calculated to represent weights. Advantages of the AHP method include the capability of assessing the consistency of the decision makers' ratings, and it allows sensitivity analysis. For these reasons, the AHP method will be used for obtaining the weighted matrix.

The formation of the unique matrix comprising the five requirements is calculated as the sum of the coefficients of each product requirement matrix, which are also multiplied by their corresponding weights. The coefficients of this matrix (a_{ij}) are calculated with expression 9.7:

$$a_{ij} = \sum_{r \in R} a_{ij_r} \sigma_r \qquad 0 \leq a_{ij} \leq 1 \tag{9.7}$$

Where R is the set of requirements, a_{ij_r} stands for the coefficients in the matrix of each requirement, and σ_r is the weight of each requirement.

Supposing that the weights obtained with the AHP methodology are: 0.15 for modularity, 0.35 for commonality and compatibility, 0.10 for reusability, and 0.05 for product demand. The coefficients for the unique matrix are calculated using expression 9.7:

$$\begin{aligned}
a_{AB} &= 0.4 \cdot 0.15 + 0.25 \cdot 0.35 + 1 \cdot 0.35 + 0.4 \cdot 0.1 + 0.5 \cdot 0.05 = 0.56\\
a_{AC} &= 0.6 \cdot 0.15 + 0.11 \cdot 0.35 + 0 \cdot 0.35 + 0.2 \cdot 0.1 + 0.5 \cdot 0.05 = 0.17\\
a_{AD} &= 1 \cdot 0.15 + 0.25 \cdot 0.35 + 1 \cdot 0.35 + 0.4 \cdot 0.1 + 1 \cdot 0.05 = 0.68\\
a_{BC} &= 0.8 \cdot 0.15 + 0.67 \cdot 0.35 + 1 \cdot 0.35 + 0.8 \cdot 0.1 + 0 \cdot 0.05 = 0.79\\
a_{BD} &= 0.4 \cdot 0.15 + 0.25 \cdot 0.35 + 1 \cdot 0.35 + 0.4 \cdot 0.1 + 0.5 \cdot 0.05 = 0.56\\
a_{CD} &= 0.6 \cdot 0.15 + 0.11 \cdot 0.35 + 0 \cdot 0.35 + 0.2 \cdot 0.1 + 0.5 \cdot 0.05 = 0.17
\end{aligned}$$

With these values, the unique matrix is shown in Table 9.8.

Table 9.8: Unique matrix.

	B	C	D
A	0.56	0.17	0.68
B		0.79	0.56
C			0.17

Application of the ALC Algorithm

The ALC methodology can be applied when the key attributes of products are displayed in one single matrix. This methodology starts by grouping products with higher coefficient of similarity. Then, a sub-matrix considering the products grouped as a family is created. Similarities between parts are recalculated as the average values using expression 9.8:

$$S_{ij} = \frac{\sum_{m\in i}\sum_{n\in j} S_{mn}}{N_i\, N_j} \tag{9.8}$$

Where i,j are families; m,n parts of family i and j, respectively; S_{ij} the coefficient of similarity between families i and j; S_{mn} the coefficient of similarity between parts m and n; and N_i, N_j the number of parts in family i and j, respectively.

This procedure is repeated until all products are grouped into a family. As a result, a dendogram is obtained. An algorithm for this method is shown in Fig. 9.7.

Applying this algorithm to the unique matrix from the above example, the maximum value of S_{ij} is 0.79 between products B and C, thus they are

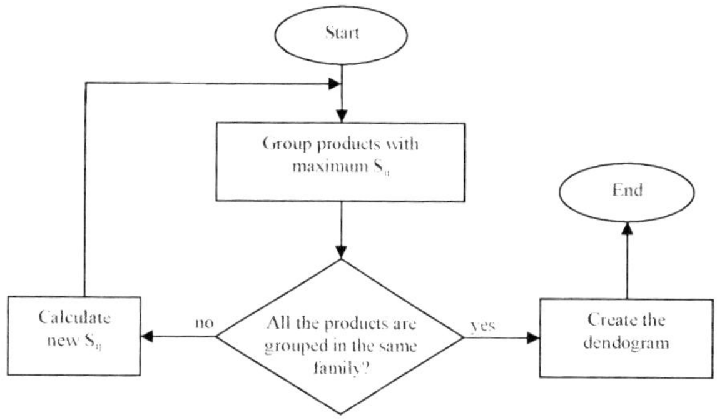

Fig. 9.7: Algorithm for the ALC method.

grouped in the family (BC) and S_{ij} are recalculated for the new sub-matrix applying expression 9.8.

$$S_{A,(B,C)} = (S_{AB} + S_{AC})/2 = (0.56 + 0.17)/2 = 0.37$$
$$S_{A,D} = 0.68$$
$$S_{(B,C),D} = (S_{BD} + S_{CD})/2 = (0.56 + 0.17)/2 = 0.37$$

These values are shown in Table 9.9.

Table 9.9: Intermediate matrix.

	BC	D
A	0.37	0.68
BC		0.37

The new matrix's maximum value is 0.68 corresponding to products A and D. Both products are grouped and the new S_{ij} are recalculated:

$$S_{(A,D),(B,C)} = (0.56 + 0.17 + 0.56 + 0.17)/4 = 0.37$$

This value is translated to the final matrix presented in Table 9.10, which indicates that the four products can be grouped together in the same family with 37% of similarity among the products.

Table 9.10: Final matrix.

	BC
AD	0.37

The resulting dendogram is shown in Fig. 9.8. From this dendogram the selection of product families can be made. There are four different levels, each with different families. Upper levels are composed of several families with few products and high similarity among them. On the contrary, families in bottom levels are composed of few families with lots of products with low similarity. For example, three families can be formed with a similarity of 79% among the products in the families (level 2). The first family is composed of products $B-C$, the second one is composed of product A, and the last family is composed of product D.

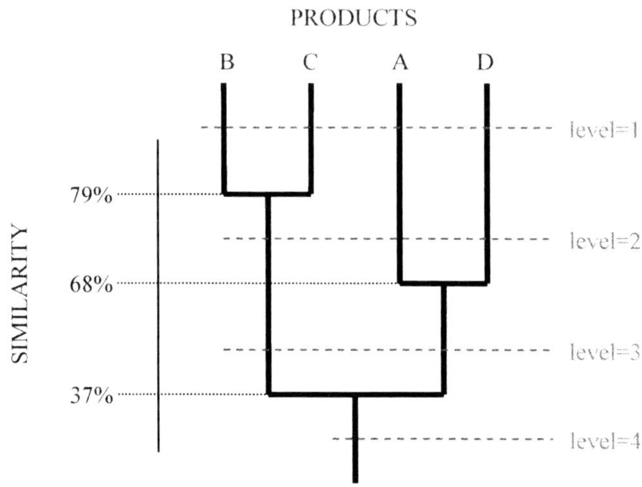

Fig. 9.8: Dendogram

In the traditional ALC method used in group technology, the selection of families was determined by the costs incurred when a product could not be manufactured within its cell, and the costs incurred when a machine within a cell could not manufacture a product of its associated family. In the design of RMS, this selection depends on other factors which are described in section 9.4.

9.4 Selection and Scheduling of Product Families for RMS

This section describes the process to select product families and their production scheduling. To achieve this goal, the main parameters required to minimize the costs to configure the production system will be identified.

The selection of families and their scheduling in the production system is a complex problem which requires the development and validation of mathematical tools. The accuracy of these tools depends on the costs used. Nowadays the implementation of RMS is in the beginning and therefore it is not possible to obtain real data of the operational costs required to configure the production system. To overcome this limitation, a methodology to estimate these costs will be developed.

9.4.1 Selection of Parameters

A reconfigurable manufacturing system focuses on manufacturing the product families at the same system, which is configured to produce each family. Once a family is manufactured, the system is reconfigured for producing the following family effectively. In each change of the configuration, the manufacturing company incurs in a changeover cost, which depends on the current configuration and the destination configuration [2].

About the formation of product families, three different situations are possible. First, it is possible to join all products in a single family, being the manufacturing system composed of the machines required to manufacture the products. In this situation, the company does not incur in changeover costs, but idle machines and/or machines whose functionality is not fully used exist. Besides, the capacity of the machines is normally higher than the one required for manufacturing each product independently.

Second, in the case of selecting a family for each product, the company has to face the costs to changeover the system, but the number of idle machines is minimized while their functionalities and capacities are both fully utilized.

Finally, the last case occurs when a smaller number of families than the number of different products is selected. In this situation, although the changeover cost is not avoided, changes are few and consequently the cost is low. Moreover, idle machines are fewer than in the first case. The functionalities and capacities of the machines are not fully used but the utilization rate is higher if compared to the first case.

Concluding, the key parameters to take into consideration for selecting product families are the changeover cost of the system (*reconfiguration cost*), cost of idle machines, cost of the under-utilization of the functionality of the machines and cost of the under-utilization of the capacity of the machines (*under-utilization costs*).

The selection of families can be solved calculating the cost of each level in the dendogram and the level with the lowest cost will be selected. Thus, all the possible solutions are evaluated. Computing time depends on the number of possible combinations to schedule the families, which may be much higher than the time available to make decisions. Therefore, a model that includes the key parameters expressed above and facilitates this selection is required.

9.4.2 Methodology to Estimate Costs

To solve the model, accuracy costs are essential. The literature does not present any methodology to estimate costs required to change modules or machines on RMS. In this section, a specific methodology to obtain those costs that takes into account the singularities of RMS is presented.

Identification of Parameters

Instead of doing a direct and global estimation for both costs, a strategy "divide and defeat" has been followed. On one hand, reconfigurability costs have been decomposed into three different parameters, each with its own cost. These parameters are:

- A single module of a machine is removed or included.
- A machine is removed.
- A machine is included.

On the other hand, under-utilization costs have been decomposed into two parameters, which are estimated. These parameters are:

- A module is not used when manufacturing a product from a certain family.
- A machine (with all its modules) is not used when manufacturing a product from a certain family.

Costs of reconfiguration deal with incurred costs to change the configuration of the system for producing a family to another configuration for producing the following family. These costs have to be calculated for each pair of families in each level of the dendogram. The last level contains one family and therefore the reconfiguration of the system is not required. This is shown in the objective function of the model, which the first term indicates the cost of reconfiguration until level $L-1$, being L the total number of levels.

Costs estimation starts with the identification of the components which compose the products that form both families. Those components are produced by a specific module belonging to certain machines. Machines in RMS are composed of different modules each of them develops a specific functionality.

The same example with four products is presented for costs estimation. Data required for the estimation are relationships among products and components, among components and the modules that produce them, and among modules and its machines. Moreover, parameters in which are decomposed the cost of reconfiguration and cost of under-utilization of resources are estimated.

Modules for manufacturing the components, together with their machines, are presented in Table 9.11:

The dendogram shown in Fig. 9.8 presents the different families that can be formed depending on the similarities among products. From level 1, four families are obtained, each of which composed of one product. The name of each family is the name of the products that compose that family. Therefore, the families in level 1 are: A, B, C, and D. From level 2, three families are obtained, one composed of two products (family BC, and the rest composed

Table 9.11: Machines, modules and components.

Machine	Modules	Components
M1	M11	1
M2	M21	2,4,10
	M22	3
	M23	11
M3	M31	5,6
	M32	7,9
	M33	12
M4	M41	8

of one product (families A and D). From level 3 two families appear, both composed of two products (families BC and AD). Finally, from level 4 one only family is obtained composed of the four products (family $BCAD$).

Cost of parameters regarding reconfigurability must be estimated based on experience. The first parameter controls if a module from a machine is used to produce one product but not other products. The second and third parameters deal with if a machine is required to produce one product but not other products. Their estimations are presented in Table 9.12:

Table 9.12: Cost estimation for reconfigurability parameters.

Parameter	Symbol	Cost
A single module of a machine is removed or included	α	1
A machine is removed	β	5
A machine is included	γ	6

For a certain scheduling of families, the cost of reconfiguration (monetary unit) is calculated as the sum of multiplying the cost of each parameter (monetary unit/change) by the number of changes or removal/inclusion of modules of machines. For example, if reconfiguration in a sequence of three families requires three removals of modules, one removal of a machine and two inclusions of machines, the reconfiguration cost is $R = 3\alpha + 1\beta + 2\gamma$.

An example of cost estimation for reconfiguration from family A to D is presented in Table 9.13. This table shows that machine $M1$ is used when producing A, but not when producing D. Therefore, $M1$ has to be removed and its corresponding cost (β) is incurred. On the contrary, machine $M4$ is not required for producing A but it is required for producing D and therefore it must be included in the new system configuration together with its cost (γ). Machine $M3$ is required for both families, so no cost is incurred because it remains in the new system. Finally, machine $M2$ is composed of three different modules. Note that the machine is required for both families, so it is not removed. Module $M21$ is necessary for both families, and it remains in the machine (without cost). Modules $M22$ and $M23$ are required only for one family and they must be removed and included, respectively.

Table 9.13: Cost of reconfiguration from family A to family D.

Module	Machine	Family A	Family D	Parameter	Cost
M11	M1	1	0	β	5
M21	M2	1	1	-	0
M22		1	0	α	1
M23		0	1	α	1
M31	M3	1	1	-	0
M41	M4	0	1	γ	6
					13

Costs of under-utilization of resources deal with incurred costs due to not using machines or modules. These costs have to be calculated for each family formed in any level of the dendogram. The first level of the dendogram is composed of families containing one single product. As the RMS is configured for producing a specific family, when producing a family composed of one product, the system is installed with the capacity and functionality needed [1] allowing full utilization of the resources installed. Consequently, the cost of under-utilization for families composed of one product is zero.

Parameters about costs of under-utilization are estimated based on experience. The first parameter deals with if a module of a machine is not used when producing a product of the family. The second parameter deals with if a machine is not used when producing a product of the family. Their estimation can be seen in Table 9.14:

Table 9.14: Cost estimation for parameters regarding under-utilization of resources.

Parameter	Symbol	Cost
A module is not used when manufacturing a product from a certain family	δ	1
A machine (with all its modules) is not used when manufacturing a product from a certain family	ϵ	7

Costs of under-utilization for a certain sequence (monetary unit) is calculated as the sum of multiplying each parameter (monetary unit/time of under-utilization) by the time of under-utilization of each module or machine. For example, for a sequence of three families, the time that a certain module has been idle for 220 minutes and a machine has been idle for 100 minutes, the cost of under-utilization is $H = 220\delta + 100\epsilon$.

As manufacturing times in reconfigurable machines are unavailable, an approximation to calculate the under-utilization costs has been developed. Thus, the number of times that a module or machine has not been used to manufacture a product is taken into account instead of time.

Table 9.15 shows the estimation of family $ADBC$. For example, focusing in machine $M2$, when producing A the module $M23$ is not used, and the cost (δ) is incurred. When producing D is now the module $M22$ which is not used and another cost δ is incurred. However, when producing B two modules are not used ($M22$ and $M23$) and a cost 2δ must be added. Finally, to produce C the machine $M2$ is not necessary, so cost ϵ is required. The total cost is $H = \delta + \delta + 2\delta + \epsilon = 4\delta + \epsilon$.

Table 9.15: Cost of under-utilization in family ADBC.

		Product					
Module	Machine	A	D	B	C	Parameter	Cost
M11	M1	1	0	0	0	3ϵ	21
M21	M2	1	1	1	0	$4\delta+\epsilon$	11
M22		1	0	0	0		
M23		0	1	0	0		
M31	M3	1	1	1	1	-	0
M32		0	0	1	1	2δ	2
M33		0	0	0	1	3δ	3
M41	M4	0	1	0	0	3ϵ	21
							58

Cost Sensitivity

Values assigned to each parameter determine the level of the dendogram that presents the minimum cost and therefore it determines the set of families to form. Thus, a sensitivity analysis for each parameter is required.

It sounds logic that, when increasing any parameter associated to the reconfiguration costs keeping constant the under-utilization costs, the minimum cost is achieved when forming few product families, or in other words, in the lower levels of the dendogram. The reason is that in the lower levels, with few families, the number of reconfigurations is low and therefore it is less sensitive to parameters of reconfiguration (α, β, γ).

Figures 9.9, 9.10 and 9.11 show the sensitivity of parameters of reconfiguration costs for the selection of different levels of the dendogram. It can be seen that when increasing the cost of reconfiguration, the minimum cost goes to lower levels of the dendogram.

Similarly, when increasing any parameter associated to under-utilization cost being constant the reconfiguration ones, the minimum cost is obtained

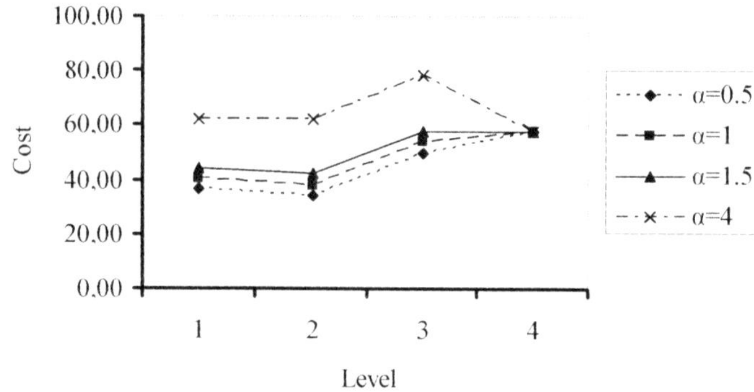

Fig. 9.9: Total costs in each level for different values of parameter α.

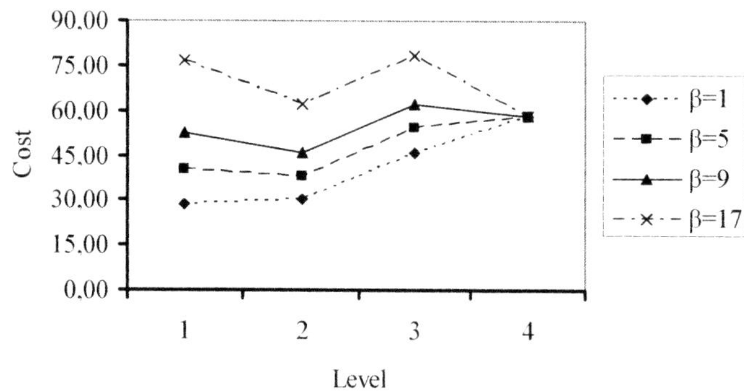

Fig. 9.10: Total costs in each level for different values of parameter β.

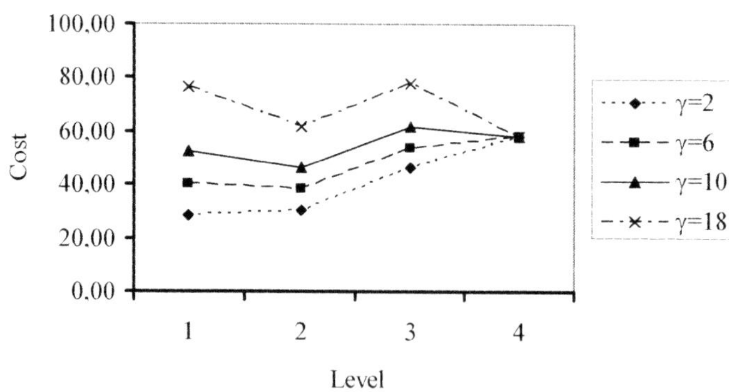

Fig. 9.11: Total costs in each level for different values of parameter γ.

when forming several product families, or in other words, in the upper levels of the dendogram. The reason is that in those levels a system with lots of families (reconfigurations) is formed and the total cost is less sensitive to under-utilization parameters.

Figures 9.12 and 9.13 show the sensitivity of parameters of under-utilization costs for the selection of different levels of the dendogram. It can be seen that when increasing the cost of reconfiguration, the minimum cost goes to upper levels of the dendogram.

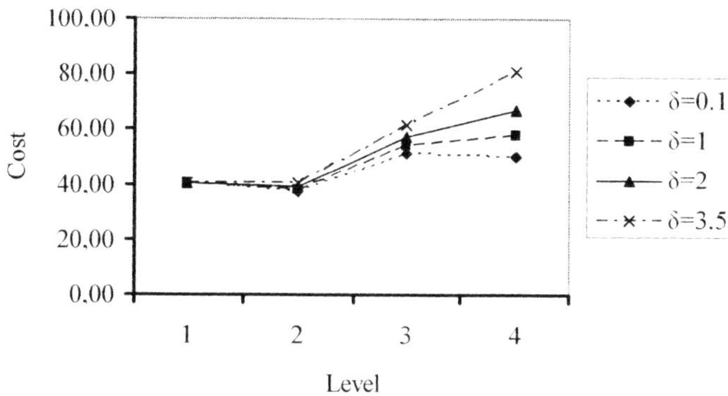

Fig. 9.12: Total costs in each level for different values of parameter δ.

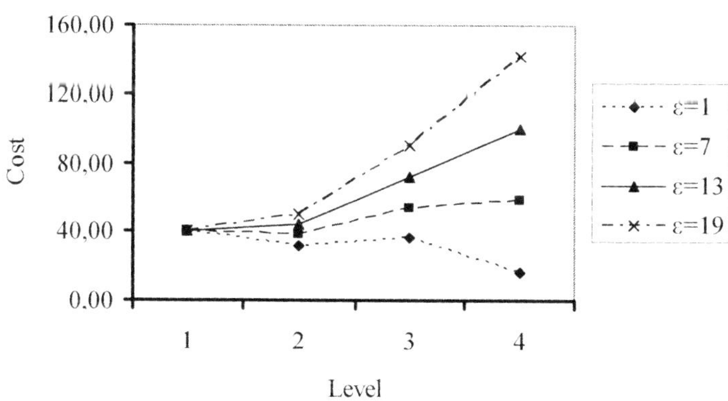

Fig. 9.13: Total costs in each level for different values of parameter ϵ.

9.4.3 Mathematical Model

For the effective working of the RMS, a family of products is selected to be produced. When finishing, the following family is ready to be produced within a specifically-designed production system. This process is repeated until the production of each product family. New orders for launching products to the market and the limitations of the storage capacity of the companies are two reasons that require the production of the same products again. Therefore, when the production of the last product family has been completed, the system is reconfigured for producing the first product family again, and the process is repeated. Only when the companies add or remove products from their portfolios, or their demands are very different, the methodology may selects different product families and different sequences of production.

This problem is very close to the Travelling Salesman Problem (TSP), which seeks to identify an itinerary that minimizes the total distance travelled by a salesman who has to visit a certain number of cities once, leaving from one of them (base city) and returning to this one. Therefore, some similarities may be highlighted. First, cities in the TSP are product families in RMS. Second, the goal in the TSP is to minimize the total distance travelled. In RMS, the goal is to minimize the total cost. Finally, in the TSP the salesman has to arrive to the base city, and in RMS when the last family has been produced the system is reconfigured for the first one.

The RMS problem requires the development of a model that solves a TSP in each level and selects the level which presents the minimum cost. Therefore, the problem to solve is a *multilevel-TSP*. Under-utilization costs for each family of a certain level are the same (they do not depend on the production scheduling) and therefore only reconfiguration costs must be taken into consideration. With the aim of calculating the total cost, the under-utilization cost must be added to the sequence that presents the minimum reconfiguration cost. As the TSP is a NP-complete problem [25], the selection and sequence of product families is NP-complete too.

The proposed model supposes that machines present infinite capacity. It is assumed that only one manufacturing routing exists for each product, so only one process plan exists for each product component and each manufacturing operation is carried out in one machine or module of machine. Each machine has its own modules which are not shared with other machines.

The notation used for model development (indices, parameters and variables) is shown in Table 9.16. The objective is to select the set of families to produce and its manufacturing sequence that minimize the reconfiguration and under-utilization costs while producing the families.

The constraints to take into account are:

1. Only one level in the dendogram is selected (const. 9.10).
2. All the families from the selected level will be produced, and none of other levels (const. 9.11). Besides, the number of variables for the changeover of the production system (T) must be equal to the number of families (const. 9.12).

Table 9.16: Indices, parameters and variables.

Indices	
i, j	Families
l	Level
Parameters	
L	Number of levels
F_l	Set of families to produce in level l ($l = 1 \dots L$)
N_l	Number of families to produce in level l, so $N_l = \lvert F_l \rvert$
R_{ij}	Cost of reconfiguration from family i to family j ($i \in F_l; j \in F_l; i \neq j; l = 1 \dots L-1$)
H_i	Cost of the under-utilization of machines' resources while producing family i
Variables	
T_{ijl}	1 if family i is produced just before family j, in level l ($i \in F_l; j \in F_l; i \neq j; l = 1 \dots L-1$)
Q_{il}	1 if family i within level l is produced ($i \in F_l; l = 1 \dots L$)
K_l	1 if all the families within level l are produced ($l = 1 \dots L$)
U_{il}	≥ 0, auxiliary variables for avoiding subtours ($i \in F_l - 1; l = 1 \dots L-2$)

3. In each level l, the families are manufactured one by one following a production order, and at the end the system is configured for manufacturing the initial family. Variables T can exist before one family (const. 9.13) and after one family (const. 9.14) only. The same occur in the TSP and Assignment problems.
4. Subtours within each level are not feasible. Therefore, the Miller-Tucker-Zemlin constraints are introduced [26]. This set of constraints avoids that, as for example, five families were produced in two different ways, such as: $1 \rightarrow 2 \rightarrow 3 \rightarrow 1 \rightarrow \dots$ and $4 \rightarrow 5 \rightarrow 4 \rightarrow \dots$ (const. 9.15).
5. Sign constraints of binary variables (T, Q, and K) and auxiliary variables (U).

Therefore, the mathematical model is the following:

$$Min \sum_{i \in F_l} \sum_{j \in F_l} \sum_{l=1}^{L-1} R_{ij} T_{ijl} + \sum_{i \in F_l} \sum_{l=1}^{L} H_i Q_{il} \tag{9.9}$$

Subject to:

$$\sum_{l=1}^{L} K_l = 1 \tag{9.10}$$

$$\sum_{i \in F_l}^{L} Q_{il} = N_l K_l \qquad \forall l = 1, \dots, L \tag{9.11}$$

$$\sum_{i \in F_l} \sum_{j \in F_l} T_{ijl} = N_l K_l \qquad \forall l = 1, \ldots, L-1 \qquad j \neq i \tag{9.12}$$

$$\sum_{j \in F_l} T_{ijl} \leq l \qquad \forall i \in F_l \qquad \forall l = 1, \ldots, L-2 \qquad j \neq i \tag{9.13}$$

$$\sum_{i \in F_l} T_{ijl} \leq l \qquad \forall j \in F_l \qquad \forall l = 1, \ldots, L-2 \qquad i \neq j \tag{9.14}$$

$$N_l T_{ijl} + U_{il} - U_{jl} \leq N_l - l \quad \forall i \forall j \in F_l - 1 \quad j \neq i \quad \forall l = 1, \ldots, L-2 \tag{9.15}$$

$$T_{ijl} = \{0, 1\} \qquad \forall i \in F_l \qquad \forall j \in F_l \qquad j \neq i \qquad \forall l = 1, \ldots, L-1 \tag{9.16}$$

$$Q_{il} = \{0, 1\} \qquad \forall i \in F_l \qquad \forall l = 1, \ldots, L \tag{9.17}$$

$$K_l = \{0, 1\} \qquad \forall l = 1, \ldots, L \tag{9.18}$$

$$U_{il} \geq 0 \qquad \forall i \in F_l - 1 \qquad \forall l = 1, \ldots, L-2 \tag{9.19}$$

The families are manufactured in cyclical order, pointing out that it does not matter which family is manufactured first, as long as the order remains the same. Costs associated with the manufacturing order do not exist at all.

The parameters required to solve the model are described in Table 9.16. Most of them (L, F_l, and N_l), can be taken directly from the dendogram, but the rest (R_{ij} and H_i) are unknown and they must be estimated. Once calculated both costs, the mathematical model can be solved.

The input information can be easily automated, with minor human intervention. The cost methodology requires information from the bill of materials (BOM), which is stored in the information system of the company. The BOM informs about both the components of a specific product and the machine that performs the required operations. The human intervention consists of the estimation of the five parameters identified above on reconfiguration and under-utilization costs. The values of these parameters are introduced in the information system and linked with the BOM to obtain the costs required for the model.

The proposed approach will be illustrated on the resolution of some instances taken from the literature. Due to the lack of existing reconfigurable systems, some instances from the literature regarding cellular manufacturing systems have been modified in order to be taken as RMS problems. Data in CMS instances are machines and parts. Machines have been converted into products, and parts into product components. Besides, the same number of reconfigurable machines than products has been selected, and the number of machine modules is twice than machines.

Table 9.17 shows the set of instances to test, their references and comments. The number of product families varies among 1 (all the products grouped in one family) and the number of products (each family composed of one product). The CPU time required to solve the proposed model increases with the number of products. Those instances have been solved using linear programming software, and a CPU at 3.06 GHz and 256 MB PC.

Table 9.17: Batch of instances.

Instance	Reference	Comment	Instance	Reference	Comment
ADI97	[27]		CHE96C	[34]	Problem 3
AKT96	[28]		COA88	[35]	
ASK87	[29]		CRA96	[36]	
AST91	[30]		KIN80	[37]	
BOC91A	[31]	Problem 1	KUM86	[38]	
BOC91B	[31]	Problem 2	MCC72A	[39]	Problem 1
BOC91C	[31]	Problem 3	MCC72B	[39]	Problem 2
BOC91D	[31]	Problem 4	NG96	[40]	
BOC91E	[31]	Problem 5	SEI89	[41]	
BOC91F	[31]	Problem 6	SHA95A	[42]	Problem 1
BOC91G	[31]	Problem 7	SHA95B	[42]	Problem 2
BOC91H	[31]	Problem 8	SHA95C	[42]	Problem 3
BOC91I	[31]	Problem 9	SHA95D	[42]	Problem 4
BOC91J	[31]	Problem 10	SRI90	[43]	
CHA82	[32]		VAK90	[44]	
CHE95	[33]		VEN90A	[45]	Problem 1
CHE96A	[34]	Problem 1	VEN90B	[45]	Problem 2
CHE96B	[34]	Problem 2			

Fig. 9.14 presents the variation of CPU time with the number of products. It can be seen the exponential tendency of the CPU time required to solve the instances.

As can be observed, the tendency of CPU time with the number of products is increased hardly when the number of products is higher than 25. Therefore, the model is appropriated for problems with 25 products or less. This result shows that an approach based on heuristics must be developed to solve problems with more than 25 different products.

9.5 Production Planning with Heuristics

Due to the existing difficulty to solve combinatorial problems in an optimal way, the development and implementation of heuristic procedures able to provide acceptable solutions within a reasonable computing time has become

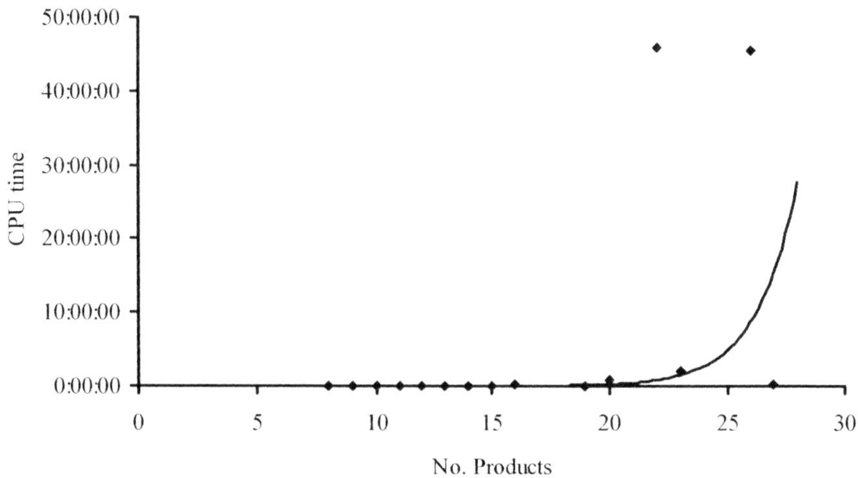

Fig. 9.14: Tendency of CPU time with the number of products.

essential [46]. In order to build a solution or to improve an existing one, heuristic algorithms frequently use any kind of specific knowledge of the problem to solve. As commented before, the problem to face is similar to the TSP. As reconfiguration from a family to another one is different than the opposite way, the problem is an asymmetric TSP (ATSP), and the heuristics that solve it may be divided in the following categories:

- Specific heuristics for the ATSP.
- General heuristics or meta-heuristics applied to the ATSP.

Some of them offer solutions near to the optimal one. Therefore, if a small deviation from the optimal solution is acceptable, these techniques can be used as solution methods [47].

The specific heuristics are simple algorithms that usually require relatively short computational times. Generally speaking, they may be divided into the following four categories [47,48]: (a) Hamiltonian cycle construction heuristics, (b) Hamiltonian cycle improving heuristics, (c) Heuristics based on patching cycles together, and (d) Hybrid heuristics. Considering a graph composed of nodes (cities) and arcs (distances among cities), a cycle is a path in which the same arc is not travelled twice, and it finishes at the base city. A Hamiltonian cycle, besides the above requirements, has to cover all the nodes only once.

Tour construction heuristics add a city in each iteration of the algorithm until a Hamiltonian cycle has been covered. In RMS, the problem is a multilevel-TSP and therefore each step of the algorithm adds one product family to sequence, in each level of the dendogram. They are very fast algorithms, so frequently are used to generate an approximate solution when time is limited, obtaining a starting point for the application of other algorithms or even an upper bound for exact algorithms. A famous heuristic is the *Nearest Neighbor* [49] which starts choosing a city randomly and then the salesman goes to the nearest unvisited city. Some other heuristics are Greedy [50], random insertion [51], or Clark-Wright heuristic [52].

Tour improvement heuristics try to improve a previously constructed tour through the implementation of several solutions which are very similar to the original one. Thus, these algorithms are known as local search algorithms. For the RMS problem, these heuristics are intended to find a production scheduling of product families which are similar to the first one but with lower costs. Some examples are the 2-Opt, 3-Opt and the Lin-Kernighan heuristics [53].

Patching heuristics start by combining two different cycles into the shortest one possible by breaking one arc in each cycle and patching together the two resulting paths. The process is repeated until one unique cycle remains. Examples are the Karp-Steele patching [54], the recursive path contraction algorithm, and the contract or path heuristic [51].

Finally, hybrid algorithms combine some of the above features. They usually start with a tour construction process improved by a local search. Some examples are the GENI and GENIUS heuristics [48].

In the last years, the development of a specific class of algorithms called meta-heuristics has gained research attention. They are based on general frameworks which can be applied to diverse optimization problems with little modifications [55]. Some examples are Simulated Annealing [56], Genetic Algorithms [57], Tabu Search [58], GRASP [59], Neural Networks [60], and Ant Colonies Optimization [61].

To solve the problem of selecting and sequencing product families in RMS, a tabu search will be applied because it has demonstrated to be a useful optimization technique to solve different combinatorial problems. Besides, ant colony optimization presents good adaptation to the TSP, driving to consider as suitable the use of ant colony optimization models to compare the obtained results with tabu search.

9.5.1 Heuristics to Solve the RMS Problem

In order to solve the RMS problem, a specific heuristic based on the nearest neighbor method will be developed and implemented. Results have been compared to those obtained with the exact method developed in section 9.4.3. Besides, a general meta-heuristic based on ant colony optimization has been implemented to compare the quality of the solutions obtained with the specific heuristic.

Variant of the Nearest Neighbor Heuristic

This heuristic, specifically developed for RMS, starts by evaluating the reconfiguration costs between each pair of product families. Then, the minimum cost is chosen to be used as starting point. Once scheduled the first pair of families, the following family to add to the sequence is the not-yet-scheduled-family with minimum reconfiguration cost. The procedure finishes when all product families are scheduled in one single sequence. For example, considering four product families to schedule A, B, C, D, the reconfiguration costs between them is presented in Table 9.18.

Table 9.18: Reconfiguration costs between families.

	A	B	C	D
A	-	5	2	7
B	2	-	3	1
C	9	5	-	8
D	6	4	6	-

The sequence that presents the minimum reconfiguration cost is $\{B \rightarrow D\}$ and it is chosen as the first pair of families to schedule. From D, the following possible family with the minimum reconfiguration cost is chosen. There are two possible families to choose $\{A, C\}$ and both have the same cost, so one of them is chosen randomly, for example family A, and the current sequence is $\{B \rightarrow D \rightarrow A\}$. As there is only one family to sequence, it is allocated at the end. Therefore, the final sequence obtained with this heuristic is $\{B \rightarrow D \rightarrow A \rightarrow C\}$.

Table 9.19 shows the results obtained with the implementation of the developed heuristic to the batch of instances used to test the mathematical model. Deviation percentages of the solutions gained with the heuristic regarding optimal solutions are shown too. In each case, computing time required to solve the instances is lower than one second.

Results showed in Table 9.19 can be considered as satisfactory because the heuristic offers good solutions to NP problems in less than one second . In 19 of the 35 instances the deviation from the optimum is lower than 5%, in 28 instances deviation is lower than 10% and all of them present a deviation lower to 18%. As an average value, the deviation of the instances from optimum is 5.85%. It must be noticed that to obtain results with the application of the optimal method, the computing time required becomes not feasible with more than 25 products to group. This heuristic can offer a solution very quickly to the RMS problem with any number of products to group.

Ant Colony Optimization

Ant colony optimization (ACO) is a general heuristic or meta-heuristic inspired in the behavior of real ants. The literature presents different algorithms based on ACO, such as Ant System [62], Ant-Q [63], Ant Colony System [61], Max-Min Ant System [64], Rank-Based Ant System [65] and Ant-Net [66].

Table 9.19: Specific heuristic and ACS meta-heuristic results.

Instance	Optimal method †	Specific heuristic	Deviation (%) from optimum	ACS values average	Deviation (%) from optimum	CPU time average
SHA95B	85	87	2.35	86.8	2.12	0:00:13
SHA95D	88	88 †	0.00	90.2	2.50	0:00:17
COA88	89	89 †	0.00	89 †	0.00	0:00:16
AKT96	126	129	2.38	133	5.56	0:00:12
CHE96A	128	138	7.81	128 †	0.00	0:00:14
SHA95C	158	167	5.70	159	0.63	0:00:19
SEI89	143	144	0.70	143 †	0.00	0:00:12
MCC72A	79	79 †	0.00	80.4	1.77	0:00:25
VAK90	166	166 †	0.00	166 †	0.00	0:00:40
CRA96	150	152	1.33	152.8	1.87	0:00:27
ASK87	159	163	2.52	163.2	2.64	0:00:32
CHA82	98	102	4.08	98 †	0.00	0:00:41
SHA95A	78	82	5.13	78 †	0.00	0:01:16
CHE95	136	148	8.82	137	0.74	0:01:01
BOC91A	340	367	7.94	340 †	0.00	0:00:12
BOC91B	242	250	3.31	243.6	0.66	0:01:33
BOC91C	211	224	6.16	220.4	4.45	0:00:59
BOC91D	302	315	4.30	329.6	9.14	0:00:51
BOC91E	240	240 †	0.00	240 †	0.00	0:01:10
BOC91F	252	260	3.17	297.8	18.17	0:00:55
BOC91G	286	297	3.85	286 †	0.00	0:00:49
BOC91H	294	301	2.38	299	1.70	0:01:13
BOC91I	241	275	14.11	256.8	6.56	0:00:35
BOC91J	222	255	14.86	246.4	10.99	0:00:40
SRI90	243	255	4.94	253.8	4.44	0:00:23
KIN80	353	368	4,25	377.8	7.03	0:00:45
ASK91	118	122	3.39	126	6.78	0:02:02
VEN90B	305	322	5.57	308.8	1.25	0:03:00
CHE96B	341	392	14.96	385.8	13.14	0:01:28
CHE96C	278	325	16.91	288	3.60	0:02:34
NG96	248	292	17.74	264.2	6.53	0:01:31
VEN90A	126	134	6.35	126 †	0.00	0:00:34
KUM86	364	412	13.19	396.8	9.01	0:02:59
ADI97	497	529	6.44	518	4.23	0:04:04
MCC72B	541	596	10.17	562.8	4.03	0:06:03

One of the most implemented methods is the Ant Colony System (ACS). In this algorithm, a set of artificial ants cooperate in the solution of a problem through exchanging information via the pheromone deposited on the paths. ACS is based on the Ant System and it improves its efficiency to solve the symmetric and asymmetric TSP.

The implemented algorithm of ACS to solve the RMS problem works as follows. An ant is located in each product family. They start to move towards other families following a state transition rule. While each ant constructs its sequence, they modify the quantity of pheromone on the visited paths through a local pheromone updating rule. Once the ants have finished their sequences, the amount of pheromone on edges is modified again by applying the global updating rule. The algorithm can be divided into four stages [61]:

1. *Initialization.* Pheromone values of each possible sequence from a family r to another family s are initialized to a positive constant value ($\tau_0 \geq 0$). Therefore, at the beginning there is no past memory and ants schedule, with the highest probability, the families with lower reconfiguration cost.
2. *Solution construction.* In this stage, the product family to schedule is selected step by step by the ants. The selection is made among the possible families probabilistically, following the pseudo-random-proportional rule presented in expression 9.20. This rule provides a balance between the exploration of new production sequences and the exploitation of sequences that offer good results.

$$s = \begin{cases} \arg\max_{u \in J_k(r)} \{[\tau(r,u)] \cdot [\eta(r,u)]^{\beta}\} & q \leq q_0 \\ S & otherwise \end{cases} \tag{9.20}$$

Where q is a random number ($0 \leq q \leq 1$), q_0 is a parameter ($0 \leq q_0 \leq 1$) and S is a random variable selected following the probability distribution presented in expression 9.21.

$$p_k(r,s) = \begin{cases} \frac{[\tau(r,s)] \cdot [\eta(r,s)]^{\beta}}{\sum\limits_{u \in J_k(r)} [\tau(r,u)] \cdot [\eta(r,u)]^{\beta}} & s \in J_k(r) \\ 0 & otherwise \end{cases} \tag{9.21}$$

Where τ is the pheromone, $\eta = 1/\delta$ is the inverse of the reconfiguration cost $\delta(r,s)$, $J_k(r)$ is the set of families not-yet-scheduled by the ant k in family r and β is a parameter that determines the relative importance of the pheromone against the reconfiguration cost ($\beta > 0$). The element $\tau(r,s)] \cdot [\eta(r,s)$ reinforces the selection of families with low reconfiguration costs but high quantity of pheromone. The relative importance of the exploitation of a good solution against the exploitation of new sequences is determined with the parameter q_0. Thus, if $q \leq q_0$ a good solution is exploited and an exploration otherwise.
3. *Local pheromone update.* When ants construct their sequence S_k, they modify the pheromone trail by applying the local pheromone updating rule, presented in expression 9.22.

$$\tau(r,s) \leftarrow (1-\rho)\cdot\tau(r,s) + \rho\cdot\Delta\tau(r,s) \quad (9.22)$$

Where ρ is a second pheromone evaporation rate ($0 < \rho < 1$) and $\Delta\tau(r,s) = \tau_0$ (initial level of pheromone). This rule tries to obtain a dynamic change on sequence preferences. If this rule is not applied, the ants will search within a little neighborhood of the best previous sequence.

4. *Global pheromone update*. This rule is applied when the ants have generated their sequences. Pheromone is deposited by the ant that generates the best sequence, with expression 9.23.

$$\tau(r,s) \leftarrow (1-\varphi)\cdot\tau(r,s) + \varphi\cdot\Delta\tau(r,s) \quad (9.23)$$

Where:

$$\Delta\tau(r,s) = \begin{cases} \frac{1}{L_{sg}} & (r,s) \in \textit{best global sequence} \\ 0 & \textit{otherwise} \end{cases} \quad (9.24)$$

Being φ the pheromone evaporation rate ($0 < \varphi < 1$) and L_{sg} is the total reconfiguration cost of the best global sequence.

The literature demonstrates that ACS parameters are independent of the problem to face. Parameters that offer the best results were presented by Dorigo and Gambardella [61], except for the number of ants, which will be set to the number of product families to schedule. This is due to the nature of the RMS problem, which solves an ATSP in each level of the dendogram, and the number of ants is set to each problem, not for each level.

The same batch of 35 instances is solved with ACS. Due to the random nature of this procedure, the experiments are repeated five times and the average values have been taken into consideration. Table 9.19 presents the average value of the obtained solutions by applying ACS and the deviation percentage regarding the optimum together with the required CPU time.

The implementation of the ACS meta-heuristic drives to acceptable results, quite similar to those obtained with the specific heuristic developed for the RMS problem. In fact, 25 of the 35 instances solved reached their solution with an average deviation from the optimum lower than 5%, 32 instances got their solutions with a deviation lower than 10%. Finally, all of them present an average deviation lower than 19%. The average instance deviation is 4.65%, better but similar than the obtained with the specific heuristic (5.85%) whereas computing time in ACS, though admissible, is higher than the required for the specific heuristic developed.

9.5.2 Hybrid Approaches to Solve the RMS Problem

The development of hybrid methods that include construction and improvement heuristics is considered as one of the most effective strategies to solve combinatorial optimization problems [48]. The problem of selecting and sequencing product families in RMS is a difficult problem, as stated in Table 9.20, which presents the growing number of possible sequences when increasing the number of products to group.

Table 9.20: Number of solutions regarding the number of products to group.

No. products	10	40	80	100	150
No. possible solutions	$4 \cdot 10^5$	$2 \cdot 10^{46}$	$7 \cdot 10^{118}$	$9 \cdot 10^{157}$	$6 \cdot 10^{262}$

This number of products to group is determined with expression 9.25:

$$\sum_{l=1}^{L-1} (N_l - 1)! \tag{9.25}$$

Where l indicates the level of the dendogram, L the total number of levels and N_l the number of families to produce in level l.

This problem can be faced by hybridization between the specific heuristic developed (nearest neighbor variant) and tabu search meta-heuristic. Once applied the specific heuristic, the results obtained will be improved with a tabu search procedure. Another option is to construct a sequence with ACS and to improve the results by applying a local search procedure.

Hybrid Approach of Nearest Neighbor Variant and Tabu Search

Tabu search selects the best possible movement in each step, allowing a solution worse than the actual one to escape from a local optimum and to continue the search for better solutions. In order to avoid the return at a former local optimum and to create a cycle, some movements are classified as "tabu" in next iterations.

The neighborhood of the solution has been generated with inserting movements, which consists of including a scheduled family in a different position of the sequence. For example, in the sequence $\{A \to B \to C \to D\}$ family A can be inserted in the third position, being the new sequence $\{B \to C \to A \to D\}$. In order to generate the neighborhood, each family is inserted in other positions of the sequence and therefore, the size of the neighborhood for a set of n product families is $n(n-1)$. As intensification rule, the algorithm searches in the neighborhood and selects as the new solution the sequence of families with the lowest cost, though it may be higher than the cost of the current sequence. The algorithm finishes after a certain number of iterations. The maximum number of iterations has been fixed according with the number of families at $n^3 + n^2$.

It is well known that tabu list size affects the performance of the heuristic [58]. Studies about optimizing tabu list size for the asymmetric TSP have not been found in the literature, but those about the symmetric TSP demonstrate that this size depends on the number of cities to visit. Therefore, tabu list size to use will be the number of product families to group (n). Tabu list does not store the whole sequence, but some attributes of it: the family that changes its position and the position to insert the family.

Note that a RMS instance is composed of several sets of families, in different levels of the dendogram. Therefore, an ATSP in each level of the dendogram must be solved by applying the above approach. As all possible sequences in the same level have the same under-utilization costs, they are not used in the above procedure but it is added at the end. Once costs have been obtained for each level, the procedure finishes by selecting the best of them.

The implemented procedure tries to improve the solutions offered by the specific heuristic. Once finished, the intensification process starts by generating the neighborhood of the solution. The new sequences are evaluated and the best one is selected in each step. The process finishes when $n^3 + n^2$ iterations have been completed. The algorithm has been storing the best solution found in each step, and gives it back when finishing. Results are shown in Table 9.21, which demonstrates that the implementation of the hybrid approach improves, or make equal in the worst case, the solution obtained with the specific heuristic for each instance. The instance average deviation from the optimum gained with this approach is 1.83%, quite better than the 5.85% obtained with the specific heuristic.

Former results have demonstrated that the hybrid approach of the specific heuristic and tabu search drives to obtain better results than the obtained without hybridization to solve the problem in RMS. Now, this approach is intended to be improved with refinements to the tabu search process: the application of a second stopping criterion and the implementation of a diversification rule.

An optimal stopping condition is compulsory to obtain satisfactory solutions with low computational effort. Studying the quality of these solutions, it has been realized that the best solutions are obtained in the early iterations. Thus, thirteen instances (composed of a total sum of 202 levels) have been selected randomly in order to study the iteration in which the best solution is found. Results have shown that the best solution has been found doing lower than 12% of the upper bound iterations number (n^3+n^2). Therefore, the batch of instances can be solved with the stopping criteria of a maximum number of iterations without improving the best solution found. For assuring more iterations than the 12% obtained experimentally, this condition has been set to 30%. Results have demonstrated that the same results are obtained, and consequently saving computing time. Figure 9.15 shows the difference in computing time with and without the stopping condition, and how it increases together with the number of products to group.

A diversification rule has been developed with the aim of searching in other areas of the solution space when it is difficult to overcome huge local optima. This rule is implemented by generating 1000 random solutions when

Table 9.21: Solution of instances applying tabu search.

Instance	Optimal method †	Specific heuristic	Tabu search	Deviation (%) from optimum	CPU time
SHA95B	85	87	85 †	0.00	0:00:01
SHA95D	88	88 †	88 †	0.00	0:00:02
COA88	89	89 †	89 †	0.00	0:00:04
AKT96	126	129	128	1.59	0:00:08
CHE96A	128	138	128 †	0.00	0:00:05
SHA95C	158	167	159	0.63	0:00:04
SEI89	143	144	143 †	0.00	0:00:07
MCC72A	79	79 †	79 †	0.00	0:00:12
VAK90	166	166 †	166 †	0.00	0:00:12
CRA96	150	152	150 †	0.00	0:00:24
ASK87	159	163	159 †	0.00	0:00:34
CHA82	98	102	98 †	0.00	0:00:44
SHA95A	78	82	78 †	0.00	0:00:45
CHE95	136	148	136 †	0.00	0:00:46
BOC91A	340	367	350	2.94	0:01:18
BOC91B	242	250	242 †	0.83	0:01:08
BOC91C	211	224	212	0.47	0:01:07
BOC91D	302	315	302 †	0.00	0:01:19
BOC91E	240	240 †	240 †	0.00	0:01:15
BOC91F	252	260	252 †	0.00	0:01:11
BOC91G	286	297	286 †	0.00	0:01:09
BOC91H	294	301	301	2.38	0:01:08
BOC91I	241	275	253	4.98	0:01:08
BOC91J	222	255	232	4.50	0:01:12
SRI90	243	255	243 †	0.00	0:01:07
KIN80	353	368	362	2.55	0:01:07
ASK91	118	122	120	1.69	0:03:19
VEN90B	305	322	305 †	0.00	0:04:32
CHE96B	341	392	364	6.74	0:04:17
CHE96C	278	325	299	7.55	0:04:10
NG96	248	292	290	16.94	0:04:08
VEN90A	126	134	128	1.59	0:07:52
KUM86	364	412	369	1.37	0:09:44
ADI97	497	529	516	3.82	0:20:49
MCC72B	541	596	559	3.33	0:25:28

the best solution has not been improved after a certain number of iterations. The heuristic evaluates all of them and selects the best one as the current solution. The stopping condition and the diversification rule can be implemented together only if the number of iterations without improving the best

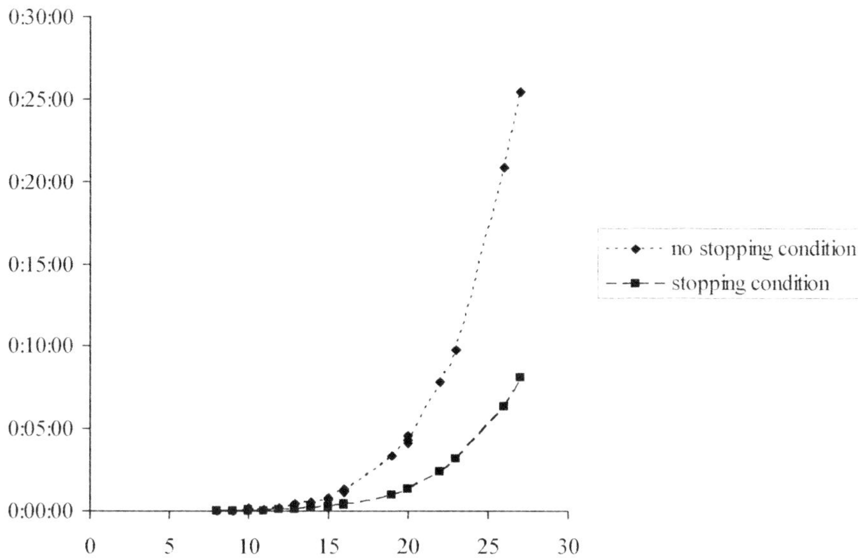

Fig. 9.15: Difference in computing time due to the second stopping condition.

solution found is smaller than the stopping condition. If the new area does not improve the best solution, new searches in other areas may be an adequate strategy.

With the aim of obtaining the best number of iterations without improving to implement the rule, some experiments have been carried out. Experiments have been applied with different numbers of iterations without improving: 10, 5, 1, and 0.5% of the maximum number of iterations. As a comparative measurement, the sum of deviation from the optmum of the previous batch of instances has been used. As computing times are similar, experimental results have shown up that the best is to use 1%. Thus, the optimum number of iterations without improving the current solution before applying the diversification rule is $0.01(n^3 + n^2)$.

The implementation of this rule improves the previous solutions in most of the cases, as it is shown in Table 9.22. Due to the random condition of the diversification rule, instances have been solved five times to obtain results.

Table 9.22 has shown successive improvements when applying the hybrid approach to solve the RMS problem, from the implementation of the specific heuristic to the implementation of the hybrid approach based on tabu search with diversification rule. In the last case, the optimal solution is obtained in 31 of the 35 instances. Instance average deviation from the optimum is 0.26%.

Hybrid Approach of ACS with Local Search

As ACS is a constructive heuristic, the development of an improvement heuristic on the generated solution may drives to a better result. This hybrid method

Table 9.22: Solution of instances applying the diversification rule.

Instance	Optimal method †	Specific heuristic	Tabu search	Tabu search with diversification rule	Deviation (%) from optimum	CPU time
SHA95B	85	87	85 †	85 †	0.00	0:00:01
SHA95D	88	88 †	88 †	88 †	0.00	0:00:02
COA88	89	89 †	89 †	89 †	0.00	0:00:02
AKT96	126	129	128	126 †	0.00	0:00:06
CHE96A	128	138	128 †	128 †	0.00	0:00:02
SHA95C	158	167	159	159	0.63	0:00:02
SEI89	143	144	143 †	143 †	0.00	0:00:04
MCC72A	79	79 †	79 †	79 †	0.00	0:00:05
VAK90	166	166 †	166 †	166 †	0.00	0:00:06
CRA96	150	152	150 †	150 †	0.00	0:00:08
ASK87	159	163	159 †	159 †	0.00	0:00:13
CHA82	98	102	98 †	98 †	0.00	0:00:16
SHA95A	78	82	78 †	78 †	0.00	0:00:16
CHE95	136	148	136 †	136 †	0.00	0:00:22
BOC91A	340	367	350	340 †	0.00	0:00:28
BOC91B	242	250	244	242 †	0.00	0:00:28
BOC91C	211	224	212	211 †	0.00	0:00:28
BOC91D	302	315	302 †	302 †	0.00	0:00:24
BOC91E	240	240 †	240 †	240 †	0.00	0:00:29
BOC91F	252	260	252 †	252 †	0.00	0:00:26
BOC91G	286	297	286 †	286 †	0.00	0:00:24
BOC91H	294	301	301	294 †	0.00	0:00:35
BOC91I	241	275	253	241 †	0.00	0:00:38
BOC91J	222	255	232	231	4.05	0:00:32
SRI90	243	255	243 †	243 †	0.00	0:00:26
KIN80	353	368	362	353 †	0.00	0:00:29
ASK91	118	122	120	120	1.69	0:01:08
VEN90B	305	322	305 †	305 †	0.00	0:01:27
CHE96B	341	392	364	341 †	0.00	0:02:08
CHE96C	278	325	299	278 †	0.00	0:01:42
NG96	248	292	290	248 †	0.00	0:02:20
VEN90A	126	134	128	126 †	0.00	0:02:23
KUM86	364	412	369	364 †	0.00	0:05:44
ADI97	497	529	516	497 †	0.00	0:10:18
MCC72B	541	596	559	555	2.95	0:10:09

has been applied on diverse combinatorial problems such as the TSP [61], the quadratic assignment problem [67] or the sequential ordering problem [55].

The local search process starts when each ant has obtained a sequence. Then, an iterative process of searching a better solution starts in the neighborhood of the current sequence and, if found, it is adopted as the new sequence. The process finishes when a local optimum has been reached. The

Table 9.23: Solution of instances applying ACS hybridized with local search.

Instance	Optimal method †	ACS	ACS with local search	Deviation (%) from optimum	CPU time
SHA95B	85	86.8	85.4	0.47	0:00:12
SHA95D	88	90.2	88 †	0.00	0:00:13
COA88	89	89 †	90.6	1.80	0:00:12
AKT96	126	133	130	3.17	0:00:01
CHE96A	128	128 †	128 †	0.00	0:00:13
SHA95C	158	159	158.8	0.51	0:00:11
SEI89	143	143 †	143 †	0.00	0:00:12
MCC72A	79	80.4	79 †	0.00	0:00:12
VAK90	166	166 †	168.8	1.69	0:00:16
CRA96	150	152.8	154.8	3.20	0:00:11
ASK87	159	163.2	163	2.52	0:00:02
CHA82	98	98 †	101.2	3.27	0:00:01
SHA95A	78	78 †	78 †	0.00	0:00:12
CHE95	136	137	136.2	0.15	0:00:13
BOC91A	340	340 †	340 †	0.00	0:00:04
BOC91B	242	243.6	242 †	0.00	0:00:07
BOC91C	211	220.4	213	0.95	0:00:13
BOC91D	302	329.6	311.6	3.18	0:00:24
BOC91E	240	240 †	247	2.92	0:00:10
BOC91F	252	297.8	259	2.78	0:00:10
BOC91G	286	286 †	286 †	0.00	0:00:05
BOC91H	294	299	306.6	4.29	0:00:06
BOC91I	241	256.8	253	4.98	0:00:20
BOC91J	222	246.4	245	10.36	0:00:12
SRI90	243	253.8	248	2.06	0:00:11
KIN80	353	377.8	356	0.85	0:00:13
ASK91	118	126	123.6	4.75	0:00:13
VEN90B	305	308.8	308	0.98	0:00:11
CHE96B	341	385.8	351	2.93	0:00:13
CHE96C	278	288	278 †	0.00	0:00:11
NG96	248	264.2	255.8	3.15	0:00:11
VEN90A	126	126 †	126 †	0.00	0:00:13
KUM86	364	396.8	372	2.20	0:00:13
ADI97	497	518	510.8	2.78	0:00:01
MCC72B	541	562.8	550	1.66	0:00:12

neighborhood is obtained through inserting movements. Results are presented in Table 9.23, which states that if applying a local search procedure to the solution obtained with ACS, it is improved in several instances, obtaining 1.93% of average deviation from the optimum using a very competitive computing time.

9.6 Conclusions

This chapter has presented a new methodology to group products into families, a central issue in the design of RMS. The importance of selecting a proper set of families lies in the fact that the production system is configured to manufacture a certain family, and the system is reconfigured to manufacture the following one.

The presented methodology takes into consideration five key requirements of products to be grouped in RMS. It is possible that those attributes do not present the same importance in different circumstances, so the application of the AHP methodology helps the designer to decide the importance of those requirements.

The output of the methodology is a dendogram that presents different sets of product families that can be selected, based on similarities among the products that compose those families. To allow a feasible reconfiguration of a production system, a mathematical model to select the families of products that minimize the costs of reconfiguration and under-utilization in RMS has been developed. The major difficulty is the value of both costs, which are unknown, so an accuracy estimation is required. This has been solved by dividing the costs into five parameters, which are easier to estimate. A methodology has been presented for a complete cost estimation based on those parameters. Finally, the model has been validated.

The selection of families among several possibilities is a highly complex problem (NP-complete), and therefore calculations for the resolution of the model grow exponentially together with the number of products. Therefore, the use of heuristic methods has been suggested. For tackling this situation, two heuristic hybrid approaches to select and sequence product families in RMS have been presented. On one hand, the procedure is based on the hybridization between a variant of the nearest neighbor heuristic and tabu search. On the other hand, between ACS and a local search procedure.

The implemented specific heuristic based on the nearest neighbor offers good solutions very quickly (in less than one second). These results have been compared to those obtained with an optimal method and a general heuristic based on ant colony optimization, showing that the developed heuristic is very competitive. Solutions to the problem are significantly improved when the specific heuristic hybridizes with tabu search, within a reasonable computing time. These results have been improved applying a second stopping criterion and a diversification rule to the tabu search procedure. Finally, the hybrid approach composed of ACS and a local search procedure has improved the solutions offered by the ACS algorithm on its own.

Therefore, it can be stated that the implementation of the hybrid approaches to solve the RMS problem drives to obtain satisfactory results within a reasonable computing time, overcoming the limitations imposed by the optimal method which was limited to offer results when there was less than 25 products to group.

References

1. Koren Y, Heisel U, Jovane F, Moriwaki T, Pritschow G, Ulsoy G, Van Brussel H (1999) Reconfigurable Manufacturing Systems. CIRP Annals 48:1–14
2. Xiaobo Z, Jiancai W, Zhenbi L (2000) A stochastic model of a reconfigurable manufacturing system, Part 1: A framework. Int J Prod Res 38:2273–2285
3. Mehrabi MG, Ulsoy AG, Koren Y, Heytler P (2002) Trends and perspectives in flexible and reconfigurable manufacturing systems. J Intell Manuf 13:135–146
4. Tu Q, Vonderembse A, Ragu-Nathan TS, Ragu-Nathan B (2004)Measuring modularity-based manufacturing practices and their impact on mass customization capability: A customer-driven perspective. Dec Sci 35:147–168
5. Jiao J, Ma Q, Tseng MM (2003) Towards high value-added products and services: mass customization and beyond. Technovation 23:809–821
6. Committee on Visionary Manufacturing Challenges (1998) Visionary Manufacturing Challenges for 2020. National Research Council, Washington DC
7. FutMan project Consortium (2003) FutMan–the future of manufacturing in Europe 2015–2020. Final Report
8. IMTR manufacturing processes & equipment group (2000) Integrated manufacturing technology roadmapping project.Integrated Manufacturing Technology Initiative, Oak Ridge
9. Pham DT, Eldukhri EE, Peat B, Setchi R, Soroka A, Packianather MS, Thomas AJ, Dadam Y, Dimov, S (2004) Innovative Production Machines and Systems (I*PROMS): a network of excellence funded by the EU sixth framework programme. Proc 2nd IEEE Int Conf Ind Inform 540–544
10. Mehrabi MG, Ulsoy AG, Koren Y (2000) Reconfigurable manufacturing systems: Key to future manufacturing. J Intell Manuf 11:403–419
11. Gershenson JK, Prasad GJ, Zhang Y (2003) Product modularity: definitions and benefits. J Eng Des 14:295–313
12. He D, Babayan A (2002) Scheduling manufacturing systems for delayed product differentiation in agile manufacturing. Int J Prod Res 40:2461–2481
13. Ulrich KT (1995) The role of product architecture in the manufacturing firm. Res Policy 24:419–440
14. Gupta S, Krishnan V (1998) Product family-based assembly sequence design methodology. IIE Trans 30:933–945
15. Martin M, Ishii K (1997) Design for Variety: Development of complexity indices and design charts. ASME Des Eng Tech Conf DETC97/DFM-4359
16. Rai R, Allada V (2003) Modular product family design: agent-based Pareto-optimization and quality loss function-based post-optimal analysis. Int J Prod Res 41:4075–4098
17. Abdi MR, Labib AW (2004) Grouping and selecting products: the design key of Reconfigurable Manufacturing Systems (RMSs). Int J Prod Res 42:521–546
18. Selim HM, Askin RG, Vakharia AJ (1998) Cell formation in group technology: review, evaluation and directions for future research. Comput & Ind Eng 34:3–20
19. Mahesh O, Srinivasan G (2002) Incremental cell formation considering alternative machines. Int J Prod Res 40:3291–3310
20. Sarker BR, Saiful Islam KM (1999) Relative performances of similarity and dissimilarity measures. Comput & Ind Eng 37:769–807
21. Gupta T (1991) Clustering algorithms for the design of a cellular manufacturing system—an analysis of their performance. Comput & Ind Eng 20:461–468

22. Vakharia A, Wemmerlov U (1995) A comparative investigation of hierarchical clustering techniques and dissimilarity measures applied to the cell formation problem. J Operat Mgment 13:117–138
23. Singhal J, Singhal K (2002) Supply chains and compatibility among components in design. J Oper Manage 20:289–302
24. Saaty TL (1980) The Analytic Hierarchy Process. McGraw-Hill, New York
25. Rinnooy Kan AHG (1976) Complexity theory. Machine scheduling problems: classification, complexity and computations. Martinus Nijhoff, The Hague
26. Miller CE, Tucker AW, Zemlin RA (1960) Integer programming formulations and the travelling salesman problem. J Assoc Comput Mach 7:326–329
27. Adil GK, Rajamani D, Strong D (1997) Assignment allocation and simulated annealing algorithms for cell formation. IIE Transactions 29:53–67
28. Akturk MS, Balkose HO (1996) Part-machine grouping using a multi-objective cluster analysis. Int J Prod Res 34:2299–2315
29. Askin RG, Subramanian P (1987) Cost-based heuristic for group technology configuration. Int J Prod Res 25:101–113
30. Askin RG, Cresswell SH, Goldberg JB, Vakharia AJ (1991) Hamiltonian path approach to reordering the part-machine matrix for cellular manufacturing. Int J Prod Res 29:1081–1100
31. Boctor FF (1991) A linear formulation of the machine-part cell formation problem. Int J Prod Res 29:2601–2614
32. Chan HM, Milner DA (1982) Direct clustering algorithm for group formation in cellular manufacture. J Manuf Syst 1:65–75
33. Chen SJ, Cheng CS (1995) Neural network-based cell formation algorithm in cellular manufacturing. Int J Prod Res 33:293–318
34. Cheng CH, Goh CH, Lee A (1996) Solving the generalized machine assignment problem in group technology. J Operational Res Soc 47:794–802
35. Co HC, Araar A (1988) Configuring cellular manufacturing systems. Int J Prod Res 26:1511–1522
36. Crama Y, Oosten M (1996) Models for machine-part grouping in cellular manufacturing. Int J Prod Res 34:1693–1713
37. King JR (1980) Machine-component grouping in production flow analysis: an approach using a rank order clustering algorithm. Int J Prod Res 18:213–232
38. Kumar KR, Kusiak A, Vannelli A (1986) Grouping of parts and components in flexible manufacturing systems. Eur J Operational Res 24:387–397
39. McCormick WT, Schweitzer PJ, White TW (1972) Problem decomposition and data reorganisation by a clustering technique. Operations Res 20:993–1009
40. Ng SM (1996) On the characterization and measures of machine cells in group technology. Operations Res 44:735–744
41. Seifodini H (1989) Single linkage versus average linkage clustering in machine cell formation applications. Comp Ind Eng 16:419–426
42. Shargal M, Shekhar S, Irani SA (1995) Evaluation of search algorithms and clustering efficiency measures for machine-part matrix clustering. IIE Trans 27:43–59
43. Srinivasan G, Narendran TT, Mahadevan B (1990) An assignment model for the part-families problem in group technology. Int J Prod Res 28:145–152
44. Vakharia AJ, Wemmerlov U (1990) Designing a cellular manufacturing system. A materials flow approach based on operation sequences. IIE Trans 22:84–97
45. Ventura JA, Chen FF, Wu CH (1990) Grouping parts and tools in flexible manufacturing systems production planning. Int J Prod Res 28:1039–1056

46. Diaz A, Glover F, Ghaziri H, Gonzalez JL, Laguna M, Moscato P, Tseng F (1996) Heuristic optimization and Neural Networks in Operations Management and Engineering. Paraninfo, Madrid
47. Helsgaun K (2000) An Effective Implementation of the Lin-Kernighan Travelling Salesman Problem. Eur J Oper Res 126:106–130
48. Johnson DS, McGeoch LA (1997) The Travelling Salesman Problem: A Case Study. In: Aarts EHL, Lenstra JK (eds) Local Search in Combinatorial Optimization. Wiley, New York
49. Rosenkrantz DJ, Stearns RE, Lewis II PM (1977) An Analysis of several heuristics for the travelling salesman problem. SIAM J Computing 6:563–581
50. Bentley JL (1992) Fast algorithms for geometric travelling salesman problems. ORSA J Computing 4:387–411
51. Glover F, Yeo G, Gutin A, Zverovich A (2001) Construction heuristics for the asymmetric TSP. Eur J Operational Res 129:555–568
52. Clarke G, Wright JW (1964) Scheduling of vehicles from a central depot to a number of delivery points. Operations Res 12:568–581
53. Lin S, Kernighan BW (1973) An Effective Heuristic Algorithm for the Travelling Salesman Problem. Operations Res 21:498–516
54. Karp RM, Steele JM (1985) Probabilistic analysis of heuristics. In: Lawler EL et al (eds) The Travelling Salesman Problem. Wiley, New York
55. Gambardella LM, Dorigo M (2000) An Ant Colony system hybridized with a new local search for the sequential ordering problem. INFORMS J Computing 12:237–255
56. Kirkpatrick S, Gelatt CD, Vecchi MP (1983) Optimization by Simulated Annealing. Sci 220:671–680
57. Holland JH (1975) Adaptation in Natural and Artificial Systems. University of Michigan Press, Ann Arbor
58. Glover, F (1989) Tabu Search—Part I. ORSA J Computing 1:190–206
59. Feo TA, Resende MGC (1989) A probabilistic heuristic for a computationally difficult set covering problem. Operational Res Letters 8:67–71
60. Hopfield JJ, Tank DW (1985) Neural Computation of Decisions in Optimization Problems. Biol Cyber 52:141–152
61. Dorigo M, Gambardella LM (1997) Ant Colony System: a cooperative learning approach to the travelling salesman problem. IEEE Trans Evolut Computation 1:53–66
62. Colorni A, Dorigo M, Maniezzo V (1991) Distributed optimisation by ant colonies. Proc Eur Conf Artificial Life 134–142
63. Gambardella LM, Dorigo M (1995) Ant-Q: a reinforcement learning approach to the travelling salesman problem. Proc 12th Int Conf Mach Learning 252–260
64. Stutzle T, Hoss HH (2000) MAX-MIN Ant System. Fut Gen Comp Sys 16: 889–914
65. Bullnheimer B, Kotsis G, Strauss C (1999) A new rank-based version of the Ant System: A computational study. Cent Eur J Op Res Econ 7:25–38
66. Di Caro G, Dorigo, M (1998) AntNet: Distributed stigmergetic control for communications networks. J Artif Intell Res 9:317–365
67. Gambardella LM, Taillard ED, Dorigo M (1999) Ant colonies for the quadratic assignment problem. J Op Res Soc 50:167–176

10

A Genetic Algorithm for Railway Scheduling Problems

P. Tormos[1], A. Lova[1], F. Barber[2], L. Ingolotti[2], M. Abril[2], and M.A. Salido[2]

[1] DEIOAC, Universidad Politecnica de Valencia, Spain
{ptormos, allova}@eio.upv.es
[2] DSIC, Universidad Politecnica de Valencia, Spain
{fbarber,lingolotti,mabril,msalido}@dsic.upv.es

Summary. This work is focused on the application of evolutionary algorithms to solve very complex real-world problems. For this purpose a Genetic Algorithm is designed to solve the Train Timetabling Problem. Optimizing train timetables on a single line track is known to be NP-hard with respect to the number of conflicts in the schedule. This makes it difficult to obtain good solutions to real life problems in a reasonable computational time and raises the need for good heuristic scheduling techniques. The railway scheduling problem considered in this work implies the optimization of trains on a railway line that is occupied (or not) by other trains with fixed timetables. The timetable for the new trains is obtained with a Genetic Algorithm (GA) that includes a guided process to build the initial population. The proposed GA is tested using real instances obtained from the Spanish Manager of Railway Infrastructure (ADIF). The results of the computational experience, point out that GA is an appropriate method to explore the search space of this complex problems and able to lead to good solutions in a short amount of time.

Key words: Scheduling, Train Timetabling Problem, Genetic Algorithms, Parameterized Regret-Based Biased Random Sampling, Real World Instances.

10.1 Introduction

Genetic Algorithms (GAs) have been successfully applied to combinatorial problems and are able to handle huge search spaces as those arising in real-life scheduling problems. GAs perform a multidirectional stochastic search on the complete search space that is intensified in the most promising areas.

The Train Timetabling Problem (TTP) is a difficult and time-consuming task in the case of real networks. The huge search space to explore when solving real-world instances of TTP makes GAs a suitable approach to efficiently solve it. A feasible train timetable should specify for each train the departure

P. Tormos et al.: *A Genetic Algorithm for Railway Scheduling Problems*, Studies in Computational Intelligence (SCI) **128**, 255–276 (2008)
www.springerlink.com

and arrival times to each dependency of the network in such a way that the line capacity and other operational constraints are taken into account. Traditionally, plans were generated manually and adjusted so as all constraints are met. However, the new framework of strong competition, privatization and deregulation jointly with the increasing of computer speeds are reasons that justify the need of automatic tools able to efficiently generate feasible and optimized timetables.

Assuming TTP is a very complex problem and GA a suitable procedure to cope with it, we have designed a GA for this train scheduling problem. Once the problem has been formally described, a Genetic Algorithm has been designed and validated through its application on a set of real-world problem instances provided by the Spanish Manager of Railway Infrastructure (ADIF). In addition, the heuristic technique described in this work has been embedded in a computer-aided tool that is being successfully used by ADIF.

The chapter is organized as follows. Section 10.2 is devoted to the formal description of the Train Timetabling Problem on a high-loaded railway line. The proposed method, based on a Genetic Algorithm (GA), is described in Section 10.3. Section 10.4 presents some examples and results of the application of the proposed GA to real-world railway timetabling problems. Finally, conclusions and directions for future research are pointed out in Section 10.5.

10.2 The Train Timetabling Problem (TTP)

Given a railway line that may have single as well as double-track sections, the Train Timetabling Problem (TTP) consists in computing timetables for passengers and cargo trains that satisfy the existing constraints and optimize a multicriteria objective function. The railway line may be occupied by other trains whose priority is higher than that of the new ones, and the new trains to be added may belong to different train operators. The locations to be visited by each train may also be different from each other. The timetable given to each new train must be feasible, that is, it must satisfy a given set of constraints. Among the constraints arising in this problem, it is possible the requirement for periodicity of the timetables. Periodicity leads to the classification of TTP as (*i*) Periodic (or cyclic) Train Timetabling and (*ii*) Non Periodic Train Timetabling.

In **Periodic Timetabling** each trip is operated in a periodic way. That is, each period of the timetable is the same. An advantage of a periodic railway system is the fact that such a system's timetable is easy to remember for the passengers. A drawback is that such a system is expensive to operate from the point of view of the use of resources such as rolling stock and crews. The mathematical model called Periodic Event Scheduling Problem (PESP) by Serafini and Ukovich [16] is the most widely used in the literature. In PESP a set of repetitive events is scheduled under periodic time window constraints. Hence, the events are scheduled for one cycle in such a way that the cycle can be repeated. The PESP model has been used by Nachtigall and Voget [13], Odijk [14], Kroon and Peeters [10], Liebchen [12].

Non Periodic Train Timetabling is especially relevant on heavy-traffic, long-distance corridors where the capacity of the infrastructure is limited due to great traffic densities. This allows the Infrastructure Manager to optimally allocate the train paths requested by the Train Operators and proceed with the overall timetable design process, possibly with final local refinements and minor adjustments made by planner. Many references consider Mixed Integer Problem formulations in which the arrival and departures times are represented by continuous variables and there are binary variables expressing the order of the train departures from each station. The non periodic train timetabling problem has been considered by several authors: Szpigel [18], Javanovic and Harker [8], Cai and Goh [1], Carey and Lockwood [4], Higgins et al. [6], Silva de Oliveira [17], Kwan and Mistry [11], Caprara et al. [3], Ingolotti et al. [7].

The main problem the Spanish Manager of Infrastructure faces is the allocation of the paths requested by transport operators and the process of designing the overall timetable. These timetables are generally non periodic and have to met a wide set of constraints and achieve a multicriteria objective function. A detailed formal description of both the constraints and the objective function is given in the following subsections. First we will introduce the notation that will be used hereafter.

10.2.1 Notation

The notation used to describe the problem is the following.

Parameters:

- T: finite set of trains t considered in the problem. $T = \{t_1, t_2, ..., t_k\}$
- $T_C \subset T$: subset of trains that are in circulation and whose timetables cannot be modified (T_C can be empty).
- $T_{new} \subseteq T$: subset of non-scheduled trains that do not have yet a timetable and that must be added to the railway line with a feasible timetable. Thus $T = T_C \cup T_{new}$ and $T_C \cap T_{new} = \emptyset$
- l_i: location (station, halt, junction). The types of locations considered are described as follows:
  - Station: Place for trains to park, stop or pass through. Each station is associated with a unique station identifier. There are two or more tracks in a station where crossings or overtaking can be performed.
  - Halt: Place for trains to stop, pass through, but not park. Each halt is associated with a unique halt identifier.
  - Junction: Place where two different tracks fork. There is no stop time.
- N_i: number of tracks in location l_i.

- NP_i: number of tracks with platform (necessary for commercial stops) in location l_i.
- $\mathrm{L} = \{l_0, l_1, ..., l_{\mathrm{m}}\}$: railway line that is composed by an ordered sequence of locations that may be visited by trains $t \in T$. The contiguous locations l_i and l_{i+1} are linked by a single or double track section.
- $\mathrm{J}_t = \{l_0^t, l_1^t, ..., l_{n_t}^t\}$: journey of train t. It is described by an ordered sequence of locations to be visited by a train t such that $\forall t \in \mathrm{T}, \exists \mathrm{J}_i : \mathrm{J}_t \subseteq \mathrm{L}$. The journey J_t shows the order that is used by train t to visit a given set of locations. Thus, l_i^t and $l_{n_t}^t$ represent the i_{th} and *last* location visited by train t, respectively.
- $\mathrm{T_D}$: set of trains travelling in the *down* direction.
  $t \in \mathrm{T_D} \leftrightarrow (\forall l_i^t : 0 \leq i < n_t, \exists l_j \in \{\mathrm{L} \setminus \{l_m\}\} : l_i^t = l_j \wedge l_{i+1}^t = l_{j+1})$.
- $\mathrm{T_U}$: set of trains travelling in the *up* direction.
  $t \in \mathrm{T_U} \leftrightarrow (\forall l_i^t : 0 \leq i < n_t, \exists l_j \in \{\mathrm{L} \setminus \{l_0\}\} : l_i^t = l_j \wedge l_{i+1}^t = l_{j-1})$. Thus $\mathrm{T} = \mathrm{T_D} \cup \mathrm{T_U}$ and $\mathrm{T_D} \cap \mathrm{T_U} = \emptyset$
- C_i^t minimum time required for train t to perform commercial operations (such as boarding or leaving passengers) at station i (commercial stop).
- $\Delta_{i \to (i+1)}^t$: journey time for train t from location l_i^t to $l_{(i+1)}^t$.
- $\lfloor \mathrm{I_L}^t, \mathrm{I_U}^t \rfloor$: interval for departure time of train $t \in \mathrm{T_{new}}$ from the initial station of its journey.
- $\lfloor \mathrm{F_L}^t, \mathrm{F_U}^t \rfloor$: interval for arrival time of train $t \in \mathrm{T_{new}}$ to the final station of its journey.

Variables:

- dep_i^t departure time of train $t \in T$ from the location i, where $i \in J_t \setminus \{l_{n_t}^t\}$.
- arr_i^t arrival time of train $t \in T$ to the location i, where $i \in J_t \setminus \{l_0^t\}$.

Planners usually use running maps as graphic tools to help them in the planning process. A running map is a time-space diagram like the one shown in Fig. 10.1 where several train crossings can be observed. The names of the stations are presented on the left side and the vertical line represents the number of tracks between stations (one-way or two-way). Horizontal dotted lines represent halts or junctions, while solid lines represent stations. On a railway network, the planner needs to schedule the paths of n_k trains going in one direction and m_k trains going in the opposite direction for trains of a given type. The trains to schedule can require (or not) a given frequency.

10.2.2 Feasibility of a Solution - Set of Constraints

In order to be feasible, a timetable has to met a set of constraints that can be classified in three main groups depending on whether they are concerning with: (*i*) user requirements (parameters of trains to be scheduled), (*ii*) traffic rules, (*iii*) railway infrastructure topology. The constraints described in this

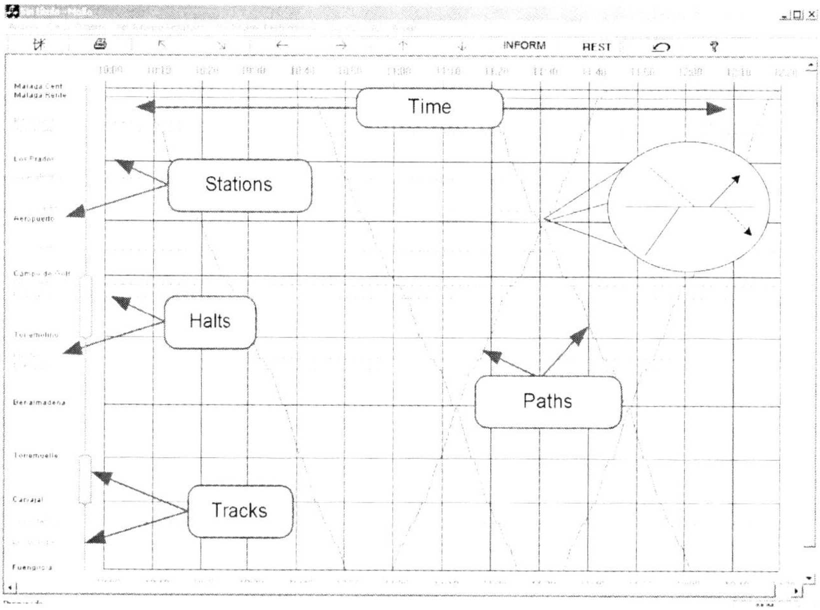

Fig. 10.1: Running Map.

work have been defined together with the Spanish Manager of Railway Infrastructure (ADIF) in such a way that the resulting timetable was feasible and practicable.

User Requirements:

- *Interval for the Initial Departure*: each train $t \in \mathrm{T}_{\mathrm{new}}$ should leave its initial station l_0^t at a time dep_0^t such that,

$$\mathrm{I}_{\mathrm{L}}^t \leq dep_0^t \leq \mathrm{I}_{\mathrm{U}}^t. \tag{10.1}$$

- *Interval for the Arrival Time*: each train $t \in \mathrm{T}_{\mathrm{new}}$ should arrive to its final station $l_{n_t}^t$ at a time $arr_{n_t}^t$ such that,

$$\mathrm{F}_{\mathrm{L}}^t \leq arr_{n_t}^t \leq \mathrm{F}_{\mathrm{U}}^t. \tag{10.2}$$

- *Maximum Delay*: a maximum delay Λ_t and a minimum journey time M_t are specified for each train $t \in \mathrm{T}_{\mathrm{new}}$; thus, the upper bound for the journey time of $t \in \mathrm{T}_{\mathrm{new}}$ is given by the following expression:

$$\frac{(arr_{n_t}^t - dep_0^t - \mathrm{M}_t)}{\mathrm{M}_t} \leq \Lambda_t. \tag{10.3}$$

Traffic constraints:

- *Journey Time*: for each train and each track section, a Journey time is given by $\Delta^t_{i\to(i+1)}$, which represents the time the train t should employ to go from location l^t_i to location l^t_{i+1}. Therefore, the following expression must be fulfilled

$$arr^t_{i+1} = dep^t_i + \Delta^t_{i\to(i+1)}. \tag{10.4}$$

- *Crossing*: according to the following expression, a single-track section ($i \to i+1$, *down* direction) cannot be occupied by two trains going in opposite directions ($t \in \mathrm{T_D}$ and $t' \in \mathrm{T_U}$).

$$dep^{t'}_{i+1} > arr^t_{i+1} \vee dep^t_i > arr^{t'}_i. \tag{10.5}$$

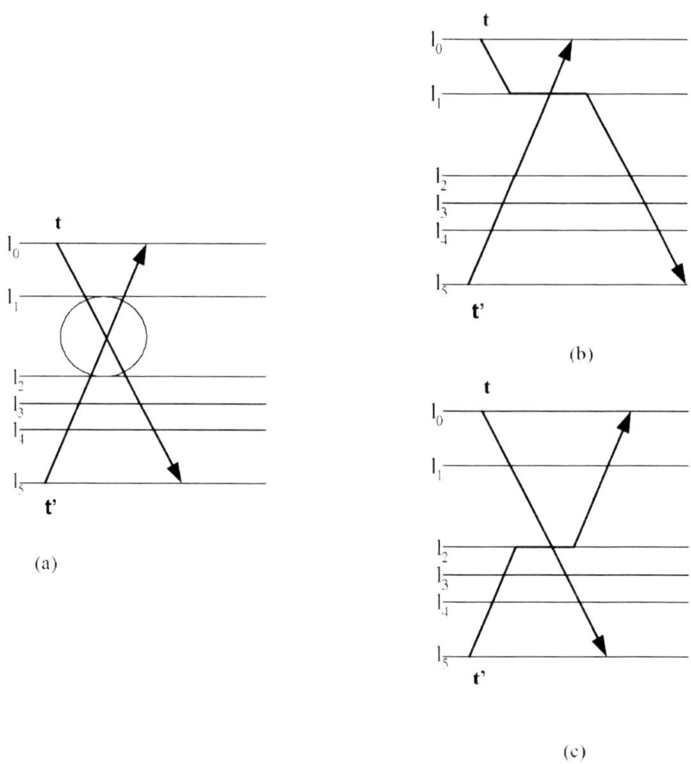

Fig. 10.2: (a) Crossing conflict. (b) Train in Down direction waits. (c) Train in Up direction waits.

- *Commercial Stop*: each train $t \in \mathrm{T_{new}}$ is required to remain in a station l^t_i at least C^t_i time units:

$$dep_i^t \geq arr_i^t + C_i^t. \tag{10.6}$$

- *Overtaking on the track section*: overtaking must be avoided between any two trains, $\{t,t'\} \subseteq T$, going in the same direction on any double-track sections, $k \rightarrow (k+1)$, of their journeys:

$$(arr_{k+1}^t > arr_{k+1}^{t'}) \leftrightarrow (dep_k^t > dep_k^{t'}). \tag{10.7}$$

- *Delay for unexpected stop*: when a train t stops in a station j to avoid conflicts with other trains (overtaking/crossing), and no commercial stop was planned ($C_j^t = 0$) in this station, the journey time of train t that corresponds to the previous ($l_{j-1}^t \rightarrow l_j^t$) and next ($l_j^t \rightarrow l_{j+1}^t$) track sections of j must be increased by Γ_t time units. This increase represents the speed reduction of the train due to the braking and speeding up in the station.

$$dep_j^t - arr_j^t > 0 \wedge C_j^t = 0 \rightarrow \Delta_{j-1\rightarrow j} = \Delta_{j-1\rightarrow j} + \Gamma_t \wedge \Delta_{j\rightarrow j+1} = \Delta_{j\rightarrow j+1} + \Gamma_t. \tag{10.8}$$

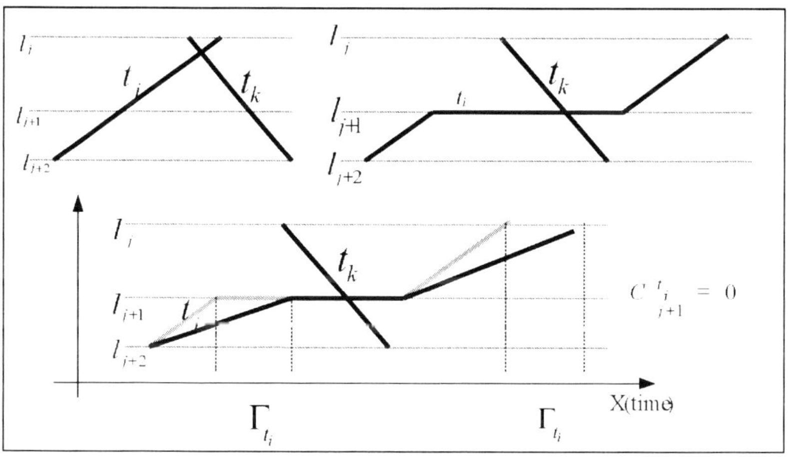

Fig. 10.3: Unexpected Stop.

- *Reception Time*: The difference between the arrival times of any two trains $\{t,t'\} \subseteq T \wedge \{t,t'\} \subseteq T_{\text{new}}$ in the same station l is defined by the expression below, where R_t is the reception time specified for the train that arrives to l first.

$$arr_l^{t'} \geq arr_l^t \rightarrow arr_l^{t'} - arr_l^t \geq R_t. \tag{10.9}$$

- *Expedition Time*: The difference between the departure and arrival times of any two trains $\{t,t'\} \subseteq T \wedge \{t,t'\} \subseteq T_{\text{new}}$ in the same station l is

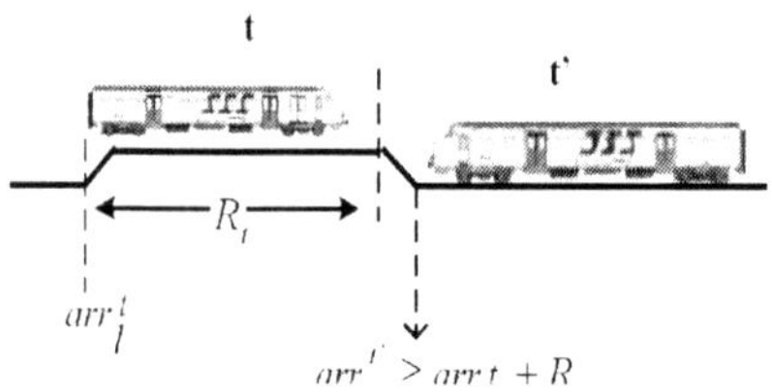

Fig. 10.4: Reception time between train t and train t'.

defined by the expression below, where E_t is the expedition time specified for t.

$$|dep_l^{t'} - arr_l^t| \geq \mathrm{E}_t. \tag{10.10}$$

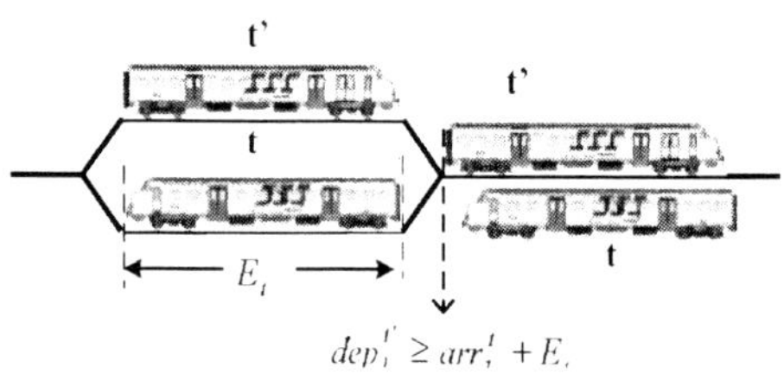

Fig. 10.5: Expedition time between train t and train t'.

- *Simultaneous Departure*: the difference between departure times from the same station of two trains going in opposite directions must be at least S, when both trains stop in the station. This constraint is formulated as:

$$\forall t, t' \in \mathrm{T}_{\text{new}} : dep_i^t - arr_i^t > 0 \wedge dep_i^{t'} - arr_i^{t'} > 0 \rightarrow |dep_i^t - dep_i^{t'}| \geq S. \tag{10.11}$$

Infrastructure constraints:

- *Finite Capacity of Stations*: a train $t \in \mathrm{T}_{\text{new}}$ could arrive to a location l_i^t if and only if it has at least one available track (with platform, if $C_i^t > 0$). In order to formulate this constraint, consider:

$$\forall x \in \mathrm{T}_{\text{new}} : T_x = \{t \in \mathrm{T} : t \neq x, J_t \cap J_x \neq \emptyset\} \text{ and}$$

$$\text{Meet}(x, t, l) = \begin{cases} 1 & \text{if } [arr_l^x, dep_l^x] \cap [arr_l^t, dep_l^t] \neq \emptyset \wedge \mathrm{C}_l^t = 0 \\ 0 & \text{else} \end{cases}$$

$$\text{Meet}_{\text{P}}(x,t,l) = \begin{cases} 1 & \text{if } [arr_l^x, dep_l^x] \cap [arr_l^t, dep_l^t] \neq \emptyset \wedge \text{C}_l^t > 0 \\ 0 & \text{else} \end{cases}$$

Hence, the constraint of finite capacity of stations is formulated as follows:

$$\forall x \in \text{T}_{\text{new}}, \forall l \in \text{J}_x : ((\sum_{t \in \text{T}_x} \text{Meet}(x,t,l) + \sum_{t \in \text{T}_x} \text{Meet}_{\text{P}}(x,t,l) < \text{N}_l) \wedge$$

$$(\text{C}_l^x > 0 \rightarrow \sum_{t \in \text{T}_x} \text{Meet}_{\text{P}}(x,t,l) < \text{N}_{\text{P}_l})). \tag{10.12}$$

- *Closing Time*: Let $[\text{H}_l^1, \text{H}_l^2]$ be the closing time for maintenance operations of station l. The closing time imposes constraints over regular operations -trains can pass but cannot stop in the station (see the next expression). And can even forbid regular operations, trains can neither pass nor stop (i.e.: the number of tracks in the station is decreased to one (see (10.12)).

$$dep_l^t < H_l^1 \vee arr_l^t > H_l^2. \tag{10.13}$$

- *Headway Time*: If two trains, $\{t, t'\} \subseteq \text{T}$, travelling in the same direction leave the same location l_k towards the location l_{k+1}, they are required to have a difference in departure times of at least φ_k^d and a difference in their arrival times of at least φ_k^a. When the blocking type in the track section is *Automatic*, then $\varphi_k^a = \varphi_k^d$. Consider the following expression

$$|dep_k^t - dep_k^{t'}| \geq \varphi_k^d. \tag{10.14}$$

$$|arr_{k+1}^t - arr_{k+1}^{t'}| \geq \varphi_k^a. \tag{10.15}$$

According to the company requirements, the method proposed should obtain the best available solution so that all the above constraints are satisfied. As we previously pointed out, the network could be previously occupied by other trains whose timetable have not been changed. That is to say $\forall t \in \text{T}_{\text{C}}$, the variables arr_i^t and dep_i^t, have been previously instantiated with given values. This means $\forall t \in \text{T}_{\text{C}}, \forall i \in J_t, arr_i^t \in \text{CONSTANT}, dep_i^t \in \text{CONSTANT}$ and the process generates the constraints so that the arrival and departure time of trains in circulation are constants, and it does not generate constraints that only involve variables corresponding to trains in circulation. Next, the process verifies that each new train satisfies each constraint taking into account the remaining new trains as well as all the trains already in circulation. In other words, if a constraint is violated and it relates new trains with trains in circulation, the only timetables that should be modified are those corresponding to new trains.

The set of constraints just described corresponds to Spanish Railway Company requirements and do not match exactly with other published works.

Several criteria can exist to assess the quality of the solution, for example: minimize travel time, minimize the passenger waiting time in the case of changeovers, balance the delay of trains in both directions, etc. The objective function considered in this work is described in the next section.

10.2.3 Optimality of a Solution - Objective function

In order to assess the quality of each solution, we obtain the *optimal solution* (optimal traversal time) for each specific train $t \in \mathrm{T_{new}}$. The optimal solution of train t is computed by the scheduling of the new train t (verifying all problem constraints) on the network being occupied only by trains in circulation ($\mathrm{T_C}$). This optimal solution for train t (Γ^t_{opt}) is the lowest time required by t to complete its journey. The other trains to be scheduled in $\mathrm{T_{new}}$ are ignored.

Once the optimal time for each new train to be scheduled has been computed, the criterion to measure the quality of each solution will be the average delay of new trains with respect to their optimum (δ). That is:

$$\delta_t = \frac{(arr^t_{n_t} - dep^t_0) - \Gamma^t_{\mathrm{opt}}}{\Gamma^t_{\mathrm{opt}}};\ \delta_{\mathrm{U}} = \frac{\sum_{t \in \mathrm{T_U} \cap \mathrm{T_{new}}} \delta_t}{|\mathrm{T_U} \cap \mathrm{T_{new}}|};\ \delta_{\mathrm{D}} = \frac{\sum_{t \in \mathrm{T_D} \cap \mathrm{T_{new}}} \delta_t}{|\mathrm{T_D} \cap \mathrm{T_{new}}|};\ \delta = \frac{\delta_{\mathrm{U}} + \delta_{\mathrm{D}}}{|\mathrm{T_{new}}|}$$

Finally, assuming TTABLE being one problem solution and therefore the timetable for all new trains, the objective function of this problem is formulated as:

$$f\,(TTABLE) = \mathrm{MIN}(\delta) \tag{10.16}$$

If there are no trains in circulation in L ($\mathrm{T} = \mathrm{T_{new}}$), the optimal time of a new train t would be:

$$M_t = \Gamma^t_{\mathrm{opt}} = \sum_{i=0}^{n_t-1} \Delta^t_{i \to (i+1)} + \sum_{i=0}^{n_t-1} C^t_i \tag{10.17}$$

10.3 Solving Process: A Genetic Algorithm Approach

The solving method proposed in this work for the TTP is based on a Job-Shop approach. In short, the General Job-Shop Scheduling Problem arising in many companies implies the execution of a set of orders consisting of jobs, satisfying a set of time and resource constraints. Precedence relationships of jobs of each order means time constraints and resource constraints that occur when one resource can not be simultaneously used by two jobs, that is, a manufacturing machine can not simultaneously perform two jobs. The main objective is to generate a schedule (job timetable) where each job satisfies time and resource constraints with the lowest computational effort and optimizing a measure of performance (usually makespan).

In the case of the TTP, each train can be decomposed in a set of ordered Train-track_section (T-ts) that has to verify a set of time and resource constraints: precedence relations of track sections for each train as time constraints and resource constraints when one-way track section can not be simultaneously occupied by two trains. The objective is to build a schedule (train timetable) where each Train-track_section satisfies time and resource constraints and optimizes a measure of performance with the lowest computational effort.

In this context, if limitation of resources is not taken into account, the problem is reduced to propagating the initial departure time, dep_0^t, from l_0^t until $l_{n_t}^t$ using the journey time $\Delta_{i\rightarrow(i+1)}^t$ that corresponds to each train t and each track section $l_i^t \rightarrow l_{i+1}^t$. However, in this problem resources have a finite capacity. A single line contains a set of single-track sections that cannot be occupied by more than one train at a given time, just as machines in a job-shop can process only one job at a given time. The TTP is an optimization combinatorial problem whose search space grows exponentially when the number of conflicts increases. This may be due to an increase in the number of trains, or to a decrease in the capacity of stations (number of tracks with platform if commercial stop exists), etc. It is well known that this problem is NP-Hard [2], therefore, heuristic methods are common approaches, able to obtain "good" solutions with a reasonable computational time. These heuristics have to be able to explore the large search spaces arising in this type of problems and to achieve a good solution in a short computational time. Therefore, the GA approach is suitable to cope with the TTP.

Assuming the parallelism just established between both problems, the GA developed to solve the TTP assumes each train as an order t that is split into specific jobs t_{jk} and where each job implies that train t uses one track to go from location j to location k. We represent a job by a pair (t, s_i^t), where t is the train and s_i^t is the i_{th} track section of its journey.

10.3.1 Basic Scheme of the GA

Fig. 10.6 shows the general scheme of a generic genetic algorithm. First, the initial population (P in Fig. 10.6), whose size is POP_SIZE, is generated and evaluated following a scheduling scheme that is described in both Fig. 10.8 and subsection *Initial Population*. The following steps are repeated until the terminating condition *end_cond* (execution time, number of feasible solutions or number of generations), is reached. Some individuals that compose the population P in Fig. 10.6, are modified by applying the procedures `Selection()`, `Crossover()`, and `Mutation()`. Thus, a new population P is obtained in each generation. Each iteration of Fig. 10.6 corresponds to a new generation of individuals.

```
Function Genetic_Algorithm(POP_SIZE, end_cond) As Timetabling

begin
    P=Generate_Initial_Population(POP_SIZE)
    while NOT (end_cond) do
    begin
        P=Selection(P)
        P=Crossover(P)
        P=Mutation(P)
        BEST_L=Evaluate_Population(P)
    end
    return BEST_L
end
```

Fig. 10.6: General Genetic Algorithm.

10.3.2 Definition of Individuals: Solution Encoding

In order to apply a GA to a particular problem, an internal representation for the solution space is needed. The choice of this component is one of the critical aspects for the success/failure of the GA for the problem under study. In the literature, we have found different types of representations for the solution of different scheduling problems. In this work, we have used an activity list as representation of a solution. This solution representation has been widely used in project scheduling. The solution is encoded as a precedence feasible list of pairs (t, s_i^t), that is, if (t, s_x^t) and (t, s_y^t) are the j_{th} and k_{th} gene of a chromosome of the same individual and $x < y$, then $j < k$.

The corresponding train schedule is generated applying a modified version of the Serial Schedule Generation Scheme used in Project Scheduling [9]. In the list of pairs all trains are merged in order to obtain a feasible solution. In our implementation, the new trains are scheduled following the order established by the list. Each individual in the population is represented by an array with as many positions as pairs (t, s_i^t) exist in the railway scheduling problem considered. Fig. 10.7 shows the activity list representation for a problem with N pairs (t, s_i^t).

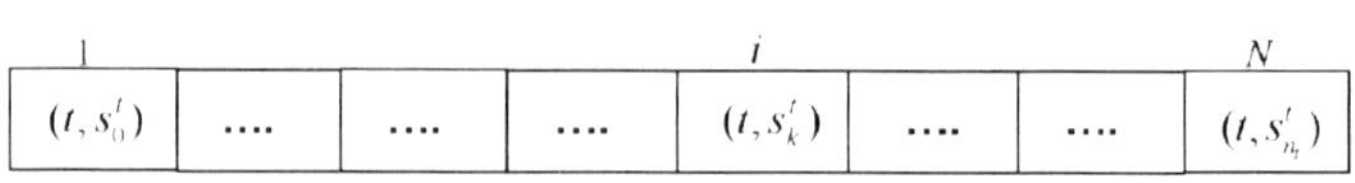

Fig. 10.7: Activity List Representation.

According to this list, (t, s_k^t) is the i_{th} pair to be scheduled. Considering that $s_k^t = l_k^t \rightarrow l_{k+1}^t$, the departure time of train t from l_k^t will be the earliest feasible time from $arr_k^t + C_k^t$. Note that when applying the de-codification process described in subsection 10.3.8, one and only one schedule can be deduced from a given sequence, but different sequences could be transformed into the same schedule.

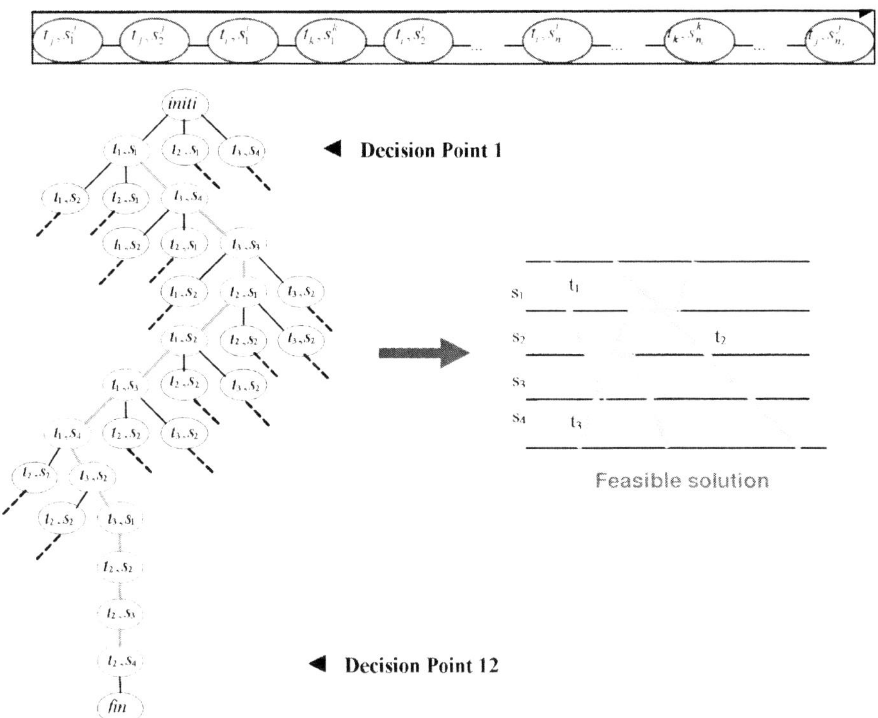

Fig. 10.8: Each activity list represents one path of the search tree.

As Fig. 10.8 shows, an activity list represents one path of the search tree. Different order of the activity list usually will imply a different solution.

10.3.3 Fitness Computation

When applying the GA, we need to define an evaluation function that determines the probability of survival of an individual to the next generation. In this chapter, the average deviation with respect to the optimal solution for the train is returned as the fitness value. In other words, an individual

χ will have the fitness value obtained from the objective function defined in Section 10.2.3.

10.3.4 Initial Population

The GA starts with the generation of an initial population, that is, a set of POP_SIZE feasible solutions. This set of feasible solutions can be obtained with different scheduling techniques. For the design of a GA, the initial population should include a variety of medium to good feasible solutions in order to increase the quality of the best solution in the evolutionary process. In this work, the value of the POP_SIZE is 50. The initial population has been obtained with an iterative heuristic based on random sampling methods and that is repeated POP_SIZE times (Fig. 10.9). N is the total number of track sections for all trains $t \in T_{NS}$. It is shown how is created the activity list by selecting a train (t) and a set of track sections $s \in J_t$ for each iteration, until all the (N) track sections have been scheduled for each new train. A solution is obtained once N iterations have been completed. Each solution gives a scheduling order, $L = (t_x, s_0^{t_x}), ..., (t_z, s_j^{t_z}), ..., (t_y, s_{n_{t_y}}^{t_y})$ that represents a new individual of the initial population.

The scheduling method developed implies the search of a path in a tree as the one shown in Fig. 10.8. At each decision point, a train with track-sections unscheduled is selected. This process obtains a feasible timetable with a value of the fitness function.

The main decision in the procedure `Select_Train()` is: which train has to be scheduled at each decision point. Even though a random decision (RANDOM) could be taken selecting the train to schedule randomly, we have applied a *Regret-Based Biased Random Sampling* (RBRS) procedure that makes the selection of a train dependent on its deviation with respect to the optimal solution (optimal running time of the train calculated as indicated in Section 10.2.3). This approach guides the scheduling process in order to obtain better solutions.

The *Parameterized Regret-Based Biased Random Sampling* (RBRS) selects trains of T_{NS} through a random device. The use of a random device can be considered as a mapping $\psi : i \in T_{NS} \rightarrow [0..1]$ where a probability $\psi(i)$ of being selected (being $\sum_{i \in T_{NS}} \psi(i) = 1$) is assigned to each $t \in T_{NS}$. The regret value (ρ_i) for each train $i \in T_{NS}$ compares the priority value of train i $-\nu(i)-$ with the worst priority value $\nu(j)$ of the trains of T_{NS} and is calculated as follows:

$$\rho_i : \max_{j \in T_{NS}} (\nu_j) - (\nu_i). \tag{10.18}$$

Therefore, the parameterized probability mapping $\psi(i)$ is calculated as:

$$\psi(i) : \frac{(\rho_j + \varepsilon)^\alpha}{\sum_{j \in T_{NS}} (\rho_j + \varepsilon)^\alpha}. \tag{10.19}$$

```
Function Generate_Initial_Population(POP_SIZE) As Population
begin
    ref=Get_Low_Bound_Opt_Sol()
    i=0
    P=""
    While (i<POP_SIZE)
    begin
        L= "" //L is a new list of chromosomes
        j=0
        While (j<N)
        begin
            t=Select_Train() //using the RBRS method
            s=Select_Track_Section(t)
            d=Get_Departure(t,s)
            Set_Timetable((t,s),d)
            L=L+(t,s) //(t,s) is inserted in L
            j=j+1
        end
        Set_Fitness_To_Individual(L, ref)
        P=P+L
        i=i+1
    end
    return P
end
```

Fig. 10.9: Procedure to obtain the Initial Population.

This parameterized Regret-Based Biased Random Sampling has been widely and successfully used in project scheduling (Schirmer and Riesenberg [15], Tormos and Lova [19]). The priority value of each train is calculated according to its current delay with respect to the scheduled timetable. Trains with higher delays have more probabilities of being selected. We have implemented the procedure `Generate_Initial_Population()` using the RBRS method to select the next train to be scheduled, with $\alpha = 1$ and $\varepsilon = 0.5$.

10.3.5 Crossover

One of the unique and important aspects of the techniques involving Genetic Algorithms is the important role that recombination (traditionally, in the form of crossover operator) plays. Crossover combines the features of two parent chromosomes to form two offspring that inherit their characteristics. The individuals of the population are mated randomly and each pair undergoes the crossover operation with a probability of P_{cross}, producing two children by crossover. The parent population is replaced by the offspring population. The crossover is one of the most important genetic operators and must be correctly designed. Crossover must combine solutions to produce new ones. Crossover must preserve and combine "good building blocks" to build better individuals [5]. Given two individuals selected for crossover, a mother **M** and a father **F**, two offspring, a daughter **D** and a son **S** are produced.

We have implemented the well-known one point crossover with $P_{cross} = 0.8$. First we draw a random crossover-point k, with k between 1 and N (number of Train-Track_Section in the problem). The first k positions in **D** are directly taken from **M**, in the same order. The rest of the activities in **D** are taken with their relative order in the father's sequence. In this way, the solution generated, the daughter, is a precedence feasible solution. Obviously, the generation of **S** is similar to the daughter's but **S** inherits the first positions directly from **F**, and the rest of the Train-Track_Section from **M**. The pseudocode for this crossover technique is shown in Fig. 10.10.

```
Function Random_Crossover_Point()

begin
    //Draw a random integer k, with 1<= k<= N
    //k is the random crossover-point
    //Generation of the daughter
    for i=1 to k do
        Di = Mi
    for i=k+1 to N do
    begin
        I = lowest index  1<= I<= N and Fi not in {D1, ..., D(i-1)}
        Di = Fi
    end
    //Generation of the son
    ........
end
```

Fig. 10.10: Crossover Procedure.

10.3.6 Mutation

Once the crossover operator has been applied and the offspring population has replaced the parent population, the mutation operator is applied to the offspring population. Mutation alters one or more genes (positions) of a selected chromosome (solution) to reintroduce lost genetic material and introduce some extra variability into the population.

The mutation operator that we have implemented works as follows: for each pair (t, s_i^t) in the sequence, a new position is randomly chosen. In order to generate only precedence feasible solutions, this new position must be higher than its predecessor and lower than its successor. The chromosome is inserted in the new position with a probability P_{mut}. In our implementation $P_{mut} = 0.05$.

10.3.7 Selection

Selection is an artificial version of the natural phenomenon called the survival of the fittest. In nature, competition among individuals for scarce resources and for mates results in the fittest individuals dominating over weaker ones. Based on their relative quality or rank, individuals receive a number of copies. A fitter individual receives a higher number of offspring and, therefore, has a higher probability of surviving in the subsequent generation. There are several ways of implementing the selection mechanism.

We have implemented *2-tournament selection*. This selection mechanism implies that two individuals are randomly chosen from the population and compete for survival. The best of them (the one with the best fitness value) will appear in the subsequent population. This procedure is repeated POP_SIZE times, until POP_SIZE individuals are selected to appear in the next population.

10.3.8 Decodification Process

In this subsection, we detail the procedure `Evaluate_Population()` of Fig. 10.6. This procedure receives a population P from which it should obtain POP_SIZE solutions. Each solution will be evaluated according to the objective function that is defined in Section 10.2.3 (`Set_Fitness_Individual()` in Fig. 10.11).

For each pair $p = (t, s_i^t) \in L$, a departure time is computed by means of the function `Get_Departure()`, which returns $d = arr_i^t + C_i^t$ if $i > 0$, otherwise $d = m$ such that m is the initial departure time given by the user.

Considering that s_i^t starts at station l_i^t and ends at station l_{i+1}^t, the procedure `Set_Timetable()` assigns a possible departure and arrival time to train t in each location between l_i^t and l_{i+1}^t, according to the journey time $(\Delta_{i \to (i+1)}^t)$ defined for this train from l_i^t to l_{i+1}^t. The next step consists in verifying whether all the constraints defined in 10.2.2 are satisfied by the timetable given for t in s_i^t. If any constraint is not satisfied, the departure time in l_i^t is increased until that constraint is satisfied. This increment in the departure time in l_i^t causes a technical stop of the train t at this station. A backtracking may occur if the station is closed for technical operations or if the station does not have enough tracks.

Once a feasible timetable has been found in this track section for train t, the same procedure is repeated with the next chromosome $(t', s_k^{t'})$.

The priority of the trains in each track section, that is, which train should be delayed if a conflict appears, is determined by the order in which each gene is numbered in the activity list. When a conflict occurs between two trains in the same track section, the priority is for the train whose timetable in this track section was assigned first. As different individuals define different priorities among the trains different solutions may be obtained.

```
Procedure Evaluate_Population(P)

begin
    i=0
    While(i<|P|) do
    begin
        L=Get_Individual(i,P)
        k=0
        while(k<|L|)
        begin
            p=Get_Chromosome(L,k)
            d=Get_Departure(p)
            Set_Timetable(p,d)
            k=k+1
        end
        i=i+1
        Set_Fitness_Individual(L)
    end
end
```

Fig. 10.11: De-codification Process.

The parameter setting of the proposed GA results from previous computational experiments.

10.4 Solving Real Cases with GA

The application of the GA described is illustrated using a real-world problem. The line considered is the railway line between Madrid and Jaen which covers 54 locations (stations, halts, siding, etc); it is 369.4 kilometers long and has 30/23 two/one-way track sections. Each horizontal line in Fig. 10.12 shows the position of each location.

The timetabling shown in Fig. 10.12a corresponds to the traffic in the line Madrid-Jaen and in the time window [12:00, 24:00] there are 103 trains in circulation, before adding the new trains. Each point of the oblique lines corresponds to the position of one train (axis Y) at a given time (axis X).

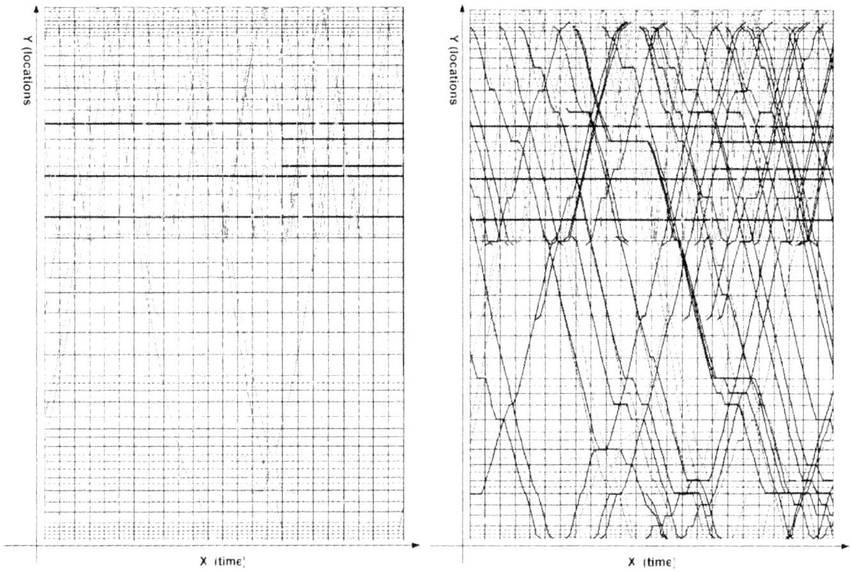

(a) Trains in Circulation (b) Solution

Fig. 10.12: Graphical Representation of a Problem Example and its Solution.

We have to add 59 new trains to the line in the time window [12:00, 24:00] satisfying the constraints, taking into account the trains in circulation and minimizing the average delay of the new trains. Usually, the planner works only assisted by a graphical tool and schedules trains in a lexicographic order (one train after another). The solution obtained is a feasible solution but not necessarily a good solution due to the multiple combinations that must be considered simultaneously. Fig. 10.12b shows the solution obtained using the GA approach in 300 seconds using a Pentium IV 3.6 Ghz processor. The value of the objective function for this solution is 7.2% (see Section 10.2.3).

10.4.1 Results

The performance of the GA developed has been tested using a set of real-world problems provided by the Spanish Manager of Railway Infrastructure (ADIF). The description of the instances is given in Table 10.1 (columns 2 to 10) by means of: length of the railway line, number of single/double track sections, number of locations and stations, number of trains and track sections (T-ts) corresponding to all these trains, considering trains in circulation and new trains, respectively.

Table 10.1: Real railway problem instances provided by ADIF.

Problems	Infrastructure Description					Trains in Circulation		New Trains	
	Km	1-Way	2-Way	Loc	Stat	Trains	T-ts	Trains	T-ts
1	96	16	0	13	13	47	1397	16	180
2	129	21	0	22	15	27	302	30	296
3	256	38	0	39	28	81	1169	16	159
4	401	37	1	39	24	0	0	35	499

Each problem has been solved using the two constructive methods used to generate the Initial Population that differs in the criterion to select the trains:

- Random selection of each train to be scheduled in each iteration (RANDOM).
- Selection of each train using the Parameterized Regret Biased Based Random Sampling method (RBRS).
- The results obtained by means of these constructive methods are compared against those achieved by the GA with the same computational time.

Table 10.2 summarizes the results for each solving method with respect to the number of solutions generated (# of Solutions) and Average Deviation with respect to the Optimal Solution (ADOS) according to expression (10.16). The tests have been carried out in a Pentium IV 3.6 Ghz processor and the running time was of 300 seconds for all the problems. $\alpha = 1$ and $\epsilon = 0.5$. The different number of solutions generated depending on the method used is because a prune procedure is applied when the RANDOM and RBRS approaches are used. That is, when a partial schedule produces a value of the objective function worse than the best value obtained at the time, then the current iteration is interrupted and the construction of a new one starts. However with the GA approach, the prune is not possible because each iteration must be completed to obtain a fitness value for the solution. This fitness value is necessary to obtain the next generation of individuals.

Table 10.2: Results of the RANDOM, RBRS and the GA scheduling methods.

Problems	RANDOM		RBRS		GA	
	# of Solutions	ADOS	# of Solutions	ADOS	# of Solutions	ADOS
1	267	18	263	15.4	255	15.1
2	611	10.1	608	10.0	313	9.6
3	424	14.7	521	14.1	382	12.4
4	405	19.2	397	17.9	285	16.0

Results of Table 10.2 show that the GA proposed outperforms both RANDOM and RBRS methods for all problem instances considered. These results demonstrate the efficiency of the GA to solve Railway Scheduling problems against other constructive algorithms and support the idea of developing more sophisticated and powerful GAs to solve complex problems such as Train Timetabling Problem.

10.5 Conclusions

Optimizing a train schedule on a single line track is known to be NP-Hard. This makes it difficult to determine optimum solutions to real life problems in reasonable time and raises the need for good scheduling techniques. The Train Timetabling Problem considered in this work implies the optimization of new trains on a railway line that is occupied (or not) by other trains with fixed timetables. The schedule for the new trains is obtained with a Genetic Algorithm that includes a guided process to build the initial population.

The GA developed has been used to solve real-world instances and its performance has been compared against constructive approaches. The results of the computational experience, point out that GA is an appropriate method to explore the search space of this complex problems and that further research in the design of efficient GA is justified. The GA approach proposed in this work might be improved with the use of local search able to intensify performance around promising regions of local optima. An added value of the proposed GA is that its main concepts are embedded in a computer-aided tool that is being successfully used by the Spanish Manager of Railway Infrastructure.

Acknowledgments

This work has been partially supported by the research projects Future and Emerging Technologies Unit of EC (IST priority - 6th FP), under contract nr. FP6-021235-2 (project ARRIVAL), TIN2004-06354-C02-01 (Min. de Educación y Ciencia, Spain-FEDER), FOM-70022/T05 (Min. de Fomento, Spain) and GV/2007/274 (Generalidad Valenciana). We are grateful to the Spanish Manager of Railway Infrastructure (ADIF) and to Jose Estrada Guijarro for his continuous support throughout this research.

References

1. X. Cai and C. J. Goh. A fast heuristic for the train scheduling problem. *Computers and Operations Research*, 21(5):499–510, 1994.
2. A. Caprara, M. Fischetti, and P. Toth. Modeling and solving the train timetabling problem. *Operations Research*, 50:851–861, 2002.
3. A. Caprara, M. Monaci, P. Toth, and P. Guida. A lagrangian heuristic algorithm for a real -world train timetabling problem. *Discrete Applied Mathematics*, 154:738–753, 2006.

4. M. Carey and D. Lockwood. A model, algorithms and strategy for train pathing. *The Journal of the Operational Research Society*, 46(8):988–1005, 1995.
5. D.E. Goldberg. *Genetic Algorithms in Search, Optimization and Machine Learning*. Adison Wesley, 1989.
6. A. Higgins, E. Kozan, and L. Ferreira. Heuristic techniques for single line train scheduling. *Journal of Heuristics*, 3(1):43–62, 1997.
7. L. Ingolotti, A. Lova, F. Barber, P. Tormos, M.A. Salido, and M. Abril. New heuristics to solve the csop railway timetabling problem. *Advances in Applied Artificial Intelligence. LNAI, Subseries of Lecture Notes in Computer Science*, 2006.
8. D. Jovanovic and P. T. Harker. Tactical scheduling of rail operations: The scan i system. *Transportation Science*, 25(1):46–64, 1991.
9. J. Kelley. The critical-path method: Resources planning and scheduling. *Industrial Scheduling*, 1963.
10. L. Kroon and L. Peeters. A variable time model for cycling railway timetabling. *Transportation Science*, 37(2):198–212, 2003.
11. R. K. S. Kwan and P. Mistry. A co-evolutionary algorithm for train timetabling. In IEEE Press, editor, *Congress on Evolutionary Computation*, pages 2142–2148, 2003.
12. C. Liebchen. *Periodic Timetable Optimization in Public Transport*. dissertation.de - Verlag im Internet GmbH 2006, 2006.
13. K. Nachtigall and S. Voget. A genetic algorithm approach to periodic railway synchronization. *Computers and Operations Research*, 23:453–463, 1996.
14. M. Odijk. A constraint generation algorithm for the construction of periodic railway timetables. *Transportation Research Part B*, 30(6):455–464, December 1996.
15. A. Schirmer and S Riesenberg. Parameterized heuristics for project scheduling-biased random sampling methods. Technical Report 456, Institute fr Betriebswirtschaftslehre der UNIVERSITT KIEL, 1997.
16. P. Serafini and W. Ukovich. A mathematical model for periodic scheduling problems. *SIAM J. on Discrete Mathematics*, 2:550–581, 1989.
17. E. Silva de Oliveira. *Solving Single-Track Railway Scheduling Problem Using Constraint Programming*. PhD thesis, The University of Leeds, School of Computing, September 2001.
18. B. Szpigel. Optimal train scheduling on a single track railway. In M. In Roos, editor, *Proceedings of IFORS Conference on Operational Research'72*, pages 343–352, 1973.
19. P. Tormos and A. Lova. A competitive heuristic solution technique for resource-constrained project scheduling. *Annals of Operations Research*, 102(1-4):65–81, 2001.

11

Modelling Process and Supply Chain Scheduling Using Hybrid Meta-heuristics

Soumya Banerjee[1], G.S.Dangayach[2], S.K.Mukherjee[3], and P.K. Mohanti[4]

[1] Dept. of Computer Science and Engineering, Birla Institute of Technology, Mesra, India soumyabanerjee@bitmesra.ac.in
[2] Department of Mechanical Eng., Malviya National Institute of Technology, Jaipur, India
[3] Vice Chancellor, Birla Institute of Technology, Mesra, India
[4] University of New Brunswick, New Brunswick, Canada

Summary. This chapter proposes a natural stigmergic computational technique *Bee Colony* for process scheduling and optimization problems developed by mimicking social insects' behavior. The case study considered in the chapter is a milk production center, where process scheduling, supply chain network etc. are crucial, as slight deviation in scheduling may lead to perish out the item causing financial loss of the plant. The process scheduling of such plants extensively deals with multi-objective conflicting criteria, hence the concept of *Pareto Dominance* has been introduced in the form of *Pareto Bee Colony Optimization*. Some facts about social insects namely bees are presented with an emphasis on how they could interact and self organized for solving real world problems. Finally, a performance simulation and comparison has been accomplished envisaging other similar bio-inspired algorithms.

Key words: Supply Networks, Scheduling, Bee Colony Optimization, Multi-objective Optimization, Pareto Bee Colony.

11.1 Introduction and Background

Supply networks are organizations of partially autonomous production and distribution centers through which goods are processed and delivered to customers. Optimizing the activities of a supply network to improve production throughput and timeliness of the delivery requires dealing with a number of large-scale, interrelated assignments, scheduling and routing problems. The optimization is especially challenging for a supply network that delivers rapidly perishable goods, such as raw materials used for manufacturing foods and beverages. The perishable goods are only used within a period of restricted time limit, so it is expected that their production and delivery are made only on stipulated demand and even their routing through proper

S. Banerjee et al.: *Modelling Process and Supply Chain Scheduling Using Hybrid Meta-heuristics*, Studies in Computational Intelligence (SCI) **128**, 277–300 (2008)
www.springerlink.com

channel also becomes an issue. Naturally the problem of such type requires multi optimization with large number of constraints at different stages. The specialty of such optimization is not only about their costs but also flexibility and robustness of the solution could also be considered. It is observed that any small deviation over local activity may inject a cascaded delay while deploying the common resources. Therefore, referring the aspect of perishable material for food industry, this may turn into substantially significant financial loss and even degrading the brand value of the company, and all the conventional predefined optimal solutions could become impractical due to various other constraints. Professionals find it as trade off between the minimizing the costs of operations and providing shock absorption over these disruptions.

There are certain commercial tools (like CPLEX) available to address these issues. A considerable number of works that demonstrate the approach of supply network logistics issues, but comparatively less versions of multi optimization scheduling based on meta-heuristics have been established. We propose a novel hybrid meta-heuristics optimally in a supply network for perishable material concerned with food industry. The genesis of the search strategy is based on evolutionary computation method, mainly to exploit its efficient exploration/exploitation capability in the large search space of the main decision variables characterizing our scheduling problem. The solution integrates the followings:

- A detailed mathematical model of the logistic problem that unambiguously specifies the free decision variables.
- A set of fast heuristics organized in a hierarchical structure that is able to construct a fully feasible solution starting from an initial assignment of a subset of decision variables.
- A multi-objective bee colony based rough set algorithm that searches for the set of best tradeoff solutions considering both the costs and the robustness of the corresponding schedules.

The proposed solution is presented in the context of reducing the risk and uncertainty in optimization. The optimization algorithm returns a set of solutions with different cost and risk tradeoffs, allowing the analyst to adapt the planning depending on the attitude to risk. The subsequent section presents an overview of supply chain modelling through diversified methodologies especially envisaging adaptive and intelligent techniques.

Background

The globalization, dynamics, and frequent variations of customer demands on today's markets increase the needs of companies to form supply chains (SCs) and cooperative business partnerships that enable them to survive on today's competitive market [6, 7]. SCs are networks of autonomous business entities that collectively procure, manufacture, and distribute certain products. The objective of a SC is to respond efficiently to customer demands and at the same time, it must minimize the cost of all participating business entities. To achieve this objective the supply chain management (SCM) system must coordinate and optimize the procurement, production, and distribution of goods. In reality, however, SCs are often operating in dynamic and non-homogenous cultural environments. Therefore, current SCM systems need to adopt adaptive learning features and reasoning of theory of evidence to reflect the changes in the dynamic cultural environment. In practical research, there are couples of supply chain works primarily concerned with adaptive reasoning and seems to be hybrid intelligent. For example, Several systems were developed to model the SCs complexity using a GA, for examples Truong and Azadivar [8] have integrated GAs, mixed integer programming methods, and simulation techniques into a hybrid optimization model, while other researchers use GAs and Pareto Optimal techniques [9]. Furthermore, Al-Mutawah, Lee, and Cheung have developed a Distributed Multi-objective Genetic Algorithm (DMOGA) to solve the SC optimization problem in [10]. One common limitation of DMOGA and other typical genetic based implementation of multi-objective optimization is the inheritance process of GA, which restricts the parents to transfer experiences only to their offspring, ignoring the influence of other external sources. In real world, particularly in a distributed environment, SC applications data are collected from heterogeneous sources, implying the need to co-opt other sources of influence as well.

11.2 Related Works on Meta-heuristics

As it has been discussed in previous sections, the supply chain is a complex network of facilities and organizations with interconnected activities but different and conflicting objectives. Many companies are interested in analyzing their supply chain as an entire and unique system to be able to improve their business. However, in most cases the task of designing, analyzing and managing the supply chain has been done based on experience and intuition; very few analytical models and design tools have been used in the process. This implies that finding the best supply chain strategies for a particular firm, group of firms or sector poses significant challenges to the industry and academia. The optimization literature focuses on algorithms for computing solutions to constrained optimization problems.

Meta-heuristics have many desirable features to be an excellent method to solve very complex SCM problems: in general they are simple, easy to implement, robust and have been proven highly effective to solve hard problems. Several other aspects are worth to mention. The first one is the meta-heuristicsc modular nature that leads to short development times and updates, given a clear advantage over other techniques for industrial applications. This modular aspect is especially important given the current times of implementing a Decision Support System (DSS) in a firm and the rapid changes that occurs in the area of SCM. The next important aspect is the amount of data involved in any optimization model for an integrated supply chain problem, which can be overwhelming. The complexity of the models for the SCM and the incapacity of solving in real time some of them by the traditional techniques, force the use of the obvious technique to reduce this complex issue by data aggregation [11]. However this approach can hide important aspects that impact the decisions. Other reports published presented a complex vehicle routing model to the distribution in the food and beverages industries [12].

A literature search in food science and technology databases reveals that optimization using response surface modelling (RSM) has been, and continues to be, the most common approach. RSM techniques were introduced in the 1950s associated with design of experiments methods [1,2]. Although the usefulness of RSM in certain conditions must be recognized, this approach has a number of important drawbacks due to the empirical, local and stationary nature of the simple algebraic models used. In contrast, a number of powerful model-based optimization methods have been developed during the last decades, which use more rigorous, time-dependent models. Primarily, at the core is the problem domain from which instances were drawn. The problem domain consist twenty four distinct instance classes on the basis of twelve distinct problem specifications. Except for the 100-job single machine total weighted tardiness instance class, for each instance class, a new benchmark set of 125 instances was created. Furthermore, solutions were obtained for all 3000 instances and recorded to serve as reference for future research. In order to find these solutions, new solution representations were developed for the problems with a parallel machine environment. Moreover, well known speed up techniques for the single machine total weighted tardiness problem were adapted to the constraints posed by objective functions that ignore weights or impose a unit penalty on each late job. Finally, the speed up techniques was adapted to work in machine environments with more than one machine in parallel [3].

Mixed planning and scheduling problem was discussed in length and it has been shown how to extend a conventional scheduler by some planning capabilities during the investigation on complex process models [4]. Balanced theory and practice of planning and scheduling in supply chains also have become prominent area of implementation [5]. The project first gives an overview of the various planning and scheduling models that have been studied in the literature, including lot sizing models and machine scheduling models. It subsequently categorizes the various industrial sectors in which planning and scheduling in the supply chains are important; these industries include continuous manufacturing as well as discrete manufacturing followed by the description how planning and scheduling models can be used in the design and the development of decision support systems for planning and scheduling in supply chains and discuss in detail the implementation of such a system at the Carlsberg A/S beer-brewer in Denmark.

The trend of research and implementation found to be more practical when natural heuristics has been incepted in scheduling the activities. Many project tasks and manufacturing processes consist of interdependent time-related activities that can be represented as networks. Deciding which of these sub-processes should receive extra resources to speed up the whole network (i.e. where activity crashing should be applied) usually involves the pursuit of multiple objectives amid a lack of a *priori* preference information. A common decision support approach lies in first determining efficient combinations of activity crashing measures and then pursuing an interactive exploration of this space.

11.3 Motivation and Importance Behind the Model

Most of the raw material of food and beverages are perishable due to their complex preservation approach and thus conventional *Just in Time (JIT)* methodology doesn't hold good as it seldom fails in uncertainty and ambiguity. The distribution network of finished frozen food products also requires utmost prompt delivery over a broader geographical coverage preventing them from damage in terms of food value and nutrition. The vehicle used for dispatching have limited capacity, and so large demands require several vehicle loads to transport all the products at proper places in time. These activities have to be properly synchronized, because the unloading at the customer site must be continuous in order to prevent compromising the food value properties of the product. Each production centers of food processing aims to increase resource utilization decrease costs and ensure the timeliness of the deliveries. Hence, those centers pursue multiple, contradictory goals. At present, many companies tend either to rely on skilled operators that work out production plans based on their experience, or to plan production operations on very short time horizons, sacrificing the optimization on longer horizon to achieve a reduced risk of delayed delivery.

The plethora of different type ambiguous problems of supply network envisages the present working project deploying certain novel components of evolutionary computations. The aim is to present a smarter and easily understood model, which could assist the logistics managers of food industry for scheduling their supply network in an optimized direction both on and off the production (including distribution).

Concept of Hybrid Meta Heuristics for the Proposed Model

We consider the problem of scheduling of events in the form of Directed Acyclic Graph (DAG). Each node in the graph represents an executable task. Each directed edge represents a precedence constraint (or simply dependence) between two tasks; the sink node cannot start execution until the source node has finished and the transmission of the required amount of data from the source node to the sink node has been completed. We assume that the DAG has always a single entry node (i.e. a node with no parents) and a single exit node (i.e. a node with no children). The target environment consists of a set of heterogeneous events, which are fully connected; a data transfer cost is given for each pair of events (like controlling the procurement of more perishable raw materials, enabling the dispatch of food product and beverage over the distribution network, etc). A task can execute on any available events; the execution cost of each task on each event is also given. The task scheduling problem is to allocate tasks for execution onto events in such a way that precedence constraints are respected and the overall execution time is minimized. It is assumed that only one task can execute on an event at a time and once a task has started execution it cannot be preempted.

The heuristic consists of three phases: (a) Ranking; (b) Group creation; and (c) Scheduling independent tasks within each group. For each of the stages the proposal is combining two different techniques concerned to tackle both optimization and uncertainty in the execution of the events. We incorporate Bee colony optimization (BCO) and rough set approach for this mode.

In the first phase, a weight is assigned to each node and edge of the graph; this is based on averaging all possible values for the cost of each node (or edge, respectively) on each events (or combination of events, respectively). Using this weight, upward ranking is computed and a rank value is assigned to each node. The rank value, of a node r_i is recursively defined as follows:

$$ri = wi + max(c_{ij+rj}), \tag{11.1}$$

where $\forall j \in s_i$ and w_i is the weight of node i, S_i is the set of immediate successors of node i and c_{ij} is the weight of the edge connecting nodes i and j.

In the second phase, nodes are sorted in descending order of their rank value; using this order, they are considered for assignment to groups as follows. The first node (i.e. the node with the highest rank value) is added to a group numbered 0. Successive nodes, always in descending order of their rank value, are placed in the same group as long as they are independent with all the nodes already assigned to the group (i.e. there is no dependence between them in the DAG). If dependence is found, then the node with the smallest rank value (i.e. the sink of the dependence) is made the member of a new group; the new group's number is the current group's number increased by one. Again, subsequent tasks, in terms of their rank value, will be added to this group as long as they are not dependent to any other node which is a member of this group; if they are, a new group will be created and so on. The outcome from this process is a set of ordered groups, each of which consists of a number of *independent* tasks, and has a predetermined priority (based on the original ranking of the nodes; a smaller group number indicates higher priority).

The third phase of heuristics comprises of a schedule of the DAG can be obtained by considering each group in ascending order of its number, and using *any* heuristic for scheduling the independent tasks within each group. It is noted that the input of the latter heuristic will be a set of (independent) tasks; a set of machines; the array giving the execution cost of each node on any machine; and, another array giving the earliest time that each task may start execution on each event.

11.4 Bee Colony Optimization

Various unsocial insect colonies such as ants, wasps, termites and bees exhibit remarkable problem solving behavior. Although a single insect is quite limited in its ability, complex behavior is exhibited at the level of the colony that emerges from the interactions of the individual insects [13]. This phenomenon is called Self-Organization. The foraging behavior of honey bees has been extensively studied and is a useful example of self-organization. Computational biology and modelling of these self organized properties mediated in solving plenty of complex optimization and scheduling problems.

Mathematical Model of Foraging for Honey Bees

Foraging is an interesting property to observe for honey bees and it is complex process involving large number of individuals collecting food from many different sources. Differential equation models have shown how quite simple communication mechanisms can produce complex and functional group level foraging patterns. Here, we concentrate our focus on the mathematical aspects of foraging including waggle dance of bees during the foraging. For example, although individual honey bee foragers follow only a small number of the waggle dances advertising flower patches, the colony can nonetheless focus its foraging effort on the most profitable patches. Similarly, certain ants deploy their foragers preferentially on the shorter of two paths, despite few if any individual insects directly comparing the paths [15, 16]. The potential benefit of the existing mathematical models are to understand how population change through time. The number of bees foraging for a particular food source can be represented as single variable that changes its value as the insects are recruited to and abandon the source. These recruitment and abandonment rates can be written as functions of the number of insects foraging at a source, waiting at the nest, or scouting for new sources. There are many impressive literatures available on the different aspects of differential equation based modelling of insects encompassing the foraging [17–26].

In the mathematical model, several behavioral states could be contemplated. Colonies have access to n number of food sources. Each state has an associated variable, indexed by source (by default). Hence, the different sates in the dynamic model would like to be:

- Waiting (denoted as W): Waiting at the nest and available to start foraging.

- Searching (denoted as S): Searching for food sources.
- Exploiting (E_i): Exploiting food source i. Workers in this state do not directly recruit nest mates, although they may leave signals, such as pheromone trails, that increase the likelihood of other foragers finding the source.
- Recruiting (R_i): Attempting to recruit nest mates to food source i.
- Following (F_i): Attempting to follow recruiters to food source i.

In order to model and deploy the differential equation in these behavioral states of bees, we would like to establish a series of mathematical anomalies stated as follows:

- A waiting worker W can become an exploiting forager at source i E_i through three different routes, iff she might be activated to search (through function a), and then discover the food source (through d_i). Or, she might be led toward the food source through direct contact or communication (h_i) with another worker, arriving (s_i) only if the communication is successful. Finally, she might reach a food source by following an indirect signal, such as a pheromone trail (through j_i).
- The function j_i represents indirect recruitment, where successful foragers influence their environment in a manner that increases the chance of nest mates finding the food.
- The function f_i represents direct recruitment, where successful foragers either physically lead nest mates to the food source or directly communicate, in the nest, the location of the source [27].
- The population of workers in the nest, W, increases as searchers are deactivated (b), exploiters retire from foraging (g_i), and followers get lost and return to the nest (v_i). Conversely, the population decreases as nest workers are activated to search (a), as they are led by indirect recruitment signals to become exploiters (j_i), and as they begin to follow direct recruitment to various food sources (h_i).

Considering these dynamic conditions, we can write series of equations [14]:

$$dw/dt = b + \sum_{i=1}^{n} g_i + \sum_{i=1}^{n} v_i - a - \sum_{i=1}^{n} j_i - \sum_{i=1}^{n} h_i \tag{11.2}$$

where b is deactivated searchers as population of workers w increases in the nest, g_i denotes exploiters retire from foraging, v_i denotes return to the nest, a denotes activated workers, j_i is the recruitment signal and h_i denotes the direct recruitment of different food source.

The exhaustive mathematical treatment even prescribe to estimate the optimal investment in workers by colonies that use this foraging mechanism. A productivity function can be defined as to show how foraging efficiency depends on the maximum number of ants foraging at a food source at stable equilibrium [28]. These collective decision making of bees assist to model more complex situation very similar to group of robots taking a decisions in a group.

11.5 Multi-Objective Optimization and Standard Bee Colony Optimization Algorithm

Most real world optimization problems are naturally posed as multi-objective optimization problems. However, due to the complexities involved in solving optimization problem and due to lack of suitable and efficient solution techniques, they have been transformed and solved as single objective optimization problem. Moreover, because of the presence of conflicting multiple objectives, a multi objective optimization problem results n a number of optimal solutions, known as *Pareto optimal solutions* [29]. In standard practice of using bee's natural properties in computation it has been observed there may be substantial number of instances where selection of best and nearly best solution against very close processes comprising of conflicts or constraints associated with it. This leads to the solution of Pareto type and in this proposal, we introduce the concept of PBCO (Pareto Bee Colony Optimization) in the context of scheduling of several processes very similar to the case of Milk Production Center presented here. Primarily, the standard bee colony and its property have been considered and subsequently their affinity to the process scheduling is discussed.

Algorithm 11.1: Basic Bee Colony Optimization Algorithm-High Level Description.

```
Step 1:   Initialization. Determine the number of Bees and
          the number of iterations I. Select the set of
          Stages ST = {st1, st2, ... , stm}.
          Find any Feasible solution x of the problem.
          This solution is the initial Best Solution.

Step 2:   Forward Pass: Allow Bees to fly from the hive
          and to choose B partial solutions from the set
          of partial solution Sj at stage Stj.

Step 3:   Backward Pass: Send all bees back to the hive.
          Allow bees to exchange information about quality
          of the partial solution created (without
          recruiting nest mates) or dance.

Step 4:   Update Best Solution and return to the nest with
          incrementing the counter.
```

A general scheduling problem can be formalized as follows [30]. We consider a finite set of operations O, partitioned into m subsets $\langle M_1, \ldots, M_2 \rangle =: M(\bigcup M_i = O)$ and into n subsets $\langle J_1, ..., J_n \rangle =: J(\bigcup J_k = 0)$. Together with a partial order $p \subseteq \mathbf{O} \times \mathbf{O}$ such that $p \cap J_i \times J_j$ for $i \neq j$ and a function $p : O \rightarrow N$. A feasible solution is a refined partial order $p* \supseteq p$ for which the restrictions $p * IM_i \times M_i$ and $p * J_k \times J_k$ are total, $\forall i, k$. Most importantly, the cost of a feasible solution is defined by $C_{max}(p/ast) := max\{\sum p(o)|C$ is a chain in $(O, p*)\}$. The effort is to minimize C_{max}. Here M_i is the set of operations that have to be processed on machine i. J_k is the set of operations belong to job k (analogues to core task and sub tasks of the Milk production center mentioned in the case study). All the processes or operations must be performed sequentially and this constraint has been expressed in p^*.

Considering theses assumptions as the benchmark of the case study of the milk production center where the processes are sequential and time bound otherwise the milk core need to be perished out, the model is likely to concentrate on the optimal execution of process scheduling maintaining their intermediate time and other constraints during the makespan. The makespan of the scheduling has been modelled through proposed *Pareto Bee Colony Optimization* (PBCO), envisaging their natural property like waggle dance and foraging.

11.5.1 Waggle Dance –Computational Interpretations

A forager f_i on return to the hive from nectar exploration will attempt with probability p to perform waggle dance on the dance floor with duration D $= d_i A$, where d_i changes with profitability rating while A denotes waggle dance scaling factor. Further, it will also attempt with probability r_i to observe and follow a randomly selected dance. The probability r_i is dynamic and also changes with profitability rating. If a forager chooses to follow a selected dance, it will use the "path" taken by the forager performing the dance to guide its direction for flower patches. We term the path as "preferred path". The path for a forager is a series of landmarks from a source (hive) to a destination (nectar).

11.5.2 Forage and Combining Rough Set

For foraging algorithm, a population of l foragers is defined in the colony. The foragers move along branches from one node to another node in the disjunctive graph and so construct paths representing solutions. A forager must visit every node once and only once in the graph, starting from initial node (i.e. source) and finishing at final node (i.e. sink), so as to construct a complete solution. When a forager is at a specific node, it can only move to next node that is defined in a list of presently allowed nodes, imposed by precedence constraints of operations. It has been observed that after complete deployment and performance benchmark of the proposed BCO uncertainty and ambiguity part still exist. In order to the model more feasible and applicable, the concept of rough set is sorted to use in conjunction with BCO.

11.5.3 Process Scheduling and Optimization under Uncertainty

The scheduling problem has usually been seen as a function of known and reliable information. Modelling approaches developed are mainly deterministic, that is, they are based on nominal or estimated values for all the parameters, thus implicitly assuming that a *predictive schedule* will be executed exactly as planned. However, this assumption is somehow utopian since most plants operate in an unstable and dynamic environment, where unexpected events continually occur. Scheduling problems involve data coming from different sources, and which varies rapidly over time as customer orders, resource availabilities and/or processes undergo changes. Data may be ambiguous, outdated or inaccurately predicted before the problem is solved. Because of the dynamic and uncertain conditions of a real process system, the schedule executed in the plant will probably differ from the predicted one. The effects of the uncertainty may impact on the system's efficiency, eventually leading either to an infeasible situation, or to the generation of opportunities that improve its performance. These situations may become even more significant with the new trends towards managing the whole SC. As stated by Aytug et al. [34], internet technology enables companies within a SC to share their production schedules. In this environment, changes to the production schedule at a downstream node of the SC can cause significant disruptions in upstream operations. These variations can be amplified causing what is known as the *bullwhip effect* [35]. The consideration of the uncertainty when modelling the problem is essential for the development of reliable and effective decision-support systems. Several methodologies are available in PSE (Process Systems Engineering) for optimization under uncertainty. They are categorized, in line with the method used to represent the uncertainty, represented as follows: (a) Probabilistic data - based methods; (b) Stochastic optimization; (c) Fuzzy or Rough Set data - based methods; and (d) Fuzzy Programming.

Blackhurst et al. [36] proposed a network-based methodology to model and analyze the operation of a SC as an abstracted network, with uncertainty in variables such as requirements, capacity, material delivery times, manufacturing times, costs, due dates and priorities. The term stochastic optimization is sometimes used referred to meta-heuristics because of the probabilistic nature of these optimization methods. In general, and as differentiated by some impressive research [37], stochastic optimization involves methods specially developed to address problems with uncertain data, whereas meta-heuristics use stochastic properties in their search.

Although the present model is not deploying stochastic optimization, but broadly the uncertainty part of scheduling events is handled through Rough set based rule metaphor. Hence, the core meta heuristics is being the Pareto Bee colony, subsequently rough set assists to model the associated events with the process and could rank them as well.

11.5.4 Rough Set

Rough set theory is an extension of conventional set theory that supports approximations in decision making. It possesses many features in common (to a certain extent) with the Dempster-Shafer theory of evidence and fuzzy set theory. The rough set itself is the approximation of a vague concept (set) by a pair of precise concepts, called lower and upper approximations, which are a classification of the domain of interest into disjoint categories. The lower approximation is a description of the domain objects which are known with certainty to belong to the subset of interest, whereas the upper approximation is a description of the objects which possibly belong to the subset. For the present problem, the features of individual events have been accumulated:

> A feature x_i is *relevant* if there exists some value of that feature and a predictor output value or A feature x_i is *weakly relevant* if it is not strongly relevant, and there exists some from the set of the features forming a pattern vi, for which there exist subset of features

followed by a ranking of such evaluated features and eventually the choice of the first best m features. Thus ranking of features of all events related to scheduling process of milk food processing industry both on and off the production, including dispatch could be modelled by rough set theory. The content of large-scale data sets containing numerical and categorical information can not be easily interpreted unless the information is transformed into a form that can be understood by human users. The rule extraction algorithms are designed to identify patterns in such data sets and express them as decision rules. The rule extraction concept is illustrated next.

Rule and Data Set

Consider the data set in Table 11.1 with five objects, four features F1-F4, and the decision (outcome).

Table 11.1: Rule Snapshot of Rough Set.

RULE 1 IF (F2 = 0)	THEN	$(D = Low)$; [2, 6.67%, 100%][3, 5]			
RULE 2 IF (F1 = 0)	AND	$(F4 = High)$	THEN	$(D = 0)$;	[1, 33.33%, 100.00%][1]
RULE 3 IF (F4 = 0)	THEN	$(D = Medium)$; [1, 100%, 100%][4]			
RULE 4 IF (F1 = 1)	THEN	$(D = High)$; [1, 100%, 100%][2]			

The features denote process parameters (e.g., temperature, pressure, time) and the decision is the component performance, high, medium, low. A rule extraction algorithm transforms the data set of Table 11.1 into the decision rules of Table 11.2. The two sets of numbers in square brackets behind each

Table 11.2: Decision rules in Rough Set.

0	0	0	1	Low
0	0	1	3	Low
0	1	0	2	Low
0	1	1	0	Medium
1	1	0	2	High

rule describe its properties. The decision rules of Table 11.1 correspond to the patterns indicated by shaded cells in the matrix in Table 11.2.

11.6 Case Study of Milk Food Product Processing and Production

The raw chilled milk so received subjected to different processes like pasteurization heating and separation/standardization. The detailed process is outlined as follows:

- Raw milk in 40 liters can is received through milk routes established all over the milk shed.
- The milk is graded, weighed an sampled

Problem Statement

The milk production center (MPC) also to undergo different processes and sub processes such as first regeneration, second regeneration, third regeneration, heating, cooling, chilling, internal sub-processing for products, packaging, storage and distribution. The parameters shown in Table 11.3 are crucial for all critical processes and sub processes:

Table 11.3: Process Parameters in Milk Processing Centers.

Process	Sub-Process	Time-Temp-Pressure	Remark
Milk reception	Dispatch	<15 mts	Crucial
Pasteurization	1st Generation, 2nd generation, Heating, Cooling and Chilling	20 sec/ 75-85 o C/25kg/cm2 Stem Pressure	Very Crucial
Standardization	Other Sub processes for Milk based products	5-6 degree Celsius	Important

There are at least 10 major processes identified without split up, which need to be monitored and scheduled priori basis. The processes comprise of different process or makespan of sub processes and its time windows. The conceptual flow diagram of the case study is shown in Fig. fig:MHS-09-1 and its associated parameters are demonstrated in Table 11.3. Among all these parameters certain are influential enough to implicate scheduling of processes, etc.

The broad idea is to incorporate the proposed Pareto Bee Colony Optimization in terms of process scheduling with multi-objectives and constraints. Later on, a few uncertain and ambiguous parameters are separated in scheduling process and rough set approach is coined to address this issue.

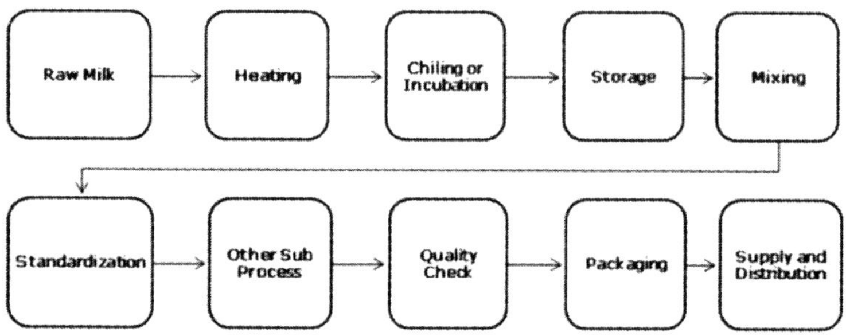

Fig. 11.1: Process Flow Diagram of MPC.

The Mathematical Model

Let us represent the given activity structure as an activity node based network in a milk production center (MPC), whose nodes can be numbered as $1, \ldots, n$. Arcs are used to indicate precedence relationship between nodes (activities). The time required for activity i is denoted by $d(i)$. For given values $d(i)$, the shortest project time δ is defined as the length of critical path. We assume that a finite set $M(i)$ of measures is to each node i. A measure $x_i \in M(i)$ is any means that influences the duration of the process based activity connected with node i. Each measures x_i is realized by certain values:

- The modified duration $d(i, x_i)$ of activity i resulting as a consequence of the measure x_i. The proposed model is concerned in speeding ups the processes and sub processes, so we assume $d(i, x_i) \leq d_i$.
- The cost associated by the process also have been accumulated.
- The model also comprises of target functions like:

$$hf_1(x) = \varphi_1(\delta(x)), \tag{11.3}$$

where, is the shortest project time after reduction of the durations $d(i)$ to the values $d(i, x_i)$ to the values $d(i, x_i)$.

The core process in the milk production center (MPC) is quite exceptional problem in the context of just in time scheduling approach.

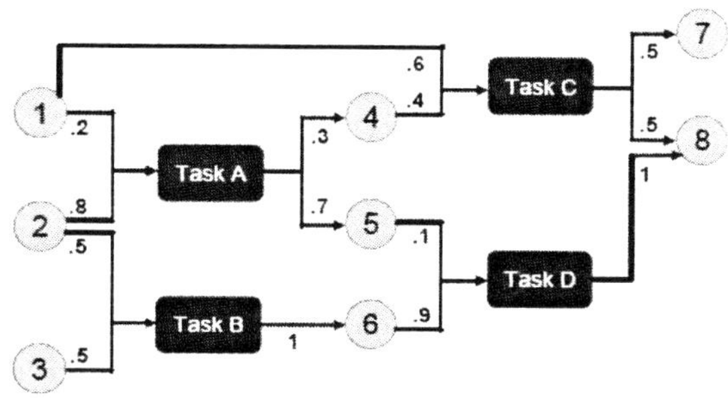

Fig. 11.2: Directed Process Graph of Milk Processing Center.

The present model of milk processing comprising four major tasks or processes and 8 sub processes (Fig. 11.2). The scenario is interpreted through directed graph. Directed graph G, consisting of two sets of nodes, V_1 and V_2, corresponding respectively to materials and tasks. Successor and predecessor nodes to a node in V_1 are always nodes in V_2 and, vice-versa, successor and predecessor nodes to a node in V_2 are always nodes in V_1. Hence the arcs in the graph always connect nodes from V_i to V_j, where $i \neq j$. An arc (r, i) from a node in $r \in V_1$ to a node $i \in V_2$ is introduced if task i requires material r as an input. The label on arc (r, i) is $\rho^{i,r}$, the fraction of input to task i due to material r. Similarly, an arc (i, r) from a node $i \in V_2$ to a node $r \in V_1$ is included in the graph if task i produces material r. The label on arc (i, r) is $\sigma^{i,r}$, the fraction of output from task i in the form of material r. Fig. 11.2 provides an example of such a network with 8 materials (numbered 1-8) and 4 tasks (labelled A-D).

11.6.1 The Proposed PBC Optimization Algorithm

The present problem of Milk Production and its sub process can be mapped with the Pareto Bee colony's characteristics. A forager f_i on return to the hive from nectar exploration will attempt with probability p to perform waggle dance on the dance floor with duration $D = d_i$ A, where d_i changes with profitability rating, while A denotes waggle dance scaling factor. Further, it will also attempt with a probability r_i to observe and follow a randomly selected dance. The probability r_i is dynamic and also changes with profitability rating. If a forager chooses to follow a selected dance it will deploy the path followed by the forager performing the dance to guide its direction for flower patch. In this model we define it as Preferred Process Path (PPP). Hence, the path for a forager is a series of landmarks from a source (hive) to a destination (nectar).

The proposed model of PBCO also implicates a direct relation with the objective function to the profitability rating. Therefore $P_{f_i} = \frac{1}{\varphi_1(\delta(x))}$, where P_{f_i} is the profitability rating for a forager. Theoretically, the average profitability rating of Bee colony is

$$P_{f_{colony}} = 1/n \sum 1/f_1(x) = \varphi_1(\delta((x)),$$

where n is the number of waggle dance just in time, (refer to the process and resources in directed process graph).

Moreover, $1/f_1(x) = \varphi(\delta(x))$ is the value of objective function, which should be maximum if a forage f_i or f_j performs waggle dance. The duration of dance is proportional to the completion of process time, which in turn just in time accomplished between all sub processes. In the process graph, forager must visit each process node i exactly once and it will follow a state transition rule to select best process path, so that no processes of milk production or its sub-process are being delayed. The state transition rule followed by the forager on the process span graph is according to the rule:

$$P_{ij} = \frac{[\rho ij(t)]^{\alpha} \cdot [1/dij]^{\beta}}{\sum[\rho ij(t)]^{\alpha} \cdot [1/dij]^{\beta}}.$$

The rating ρ_{ij} of the directed edge between process nodes i and j is given by:

$$\rho_{ij} = \begin{cases} \alpha \\ 1 - m\alpha/k - m \end{cases}$$

where α is the value assigned to the Preferred Process Path (PPP), $\alpha < 1.0$; k is the number of allowed nodes and m the number of preferred paths. The parameters α and β are the probability of the best process path which is in relation between preferred path versus heuristic distance. According to this rule, edges that are found in the preferred path and that are shorter will have a higher probability to be chosen for the solution. The heuristic distance is the processing time of the operation associated with node j. When a forager completes a full path, the edges it has travelled and the make-span (process span) of the resulting solution will be kept for the waggle dance when it returns to the hive.

Algorithm 11.2: Pareto Bee Colony Optimization -High Level Description

Initialize solution set by the empty set determine the number of bees and iterations
Select the set of processes stage $P_T = \{p_{t1}, p_{t2}, ...p_{tm}\}$
Initilaze bee pheromone matrix τ^k {Store best update solution}
Determine the weight say w_k for each objective k for different processes at random {Start Pareto}
repeat
  for bees 1 to B **do**
    Set $i = 1$
    repeat
      repeat
        Set $j = 1$
      until $j = m$
    until $i = i$
  end for
until termination criteria of each process is satisfied
Call ForwardPass ( ); {Allow bees to fly from the hive and to choose B partial solutions from the set of partial solutions S_j at stage p_{tj}}
for $i = 1$ to n **do**
  select the next process node to traverse according to $P_{ij} = \frac{[\rho_{ij}(t)]^{\alpha} \times [\frac{1}{d_{ij}}]^{\beta}}{\sum[[\rho_{ij}(t)]^{\alpha} \times [\frac{1}{d_{ij}}]^{\beta}]}$
  {$Pj \in$ allowed process node} { where ρ_{ij} is the rating of the edge between node i and j} { d_{ij} denotes heuristic distance between node i and j } { P_{ij} prior probability to traverse from node i to j}
  if the solution x is efficient solution achieved till i^{th} iteration **then**
    update the best solution
    $x := xi$
    Waggle Dance ( ) {profitability rating of Bee colony is

$$Pf_{colony} = 1/n \sum /f1(x) = \varphi 1(\delta((x))$$

    where n is the number of waggle dance through process span}
  end if
  $j = 1$
end for
for each objective k from the solution just found by B bees **do**
  identify ($R - 1$)best solution x^{tk} for object k. {τ^k is the bee pheromone matrix for each solution construct x }
end for

11.7 Implementation of PBCO as Multi Objective Optimization

The proposed algorithm has been implemented in C++ on Window XP platform (Pentium IV-2.4 GHz. processor, 256 MB RAM). The milk processing is completely just in time approach as each sub processes are also time bound and material is perishable. The time duration of waggle dance on the process graph is again dependent on the polymorphic method comprising of objective function, shortest process schedule time, and path trace iteration of milk processing. A List type data structure is maintained for and checked against the maximum number of iterations of honeybees across process graph. Practically, the model keeps the traced path in a list, which comprises consecutive operations in pairs. The effective algorithm is involved in operation scheduling and its uncertainty in supply chain process observed in a typical m time bounded milk production unit. The foraging algorithm (waggle dance and nectar exploration) has been incorporated considering bifocal approaches identical with either process or machine centric.

For any process centric approach, a list of currently eligible processes that can be scheduled is always maintained during scheduling process. In order to be viable to process span (makespan), a process's preceding sub-process (of a job) must have been scheduled. Each process planned for Milk production Center in the list is checked against the most recently scheduled sub-processes (on the same machine or parallel processes like cleaning the container, etc.) to identify if the "edge" between the two operations (the most recently scheduled process and the process under consideration from the list) is found in the preferred path. Higher rating ρ_{ij} are assigned to the process with edge found in the path. The scheduling is implemented through Pareto optimal solution, as there are conflict objective among the processes. On the other hand for machine centric approach, a discrete-event simulation and event list of events, which are in sorted order of increasing time, is maintained during scheduling process. At time $t = 0$, events relating to machine-ready status are inserted into the list. Events in the list are removed and executed one by one according to the event time. In case of tie for events having the same time, an event will be randomly picked. For the machine that is associated with the selected event, a list of currently eligible operations will be identified. Each operation in the list is checked against the most recently scheduled operation on the same machine to identify if the "edge" between the two operations is found in the preferred path. Higher rating ρ_{ij} will be assigned to the operation if the edge is found in the path. Although machine centric approach is slightly better than process centric, but the present case study has the variation of processes depending on different milk by products, hence it adopts the process centric approach to test the proposed PBCO algorithm.

11.7.1 Experimental Evaluation

A commercial snippet of data set for a typical Milk Production Center of Asian Country has been accumulated and referred in the case study. The climatic condition is extremely sensitive for Milk and its associated products

The simulation of the data set through C++ coding results some interesting process centric features (see Figs. 11.3 and 11.4. The Pareto distribution of process, optimal time and makespan of processes are presented in the bivariate polynomial form.

Month	Milk procu rement (mill. lit.)	Pasteu rized milk Temp. ^{o}C Ideal Val. 75-78	Pasteurized time in seconds Ideal Val. <20	Holding temp ^{o}C Ideal Val. 4-(-1)	Chilling Storage temp. (in Celsius) Ideal Val.5-6	Milk Milk Dis- -patch Time) (in minutes Ideal Val. < 15	Tanker time of Pasteurized milk (in hrs) Ideal Val. > 4.0	Dispatch Milk Products processing time, storage etc. Max. Val. 8-1
DEC. 05	12.6	78	18	4	5	13	3	8
JAN. 06	14	75	20	5	6	12	4.5	9
FEB. 06	12.4	75	19.5	-1	5	13	4	11
MAR. 06	11.4	76	19	-1	5	15	5	8
APR. 06	9	78	19	-1	5	11	4	9
MAY. 06	7. 4	75	17	-1	5	13	4.2	10
JUNE. 06	6.1	77	20	-2	5	14	3.8	10
JULY. 06	5.7	76	20	0	6	12	4	9
AUG. 06	6.6	75	19	3	6	13	4.5	9
SEPT. 06	7.6	77	18	3	5	11	5	10
OCT. 06	9.2	78	17	1	6	12	3	12
NOV. 06	12.6	76	20	1	5	13	4.5	11
DEC. 06	13.8	77	20	4	6	13	4	10

Table 11.4: Dairy data Evaluated through PBCO (Courtesy Jaipur Dairy, Jaipur, India).

The distribution of normal makespan of processes identified in MPC (Milk Production Center) is shown in Fig. 11.3 (Colony Multi-Objective Process Span -Cosine Series Bivariate Order 8) envisaging X, Y and Z axis span. Black Dots are glimpses of bee colony which take care of process span considering source and sink part of process. This denotes the complex multi objective scheduling in the Milk production corresponding the various processes and including time process and temperature. The approach of model is just in time, where the trade off between distribution of bee agents and reinforcement provided by Bee colony is shown by the black dots.

Another distribution map of time (Pareto Optimal Time Span) is shown in Fig. 11.4, scaling and process overlap among the overlapped processes. The peaks of the dots give predictive approach to notify the completion of process in optimal time or either in just in time.

11.7.2 Process Betterment through PBCO - A Comparative Study

To compare and evaluate the performance of the proposed bee colony algorithm, we have included two other meta-heuristics in our experimental study.

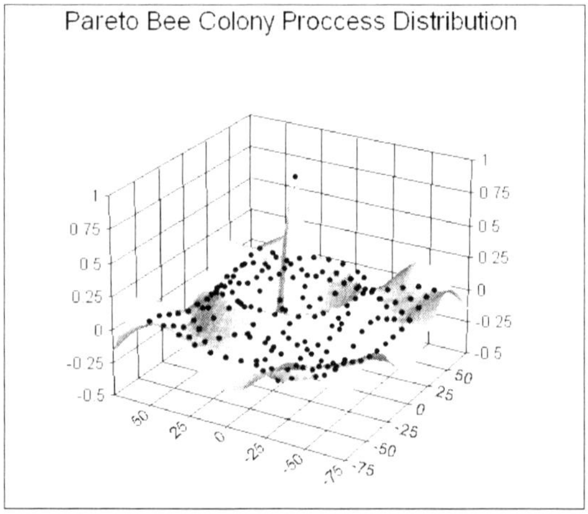

Fig. 11.3: Pareto Bee Colony Process Span, Multi-Objective and Optimal Time Distribution.

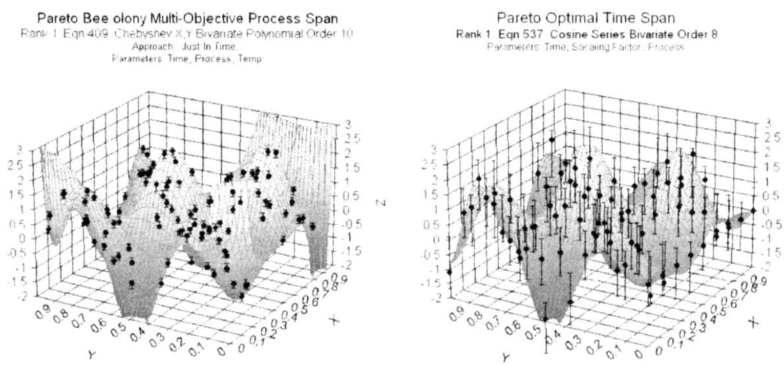

Fig. 11.4: Pareto Bee Colony Process Span, Multi-Objective and Optimal Time Distribution.

The first is an Ant Colony Optimization (ACO) algorithm [32]. The second algorithm is a Tabu Search (TS) algorithm developed by Nowicki and Smutnicki [33].

Within a multi-objective optimization approach, the solutions are compared with respect to their relative "dominance" on both cost and availability objectives. The solutions not dominated by any other are non-dominated solutions. A set (archive) of Pareto-optimal, non-dominated solutions (solutions which are not dominated by any other one) can be collected during the ACO search. Similarly, in order to improve the efficiency of the exploration process, one needs to keep track not only of local information (like the current value of

Table 11.5: Parameters.

Parameters	PBC	ACO	TS
Maximum Number of iterations	1500	1500	1000
Population Size	Process No. (MPC)	Process No No (MPC)	
Weight of Pheromone Trail α	1.0	1.0	
Weight of heuristic value β	2.0	2.0	
Parameter for local updating and profit rating ρ_{ij}	0.8		
Scale Factor	100		
Pheromone Evaporation Coefficient		0.1	
Weight of the availability in the heuristic η	0.3	0.5	
Weight of the cost in the heuristic η	0.3	0.5	
Maximum Number of Elite Solution	25		25
Maximum Size of Tabu List			10

Table 11.6: Performance Comparison of four Meta-heuristics.

Betterment	PBC	BC	AC	TS Process in MPC	Most Critical of Process
Mean percent	10.05	11.2	11.45	6.62	Pasteurization
Maximum	38	38.7	38.6	27.33	Holding and chilling
Most Close Solution schedule without failure	16	15	11	22	For all processes

the objective function) but also of some information related to the exploration process. This systematic use of memory is an essential feature of TS method.

In order to establish comparison, the parameter settings for three algorithms have been presented in Table 11.5.

Table 11.6 elaborates the brief comparison of the process span scheduling in Milk Production Center with three major perspective of meta-heuristics algorithms. Although results exhibit that TS is the smartest choice among the four, it provides most likelihood results within smallest execution time. Practically, TS has the advantage to solicit the best solution and it takes care to the Tabu list, instead of constructing the solution from source to sink applied to Bee and Ant Colony. The proposed Pareto Bee is slightly better than standard Bee Colony, where as broadly different from ACO heuristics in the context of process scheduling. The inclusion of rough set is a primary back up for the associated events with the process and could rank them as well under diversified condition.

11.8 Conclusions and Future Work

A hybrid meta-heuristic approach based on a multi-objective Bee Colony algorithm combined with constructive rough set heuristics is a valuable decision support tool for planning operations in a supply network for rapidly perishable material for food processing industry. Provided a detailed mathematical model of the supply network, our experimental investigation shows that such a hybrid approach is able to provide an effective scheduling algorithm. This work also provides a comparative platform on bio-inspired algorithms and rough set (e.g. Bee Colony, Ant Colony and Tau Search) used for scheduling in food and beverage processing industry. The involvement of multi-objective process span through Pareto scheduling also is proposed in this chapter. Further hybridization with other soft computing approach like fuzzy logic [31] could be developed on honey bee algorithm. The more extension on rough set in process scheduling is also expected. In this work, exact preferred path of the process schedule span has not been evaluated; rather we incorporated local search heuristics. This is because the processes and sub processes in the case study of Milk production Center are somewhat static in nature. It has been observed that among the all bio-inspired and stigmergic formulation of algorithm, ant colony is the most prominent one, but there are different agent based application areas where the agents work in a completely distributed environment and thus maintaining the pheromone transition table becomes slightly impractical. In those cases, Bee colony could be a better alternative.

Acknowledgement

The present case study and data has been considered from a milk food product processing *Jaipur Dairy* and its production house based on Jaipur, India. The work is partially supported by Department of Mechanical Engineering, Malviya National Institute of Technology, Jaipur, India.

References

1. Box, G.E.P., Hunter, W.G. & Hunter, J.S. (1978) Statistics for experimenters: an introduction to design, data analysis, and model building, Wiley, New York.
2. Myers, R.H. & Montgomery, D.C. (2002) Response surface methodology: process and product optimization using designed experiments, Wiley, New York
3. den Besten, L.M. Simple Meta-heuristics for Scheduling. An empirical investigation into the application of iterated local search to deterministic scheduling problems with tardiness penalties, PhD Thesis, October 2004.
4. Nareyek, A. (2000): AI Planning in a Constraint Programming Framework, in Proceedings of the Third International Workshop on Communication-Based Systems (CBS-2000)

5. Kreipl, S., Pinedo, M. Planning and Scheduling in Supply Chains: An Overview of Issues in Practice, Production and Operations Management Vol. 13, No. 1, spring 2004, pp. 77-92.
6. Chopra, S. and Meindl, P. Supply Chain Management: Strategy, Planning, and Operations: Prentice Hall College, 2001
7. Harrison, T.P., Lee, H.L. and Neale, J.J. The Practice of Supply ChainManagement: Kluwer Academic Publishing, 2003.
8. Truong T.H and Azadivar, F. Simulation based optimization for supply chain configuration design, presented at the Winter Simulation Conference, Piscataway, NJ, 2003.
9. Joines, J.A., Kupta, D., Gokce, M.A., King, R.E. Supply Chain Multi-Objective Simulation Optimization, presented at the 2002 Winter Simulation Conference, 2002.
10. Al-Mutawah, K., Lee, V., Cheung, Y. Modeling Supply Chain Complexity using a Distributed Multi-objective Genetic Algorithm, presented at The 2006 International Conference on Computational Science and its Applications ICCSA'06, Glasgow, Scotland, 2006.
11. Simchi-Levi D., Kaminsky, P. and Simchi-Levi, E. Designing and Managing the Supply Chain, McGraw-Hill, 2000.
12. Ribeiro, R. and Lourenço, H.R. A multi-objective model for a multi-period distribution management problem Economic Working papers Series, Universitat Pompeu Fabra, Spain, 2001.
13. Bonabeau, E., Dorigo, M., Theraulaz, G.: Swarm Intelligence: From Natural to Articial Systems. Oxford University Press (1999).
14. Sumpter, D., Pratt, S. A modelling framework for understanding social insect foraging. Behavioral Ecology and Sociobiology 53 (2003) 131-144.
15. Goss, S., Deneubourg, J.L. The self-organising clock pattern of Messor pergandei (Formicidae, Myrmicinae). Insectes Soc. 36:339-346, 1989.
16. Bartholdi, J.J., Seeley, T.D., Tovey, C.A., Vande Vate J.H. The pattern and effectiveness of forager allocation among flower patches by honey bee colonies. J. Theor. Biol. 160:23-40, 1993.
17. Beckers, R., Deneubourg, J.L., Goss, S., Pasteels, J. Collective decision making through food recruitment. Insectes Soc. 37:258-267, 1990.
18. Beckers, R., Deneubourg, J.L., Goss, S. Modulation of trail laying in the ant Lasius niger (Hymenoptera: Formicidae) and its role in the collective selection of a food source. J Insect Behav 6:751-759, 1993.
19. Beekman, M., Sumpter, D.J.T., Ratnieks, F.L.W. Phase transition between disordered and ordered foraging in Pharoah's ants. Proc. Natl. Acad. Sci. USA 98:9703-9706, 2001.
20. Bonabeau, E. Comment on Phase transitions in instigated collective decision making. Adapt. Behav. 5:99-105, 1997.
21. Deneubourg, J.L., Aron, S., Goss, S., Pasteels, J.M. The self-organizing exploratory pattern of the Argentine ant. J Insect Behav 3:159-168, 1990.
22. Goss, S., Deneubourg, J.L. The self-organizing clock pattern of Messor pergandei (Formicidae, Myrmicinae). Insectes Soc 36:339-346, 1989.
23. Nicolis, S.C., Deneubourg, J.L. Emerging patterns and food recruitment in ants: an analytical study. J. Theor. Biol. 198: 575-592, 1999.
24. Biesmeijer, J., de Vries, H. Exploration and exploitation of food sources by social insect colonies: a revision of the scout recruit concept. Behav. Ecol. Sociobiol 49:89-99, 2001.

25. Traniello, J.F.A. Recruitment behavior, orientation, and the organization of foraging in the carpenter ant Camponotus pennsylvanicus DeGeer (Hymenoptera: Formicidae). Behav Ecol Sociobiol 2:61-79, 1997.
26. Holdobler, B., Stanton, R.C., Markl, H. Recruitment and food retrieving behavior in Novomessor (Formicidae, Hymenoptera), I. Chemical signals. Behav. Ecol. Sociobiol 4:163-181, 1978.
27. Reuter, M., Keller, L. Sex ratio conflict and worker production. Am. Nat. 158:166-177, 2001.
28. Hall, R.W. Driving the Productivity Machine: Production Planning and Control in Japan. American Production and Inventory Control Society, Falls Church, Vancouver, 1981.
29. Deb, K. Multiobjective Evolutionary Algorithms: Introducing Bias among Pareto-Optimal Solutions. In A.Ghosh and S. Tsutsui(Eds.), Theory and Applications of Evolutionary Computation: Recent Trends, Spinger -Verlag, London, 2002.
30. Blum, C. ACO applied to Group Shop Scheduling: A case study on Intensification and Diversification. In Proceedings of the 3rd International Workshop on Ant Algorithms (ANTS 2002), 2002. Also available as technical report TR/IRIDIA/2002-08, IRIDIA, Universite Libre de Bruxelles.
31. Teodorvic, D., Dell'orco, M. Bee Colony optimization- A Cooperative Learning approach to Complex Transporation Problem, Advanced OR and AI Methods in Transportation, 2005, pp 51-60.
32. Dorigo, M., Maniezzo, Vittorio, Colorni, Alberto, Ant system: optimization by a colony of cooperating agents. IEEE Transactions on Systems, Man, and Cybernetics,Part B: Cybernetics, 1996. 26(1): p. 29-41.
33. Nowicki, E. and Smutnicki, C., A fast taboo search algorithm for the job shop problem, Management Science, Vol. 42, No. 6 (1996), pp. 797-813.
34. Aytug, H., Lawley, M. A., McKay, K., Mohan, S., Uzsoy, R. M., 2005. Executing production schedules in the face of uncertainties: A review and some future directions. European Journal of Operational Research 161, pp. 86-110.
35. Lee, H. L., Padmanabhan, V., Whang, S., 1997. The bullwhip effect in supply chains. Sloan Management Review 38, pp.93-102.
36. Blackhurst, J., Wu, T., O'Grady, P. Network-based approach to modeling uncertainty in a supply chain. International Journal of Production Research 42 (8), pp.1639-1658, 2004.
37. Fu, M. C. Simulation optimization. In: Peters, B. A., Smith, J. S., Medeiros, D. J., Rohrer, M. W. (Eds.), Proceedings of the 2001 Winter Simulation Conference.

12

Combining Simulation and Tabu Search for Oil-derivatives Pipeline Scheduling

Álvaro García-Sánchez[1], Luis Miguel Arreche[2], and Miguel Ortega-Mier[3]

[1] Universidad Politécnica de Madrid. José Gutiérrez Abascal 2, 28006. Madrid (Spain) alvaro.garcia@upm.es
[2] Universidad Politécnica de Madrid. José Gutiérrez Abascal 2, 28006. Madrid (Spain) arreche@etsii.upm.es
[3] Universidad Politécnica de Madrid. José Gutiérrez Abascal 2, 28006. Madrid (Spain) miguel.ortega.mier@upm.es

Summary. In this chapter a methodology for addressing a real-world multi-commodity pipeline scheduling problem is presented. For this problem, a Tabu Search and a Simulation model have been devised. The simulation model allows an accurate and suitable assessment of every particular schedule, whereas the Tabu Search guides the searching process and eventually succeeds in obtaining satisfactory schedules in terms of a set of relevant criteria. This methodology has been applied to a particular subsystem of the pipeline Spanish network. As a result, the proposed Tabu Search has proved to be an effective and efficient technique for solving the problem.

Key words: Multi-commodity Pipeline, Scheduling, Tabu Search, Simulation, Real World Instances.

12.1 Introduction

Among the different means of transport for delivering oil derivatives to distant destinations, pipeline transportation is the one which yields smaller variable costs along with a great degree of reliability, especially if ships are not a feasible alternative.

The main particularity of pipeline transportation arises from the way packages advance through the pipes: each package pushes the one previously pumped and so on. Since there is not any physical separation in between every two packages, some mixture occurs while they are displaced, producing an interface of contaminated product, which may not meet the specifications of either of the products that produce the contaminated interface.

Depending of the nature of the products which cause the interface, the treatment is different, ranging from pouring the interface into different tanks

Á. García-Sánchez et al.: *Combining Simulation and Tabu Search for Oil-derivatives Pipeline Scheduling*, Studies in Computational Intelligence (SCI) **128**, 301–325 (2008)
www.springerlink.com

(the cheapest case) to reprocessing the interface in a refinery (the most expensive one). Some products should by no means get mixed, and others can be allowed to get mixed to some extent.

Apart from the consideration made above, there are other elements that condition the feasibility of the schedule, such as the following ones: products should be available in the right amount at the right time to meet the customers' demand; the storage capacity is limited; refineries cannot produce over a maximum rate; pumping stations can work at a flow rate somewhere between the minimum and maximum admissible values for the pipeline configuration; besides, pipeline usually work combined with other means of transport, whose effects over the inventory level and the demand might have be considered; additionally, the time devoted to maintenance tasks might be relevant, in which case it ought to be taken into account. Finally, sometimes a long-chain polymer is added to obtain a higher flow rate. This polymer, called *flow rate enhancer*, is only needed in tiny quantities, but nevertheless it may represent a significant cost. As the flow rate enhancer advances through the pipes, it degrades into short-chain polymers not affecting at all the specifications of the different derivatives.

CLH (*Compañía Logística de Hidrocarburos)* is the most important logistic operator in Spain, with a network covering most of the Iberian Peninsula and comprising more than 3400 km of pipes. Traditionally, to address the pipeline scheduling problem, some schedulers have had to use their expertise to obtain a schedule without specific tools for this purpose. The objective of this work is to provide schedulers with useful tools to assist in their task. Schedulers from this company have to obtain a schedule on a monthly basis. For that, they are given all the necessary information five days ahead of the beginning of the scheduling horizon. With the aid of support-decision tools such as this presented here, schedulers can achieve better schedules and in a shorter period of time.

As will be described later, schedules are assessed with regard to a set of criteria. The company managers express their preferences in terms of that set instead of using a single measure of the schedule characterizing its quality. That is the reason why the problem addressed in this chapter consists in obtaining a so-called *satisfactory schedule*, which is a schedule with all its criteria being equal or better than certain satisfactory values for those criteria.

This chapter is organized as follows. Section 12.2 offers a literature review of the pipeline scheduling problem. Following, in Section 12.3, the problem under study is stated as well as the approach adopted to address it. In Section 12.4, the simulation models to represent pipelines are presented. Section 12.5 contains a brief description of the proposed Tabu Search. In Section 12.6, some computational results are offered and some final conclusions and further work are presented in Section 12.7.

12.2 Literature review

With regard to the multi-commodity pipeline scheduling problem, although the literature is relatively scarce [1–3], there are different approaches. In every work simplifications have been introduced and sometimes the resulting problems are quite unrealistic, harnessing their credibility for real-world application. Both exact and heuristics methods have been devised for the pipeline scheduling problem.

One of the earliest formalizations of the problem corresponds to Hane and Ratliff [2], whose objective is to obtain a schedule with the minimum value of an index which is related the pumping costs. As the authors themselves reckon it is not suitable for real-world problems. Milidú et al. [3] study the complexity of the problem and propose an algorithmic approach to obtain a feasible schedule which is optimum in terms of the pumping costs. Rejowski and Pinto [4] offer a Mixed Integer Linear Programming (MILP) model to solve the scheduling problem for a linear pipeline, where the objective function is the global cost (pumping, storage and interface costs). The work in [5] represents a step further, since the new MILP includes some additional constraints which improve the efficiency of the model. In [6] and [7] a continuous time MILP solves the problem presented in [4] in quite a shorter time, being a much simpler and more accurate model.

Other approaches based either on heuristics or on meta-heuristics have been developed. Campognara and De Souza [8] present a graph theory based approach to represent the problem and offer a MILP to solve the it, but since it is not efficient enough, a heuristic approach is presented that allows the schedulers to obtain a daily schedule in time to implement it. Sasikumar et al. [9] utilize a beam search heuristic to obtain a good enough schedule for a tree-topology pipeline. Ali et al. [10] have developed a model to deal simultaneously with the problem of scheduling and that of setting an appropriate pumping configuration; for that purpose they propose a two-stage methodology where stages feed-back one another. Finally, in [11] a genetic algorithm guided by a multiobjective fitness function can be found. This work is broadened in [12] where a MILP is also formulated and, finally, a hybrid technique is proposed, which yields relatively small times by feeding the genetic algorithm with schedules obtained with the MILP.

A common characteristic of most of the previous work is that the time horizon is much smaller than a month, which is a typical time frame for a schedule. This is precisely the motivation of the present work, since the developed tool is capable of addressing the task of finding a satisfactory schedule for time horizons of a month for real-world problems.

12.3 Problem Statement and Modelling Approach

12.3.1 Pipeline system characteristics

The problem addressed in this paper refers to systems that consist of a series of elements connected through pipelines of different length and radius. These elements can be of any of the following types:

- A refinery, which is the location where different derivatives are produced from oil and from which derivatives are pumped.
- A terminal, which is a set of tanks where derivatives are stored to meet the demand. Tanks are fed by a pipe which connects the terminal with some upstream nodes. There might exist as well a pipe connecting the terminal to other nodes downstream.
- A branching node, which is a location at the end of a pipe from which two o more other pipes are fed.
- A branching terminal, which is a terminal which feeds two or more pipes downstream.

As to the characteristic of the systems that can be addressed with the developed tool, the previous elements can be part of pipelines meeting the following requirements:

- Pipes only have one flow sense, not being allowed reversible pipes.
- Branchings, but not cycles, are admissible.
- There can be different sources, from which products can be pumped to one or more pipes.
- There is no limitation regarding the number of nodes in the pipeline.
- The flow rate at which products are pumped from sources into the pipeline can either be constant or depend on the contents and splitting of the packages present in branchings.

A great number of pipelines -mostly, tree topology ones- can be represented in the terms described above and, indeed, other configuration can be modelled through a set of tree topology pipelines working together [13]. One last remark is due concerning the pumping rates. Although actual pipelines have different pumping stations along their pipes to keep fluids with the appropriate pressure, for modelling purposes, a pumping station s will be a set of logical elements with which the flow is controlled downstream a particular source. Each station is linked to a particular source and feeds a set of nodes downstream that source. For example, a refinery may feed two totally different branches. Each branch will be run by a different pumping station. The total number of stations is S.

12.3.2 Problem Definition and Schedule Representation

The problem is defined when the following information is established:

- The number of nodes, N; the number of products, P.
- Relation between nodes. Each node is characterized by an identifier $1, 2, ..., N$, its type (refinery, 0; terminal, 1; or branching, 2) the number and the identifiers of its immediate downstream successors. For example, node 1 may be a terminal (node type = 1), with two nodes immediately located downstream.
- Initial contents. In order to properly represent the system state at the beginning of the horizon, the different products in each pipeline, their relative position within the pipe and their splitting downstream must be set.
- Scheduling horizon, H, which is an integer value of weeks (typically, four or five in order to analyze a month).
- Storage capacity and initial level of inventory. For every tank in the system (defined by the node, n and the product, p, within that node), its capacity, $CAP(n,p)$, and initial level of inventory, $IS(n,p)$, have to be set.
- Demands. Demands for every tank in the system for the scheduling horizon, including the withdrawal pattern during the week. By letting $D(n,p)$ be the demand of product p in node n for the complete horizon, the demand of product p in node n for the k-th hour of the j-th day of the i-th week, referred to as $D_{ijk}(n,p)$, is calculated as a product of the proportion withdrawn during the week, $\varphi_i^{week}(n,p)$, the proportion withdrawn during the j-th day within that week, $\varphi_j^{day}(n,p)$, and the proportion withdrawn during the k-th hour within that day, $\varphi_k^{hour}(n,p)$. Therefore:

$$D_{ijk}(n,p) = \varphi_i^{week}(n,p) \cdot \varphi_j^{day}(n,p) \cdot \varphi_k^{hour}(n,p). \qquad (12.1)$$

- Delivery plan. In the long run, the demand for each product-node pair has to be equal to the deliveries of that pair. For the sake of simplicity the delivery plan is built so that every tank receives an amount of product that is equal to the its total demand during the scheduling horizon.
- Available flow rates. The flow rate of every pipe can depend on the level of flow rate enhancer and on the contents of the pipeline, this dependence should be formulated to be included in the simulator. For each particular instance, this relationship is to be determined, as it is described for the case study presented in Section 12.6.

Each schedule consists of a series of K packages of different products, where the k-th package of pumping station s, referred to as $PCK(k,s)$, is defined by:

- The type of product, $P_{pck}(k,s)$; each product being referred to as an integer value between 1 and P;
- The volume, $V_{pck}(k,s)$ (any integer number);
- The level of flow rate enhancer injection, $E_{pck}(k,s)$, which can be: 1(high level), 2 (low level) and 3 (no flow rate enhancer added);

- The splitting downstream along the different terminals, given by $fr_n(k,s)$, $n = 1, ..., N$, where $fr_n(k,s)$ is the portion of the k-th package which is sent to node n. If node n is not a terminal $fr_n(k,s) = 0$. The volume of package k sent to node n from pumping station s is, therefore, $V_{pck}(k,s)fr_n(k,s)$.

12.3.3 Objective

The objective is to obtain a set of satisfactory schedules (see later) for the set of pumping stations.

To assess how good a schedule is, six criteria are defined, $F_1, F_2, ..., F_6$:

- Shortages (F_1): the amount of products which could not be withdrawn when customers required them because the incumbent tank was empty.
- Forbidden interfaces (F_2): the times two products that are not allowed to get in contact actually do.
- Interface stoppages (F_3): the time during which some interface is stuck in a pipe. When this occurs, interfaces grow too fast and, thus, interface stoppages are not desirable, but sometimes are not avoidable.
- Blockages (F_4): the time during which a package cannot be pumped into a tank at the desired flow because the tank is full.
- Interfaces cost (F_5): the cost associated with the sequence of products and their mixing.
- Non-delivered volume (F_6): the amount of the planned volume which has not been delivered within the time horizon.

A schedule is satisfactory if for the previous criteria, its values are not larger than a set of target-values, $F_1^{obj}, F_2^{obj}, ..., F_6^{obj}$. Notice that these target values can be equal to zero. For example, the number of forbidden interfaces is typically zero ($F_2^{obj} = 0$).

12.3.4 Modelling Approach

As commented above, the approaches developed so far lack some realism in terms either of the assumptions made or the horizon length for which a schedule is obtained or the problem size (too short for real-world application). Therefore, for this purpose a simulation optimization approach was adopted. By means of using a simulation model it is possible to evaluate the quality of a schedule including some complex elements difficult to model when using exact methods. A set of predefined modules were developed in Witness [14] for easily building and modifying pipeline simulation models. This simulation models offer a quick assessment of any particular schedule. Besides, a VB application was also developed, first, for helping configure and run the simulation and, second, for implementing the Tabu Search (TS). This TS generates alternative schedules and guides the searching process in order to eventually obtain satisfactory schedules.

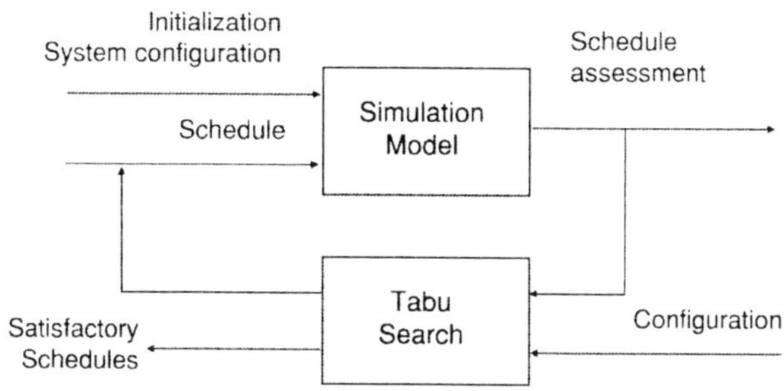

Fig. 12.1: Modelling Approach.

The scheme in Fig. 12.1 illustrates the modelling approach for this problem. First, the simulation model can be used as a stand-alone tool to evaluate the quality of a schedule. Actually, schedulers can introduce different schedule s, modify them and evaluate their quality. For that purpose, the model is to be configured and initialized: the model should properly represent the system under study and the state of the system at the beginning of the horizon. Once the model has been run, an evaluation of the schedule is obtained. In the following sections an outline of the simulation models and the Tabu Search are given. When combined with the Tabu Search, which needs to be configured and tuned, the information from the simulation model is properly exploited in order to generate new schedules. By operating iteratively, eventually, satisfactory schedules can be obtained.

12.4 The Pipeline Simulator

There are several commercial pipeline simulators (such as Stoner pipeline Simulator or pipeline Studio, from Energy Solutions) that emulate with great fidelity the behavior of the different elements in the system in fluid-dynamic terms. In particular, these simulators help the system operators assess the conditions anywhere (pressure, viscosity, etc.) However, these tools are not useful for the purpose of scheduling the pipeline. Since, in order to run the simulator, it has to be operated in the same way the real pipeline is operated, the simulated time to real time ratio is notably high (roughly, 1 to 10), which means that they are useless for assessing a great number of schedules in time to choose one. Besides, it takes a great amount of time, work and money to set one of these simulators up.

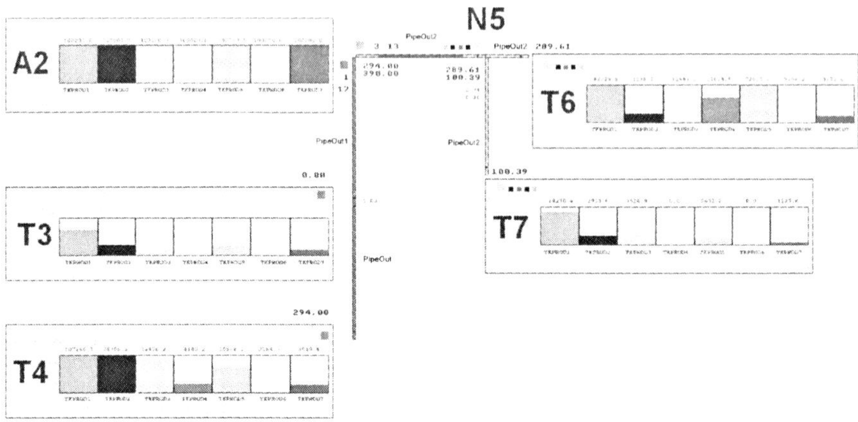

Fig. 12.2: Witness overall view.

For the specific purpose of scheduling, a simulator has been developed with the appropriate accuracy for this task. This simulator has been developed taking advantage of Witness, a commercial simulation package. In Fig. 12.2 it can be seen a screen-shot of Witness for a particular simulation model. Though any general purpose language would have allowed the creation of the models, Witness offers a series of predefined elements that highly facilitate the model building process. Additionally, to make the process even easier and faster, some Witness modules have been created which properly combined can represent some models. These models can be suitable for the system under study or can be a starting point to create an even more complex model or to add some particularities of a specific system.

There are two basic types of modules: those that represent elements of the system (refineries, storage facilities, terminals and branchings) and those designed to control the model (general, statistics, calendar and set-up).

After running a schedule, the simulator offers the following information:

- Non-met demand for every tank, which is the amount of each product that was not available when requested at the different destination points.
- Non-delivered volume, which is the amount of product which was not delivered during the time horizon. This value is not necessarily equal to the non-met demand. Actually, if the initial inventory level at a tank is

high enough it might meet all the demand without receiving the scheduled volume. Similarly, all the amount of a particular product might be sent during the time horizon but not in a timely manner, so that sometimes, customers cannot withdraw the demanded product.

- The global numbers of shortages, blockages and stopped interfaces.
- In case of shortage for a particular product in a particular terminal, it is recorded the first critical package, being the package that is traversing that terminal when the shortage occurs.
- In case of blockage, due to a package being pumped to a full tank, the package that originates the blockage is recorded. Besides, the model offers the amount of product that could actually be pumped into the tank without blocking the pipeline.
- When blockages occurs it is not irrelevant how much time this has happened. Therefore, the amount of time that each tank has been blocked is reported.
- Finally, the last package pumped for each pumping station is given, which is an indicator of how far the schedule is from being feasible with regard to the non-delivered volume.
- Besides, an estimate of the cost associated to the contaminated product is given according to an average estimation of the volume of each kind of interface. For the given sequence of products, an empirical formula provided by the CLH managers was used, specific to the pipelines under study.

Finally, to build up a model, a start-up file has been created to which modules must be added. To facilitate the configuration of the model, a Visual Basic application has been created. Fig. 12.3 illustrates the interrelations of the applications. The VB application offers several forms which properly filled out depict a particular system for a specific time horizon and automatically configures the elements in the Witness model. This Visual Basic application reads data both from the simulator itself and from the text files the latter generates after assessing a schedule. Besides, the Tabu Search (presented in Section 12.5) is embedded in the VB application. It is noteworthy that the simulator allows the introduction of several values for the flow rate for every source depending on the contents of the pipeline and the amount of flow rate enhancer in the packages sent.

12.5 Tabu Search

In this section, the main characteristics of the Tabu Search are described. First, a general outline of its logic is given, and later, in different subsections its most important elements are presented.

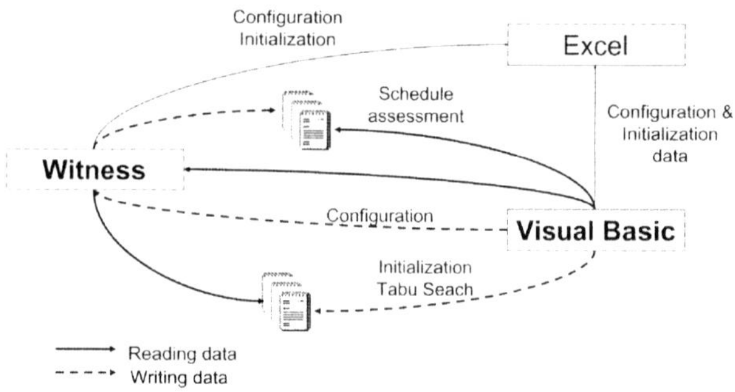

Fig. 12.3: Applications interconnection.

12.5.1 General Outline

The concept of stage is defined and used in the Tabu Search method. A total of nine stages have been identified after experimenting with different sets of stages. Although there may still be some room for improvement, theses nine stages have proved to be effective. Each stage is defined by two characteristics:

1. The type of **movements** to generate the neighbours from a given schedule and
2. What is referred to as its **main criterion**. Every stage attempts to improve a particular criterion, which is called the main criterion for that stage, without allowing the rest of them worsen more than a predefined extent (secondary criteria), therefore eligible schedules must be "within limits".

As described in Algorithm 12.1, the procedure begins with an initial schedule $\mathfrak{s}^*$, whose criteria are $F_{\mathfrak{s}^*}(c)$, $c = 1, ..., 6$. If the schedule is satisfactory, the process ends. If not, the first stage is selected to be run. The stage will stop if a schedule with a satisfactory value of the main criterion is found: $F_{\mathfrak{s}^*}(c^*) \leq F^{obj}(c^*)$. Once the stage stops, this same logic applies again until some terminating condition is met.

The stages operate similarly: from a given schedule, a neighbourhood ($\mathfrak{N}$) is generated (using the particular movements of a stage), the best not-tabu and within-limits neighbour is chosen. Sometimes, the neighbourhood may be too large to be explored. That is the reason why a parameter, Car^{max} is defined so that if the neighbourhood is larger that L only a group of L neighbours are evaluated. The neighbors to be explored are selected randomly among the original neighborhood, so that every schedule has the same probability to

Algorithm 12.1: General Logic.

```
Initializing routine
Generate 𝔰*
while Global Terminating Criterion not met do
  Select Stage (being c* its main criterion)
  while Local Terminating Criterion does not apply do
    if F_𝔰*(c*) ≥ F^obj(c*) then
      Execute stage's routines (see algorithm 12.2)
    else
      Exit while
    end if
  end while
end while
```

qualify for evaluation. Other times, there are not any eligible neighbors, in which case a diversification list is kept. This diversification is updated every time a neighborhood is explored, by randomly adding any eligible neighbor different from the chosen one. When the diversification list is needed, it is evaluated as any regularly obtained neighborhood.

The pseudo-code in Algorithm 12.2 illustrates how stages work in some more detail. In that pseudo-code:

$\mathfrak{s}$ is the current schedule

$\mathfrak{N}$ stands for the neighborhood obtained from the current schedule

$\mathfrak{T}$ stands for the global tabu list

$\mathfrak{t}$ stands for the local tabu list

$\mathfrak{D}$ stands for the diversification list

$Card^{max}$ is the maximum number of neighbors to be examined during each iteration within a stage

Both the tabu lists and the diversification lists will be presented later.

12.5.2 Types of Movements

In order to obtain new schedules from a given one, four different types of movements are considered. These movements always operate with packages within the same pumping station. Following a formal description of them is given whereas Fig. 12.4 graphically illustrates them.

1. **Package insertion.** This movement consists in inserting a package in a new position. If for station s, $PCK(i, s)$ is inserted into the j-th position, the resulting sequence for station s is the following (the apostrophes refer to the new packages, after the movement has been executed).
   If $i > j$

Algorithm 12.2: Stage logic. Main criterion: F_c.

BEGIN Stage
$s = 0$
if $F_c \geq F_c^{obj}$ **then**
  $s = s + 1$
  if $s \leq S$ **then**
    while Terminating condition does not apply **do**
      Generate $\mathfrak{N}$
      if $\mathfrak{N} \neq \emptyset$ **then**
        if $Card(\mathfrak{N}) > Card^{max}$ **then**
          Reduce $\mathfrak{N}$
        end if
        Evaluate $\mathfrak{N}$ with the simulator
        if $\exists \mathfrak{n} \in \mathfrak{N}/\mathfrak{n}$ is not tabu and $\mathfrak{n}$ is within limits **then**
          $\mathfrak{s}^* = \{\mathfrak{n}^* \in \mathfrak{N}/F_c(\mathfrak{n}^*) \leq F_c(\mathfrak{n})\forall n \in \mathfrak{N}\}$
          Update $\mathfrak{T}$, $\mathfrak{t}$ and $\mathfrak{D}$
        else
          if $\mathfrak{D} \neq \emptyset$ **then**
            Evaluate $\mathfrak{D}$ with the simulator
            if $\exists \mathfrak{n} \in \mathfrak{D}/\mathfrak{n}$ is not tabu and $\mathfrak{n}$ is within limits **then**
              $\mathfrak{s}^* = \{\mathfrak{n}^* \in \mathfrak{D}/F_c(\mathfrak{n}^*) \leq F_c(\mathfrak{n})\forall n \in \mathfrak{N}\}$
              Update $\mathfrak{T}$ and $\mathfrak{D}$
            else
              exit for
            end if
          else
            exit for
          end if
        end if
      else
        if $\mathfrak{D} \neq \emptyset$ **then**
          Evaluate $\mathfrak{D}$ with the simulator
          if $\exists \mathfrak{n} \in \mathfrak{D}/\mathfrak{n}$ is not tabu and $\mathfrak{n}$ is within limits **then**
            $\mathfrak{s}^* = \{\mathfrak{n}^* \in \mathfrak{D}/F_c(\mathfrak{n}^*) \leq F_c(\mathfrak{n})\forall n \in \mathfrak{N}\}$
            Update $\mathfrak{T}$ and $\mathfrak{D}$
          else
            exit for
          end if
        else
          exit for
        end if
      end if
    end while
  end if
end if
END stage

Table 12.1: Characterization of stages.

Stage	Main Criterion	movement type
1	F_1	Package insertion
2	F_1	Volume insertion
3	F_2	Package insertion
4	F_2	Volume insertion
5	F_3	Package insertion
6	F_3	Volume insertion
7	F_4	Package insertion
8	F_4	Volume insertion
9	F_5 and F_6	All movements

$$PCK'(m,s) = PCK'(m-1,s) \text{ if } j < m \leq i \tag{12.2}$$

$$PCK'(m,s) = PCK'(m-1,s) \text{ if } m > i \text{ or } m < j \tag{12.3}$$

If $i < j$

$$PCK'(m,s) = PCK'(m-1,s) \text{ if } j < m \leq i \tag{12.4}$$

$$PCK'(m,s) = PCK'(m-1,s) \text{ if } m > j \text{ or } m < i \tag{12.5}$$

For example, when a shortage in a particular tank of product p occurs, moving forward in the sequence a package containing this product (and being part of it directed to the incumbent tank) may reduce or even avoid that shortage.

2. **Volume insertion.** This consists in moving part of the volume (ΔV) directed to node n from a package, $PCK(s,i)$, to a different package (of the same product), $PCK(s,j)$. The resulting packages remain as follows. As to the volume of each package:

$$V_{pck}(i,s) = V_{pck}(i,s) - \Delta V \tag{12.6}$$

$$V_{pck}(j,s) = V_{pck}(j,s) + \Delta V \tag{12.7}$$

And as to the splitting of the packages:
For $PCK(j,s)$

$$fr'_n(i,s) = \frac{fr_n(i,s)V_{pck}(i,s)}{V'_{pck}(i,s)} \text{ if } n \neq n^* \tag{12.8}$$

$$fr'_{n^*}(i,s) = \frac{fr_{n^*}(i,s)V_{pck}(i,s) - \Delta V}{V'_{pck}(i,s)} \tag{12.9}$$

For $PCK(i,s)$

$$fr'_n(j,s) = \frac{fr_n(j,s)V_{pck}(j,s)}{V'_{pck}(j,s)} \text{ if } n \neq n^* \quad (12.10)$$

$$fr'_{n^*}(j,s) = \frac{fr_{n^*}(j,s)V_{pck}(i,s) + \Delta V}{V'_{pck}(j,s)} \quad (12.11)$$

Again, following the previous example, moving part of a package to another one may help reduce the shortage in a tank.

3. **Package disaggregation.** Disaggregating package $PCK(i,s)$ into two different ones, $PCK'(i,s)$ and $PCK'(i+1,s)$ consists in splitting a proportion (pr) of it, so that:

$$V'_{pck}(i,s) = V_{pck}(i,s)pr \quad (12.12)$$

$$V'_{pck}(i+1,s) = V_{pck}(i,s)(1-pr) \quad (12.13)$$

$$fr'_n(i,s) = fr'_n(i+1,s) = fr_n(i,s) \quad (12.14)$$

$$(12.15)$$

As for the rest of the packages:

$$PCK'(j,s) = PCK(j,s) \text{ if } j < i \quad (12.16)$$

$$PCK'(j,s) = PCK(j-1,s) \text{ if } j > i \quad (12.17)$$

Sometimes the previous movements may not be either possible or effective. Therefore, another movement which can be useful is disaggregating a package into two different ones, yielding two identical packages only differing in their volumes.

4. **Package aggregation.** Aggregating packages $PCK(i,s)$ and $PCK(i+1,s)$ means obtaining a new package $PCK'(i,s)$, so that:

$$V'_{pck}(i,s) = V_{pck}(i,s) + V_{pck}(i+1,s) \quad (12.18)$$

$$fr'_n(i,s) = \frac{fr_n(i,s)V_{pck}(i,s) + fr_n(i+1,s)V_{pck}(i+1,s)}{V'_{pck}(i,s)} \quad (12.19)$$

As for the rest of the packages

$$PCK'(j,s) = PCK(j,s) \text{ if } j < i \quad (12.20)$$

$$PCK'(j,s) = PCK(j+1,s) \text{ if } j > i \quad (12.21)$$

After completing every stage, all contiguous packages of the same product are aggregated. The rationale of this movement is that a set of small packages of the same product may not allow obtain any neighbour but the aggregation of them all may.

Each stage operates by using a single type of movement or serveral movements as listed in Table 12.1, so that selecting a stage determines the type of movements to be used.

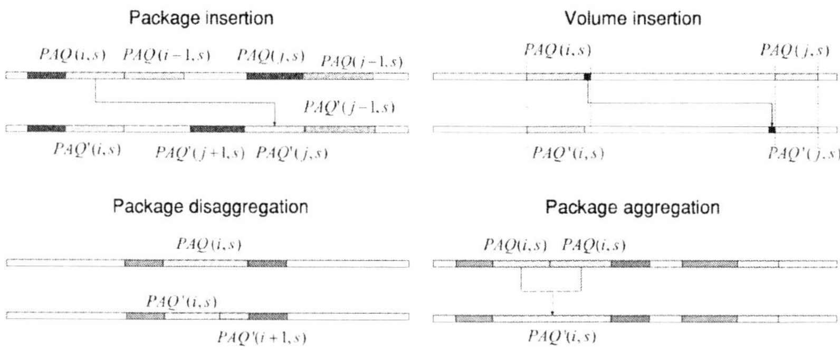

Fig. 12.4: Types of movements.

12.5.3 Stage selection

After experimenting with several rules to choose the following stage, a good performance was achieved by means of choosing a stage whose main criterion is that one which has experienced the largest relative increase over the last stages.

Since some stages share their main criterion, the stage whose movements entail a deeper structure change in the schedule is run first, followed (if necessary) by the stage with that same criterion and which introduces minor changes in the schedule structure.

Therefore, if stage 1 ends without obtaining a schedule with a satisfactory value for its main criterion (shortages, F_1), automatically, stage 2 is run. Stages 3, 5 and 7 operate likewise, according to information in Table 12.2

Table 12.2: Stage sequence.

Ending Stage	Following Stage
1	2
3	4
5	6
7	8

When either any stage entailing minor changes to the schedule (2, 4, 6 and 8) or stage 9 finish, the following stage is selected by means of choosing, first, the new main criterion and next the stage itself. For every criterion c, it is defined an index $I(c)$ (Eq. 12.22).

$$I(c) = \begin{cases} \dfrac{F_t(c) - F^{obj}(c)}{\max\limits_{x \geq t-H} \{F_x(c)\} - F^{obj}(c)} & \text{if } F_t(c) > F^{obj}(c) \\ 0 & \text{if } F_t(c) \leq F^{obj}(c) \end{cases} \quad (12.22)$$

where:

H is a parameter, so that the greater it is, the more influcence the search history has to select the following stage

t is the number of stages run so far.

$F_x(c)$ is the value of criterion c of the schedule obtained after completing the x-th stage.

$F_t(c)$ is, therefore, the value of criterion c of the schedule obtained after completing the last stage run so far. After the main criterion has been

Table 12.3: Stage sequence (II).

Main criterion	Following Stage
1, Shortages	1
2, Forbidden Interfaces	3
3, Blockages	9
4, Interface Stoppages	5
5, Interfaces Costs	7
6, Non-delivered Volume	9

selected, the next stage to be run is selected according to the correspondence in Table 12.3. For example, if the following main criterion is F_1 (shortages), the next stage will be stage 1. Likewise, if F_5 (the interface cost) is selected as the next main criterion, stage 5 would be chosen. As will be seen later (Section 12.5), this method has proved to be effective, since it does not allow criteria worsen so much that it is not possible to obtain schedules with values for the objective function close to the target values and, eventually, it obtains satisfactory schedules.

12.5.4 Tabu Lists and Aspiration Criterion

Two different Tabu lists condition and guide the selecting process of the available neighbours, one at a local level and another one at a global level.

- **Local tabu lists,** t. There are as many local tabu lists as stages, where the information regarding the packages involved in each movement are

stored, so that these packages are now allowed to be found again during some amount of iterations.
Every time a new schedule is selected, the tabu list stores the information corresponding to the package which has been modified, inserted, etc. so that for a certain number of iterations schedules with an identical package in that position does not qualify as eligible.
When a schedule is tagged as tabu, if it offers a value for criterion c greater or equal than the best obtained so far, the aspiration criterion applies and the schedule is no longer tabu

- **Global Tabu List, $\mathfrak{T}$.** Besides, a global tabu list is created containing whole schedules for a pumping station, so that no neighbour is selected if it is contained in the global tabu list. There is no aspiration criterion for this list.

12.5.5 Diversification list

Tipically, from a given schedule it is possible to obtain several neighbours. Every iteration, after a new schedule has been chosen from the neighbourhood, another schedule is randomly selected among the rest of neighbours and it is stored in the diversification list. If for a particular iteration, there are not any neighbours or all of them are tabu or not within limits, the elements contained in the diversification list are studied as if they were the neighbourhood itself.

12.5.6 Terminating criterion

Two different sets of termination criteria need to be set: at global level and locally for every stage.

- Globally, the searching procedure ends when any of the following conditions apply:
  - a certain number of satisfactory schedules have been found,
  - the total computational time has exceeded a determined value,
  - the number of stages that have been run is greater than a specified value.
- For every stage, the process stops if:
  - a maximum time has elapsed,
  - a certain number of iterations have been accomplished,
  - a different number of iterations have been done without attaining any improvement, or
  - there are not any eligible neighbours for any station -not even in the diversification list.

12.5.7 Worsening Limits

Every time a neighbourhood $\mathfrak{N}$ is explored, being c^* the main criterion of the stage and $F^0(c)$ the value of criterion c for the initial schedule of the incumbent stage, a neighbour $\mathfrak{n} \in \mathfrak{N}$ is eligible as the new schedule for the following iteration if:

$$F_{\mathfrak{n}}(c) \leq \{F^0(c)(1 - \alpha(c)), F^0(c) + \beta(c)\} \; \forall c \neq c^*, \qquad (12.23)$$

where $\alpha(c)$ and $\beta(c)$ are parameters that determine the allowance for the secondary criteria to worsen during that particular stage. The rationale behind is to let the searching process wander and explore those areas with schedules that are not attractive *a priori*, and avoid the process to get stuck in a particular region.

12.6 Case Study. Computational results

The previous method has been applied to one CLH subsystem, for which a simulation model was created. Following, the main results are shown.

The system consists of six nodes, including a refinery which feeds two branches, connecting a branching node and four terminals. The time horizon is four weeks, which is a very typical time span for the scheduling problem. In Fig. 12.5, a sketch of the system is available. The storage capacity for

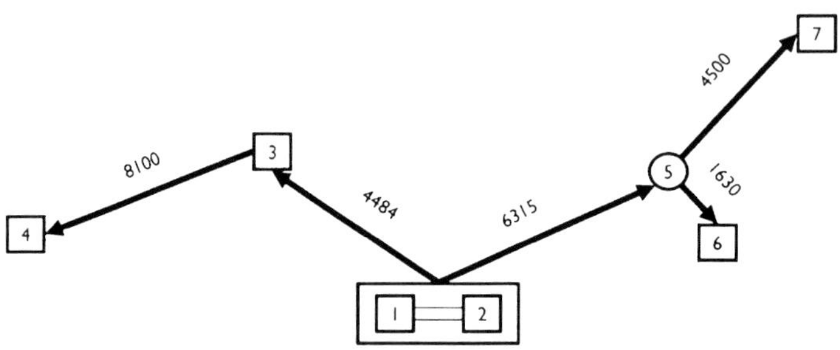

Fig. 12.5: System configuration.

every tank in this system is given in Table 12.5, where the figures indicate the capacity of the corresponding pipe. The initial contents of the tanks for the horizon are given in Table 12.5.

The pipeline contents are given in Table 12.6. The demand is characterized

Table 12.4: Storage Capacity (m^3).

	Prod.1	Prod.2	Prod.3	Prod.4	Prod.5	Prod.6	Prod.7
Node 3	20055	6784	12597	0	9127	0	4488
Node 4	229872	34619	34914	35162	18622	8555	4280
Node 6	148192	16393	18804	51295	85743	18178	11937
Node 7	15376	9245	10179	0	11451	0	4852

Table 12.5: Initial contents (m^3).

	Prod.1	Prod.2	Prod.3	Prod.4	Prod.5	Prod.6	Prod.7
Node 3	14900	1346	3660	0	4591	0	3069
Node 4	106228	27300	14779	6400	11477	2569	3549
Node 6	90815	5393	12014	11019	62916	10075	3685
Node 7	6619	4310	7781	0	3435	0	3353

Table 12.6: Initial pipeline contents.

Pipe	Product	Volume (m^3)	fr_3	fr_4	fr_6	fr_7
N1/N2-N3	1	4484	0.50	0.50	0.00	0.00
N1/N2-N5	3	895	0.00	0.00	1.00	0.00
N3-N4	1	8100	0.00	1.00	0.00	0.00
N5-N6	3	1630	0.00	0.00	1.00	0.00
N5-N7	1	3381	0.00	0.00	0.00	1.00

by the values or φ^{week},φ^{day} and φ^{hour}. The weekly demand is balanced over the horizon, so that $\varphi_i^{week} = 0.25\ \forall i$. As to the daily demand, on Sundays, no product is withdrawn from tanks, and $\varphi_j^{day} = 1/6$ for the rest of the days. The hourly pattern demand is given in Table 12.7. Finally, the amounts of

Table 12.7: Hourly demand pattern, φ^{hour}.

φ_1^{hour}	φ_2^{hour}	φ_3^{hour}	φ_4^{hour}	φ_5^{hour}	φ_6^{hour}	φ_7^{hour}	φ_8^{hour}	φ_9^{hour}	φ_{10}^{hour}	φ_{11}^{hour}	φ_{12}^{hour}
0.00	0.50	0.50	0.50	2.00	10.00	15.00	12.00	14.00	10.00	8.00	8.00

φ_{13}^{hour}	φ_{14}^{hour}	φ_{15}^{hour}	φ_{16}^{hour}	φ_{17}^{hour}	φ_{18}^{hour}	φ_{19}^{hour}	φ_{20}^{hour}	φ_{21}^{hour}	φ_{22}^{hour}	φ_{23}^{hour}	φ_{24}^{hour}
5.00	4.00	2.00	2.00	1.00	2.00	1.00	1.00	1.00	0.50	0.00	0.00

the different products to be sent are given in Table 12.8.

Table 12.8: Total demand for the horizon (m^3).

	Prod.1	Prod.2	Prod.3	Prod.4	Prod.5	Prod.6	Prod.7
Node 3	40825	6683	14451	0	7948	0	3612
Node 4	76141	0	3703	72258	5870	3658	0
Node 6	38567	16258	3974	0	27096	3341	6141
Node 7	37935	4516	3070	0	9393	0	2709

For this system the target values set for the 4-week horizon were those in Table 12.9. Basically, shortages and forbidden interfaces are not allowed in any case. All the planned volume should be delivered within the horizon. The pipeline can be blocked up to 12 hours, interfaces can be stopped no more than 12 hours. Finally, a target value was set for the interface cost according to the managers' preferences (values both in Table 12.9 and Figs. 12.6 and 12.7 are a linear transformation of the actual values, for the sake of confidentiality).

Table 12.9: Target values, F^{obj}.

Criterion	Target value
Shortages	0
Non-delivered volume	0
Forbidden interfaces	0
Interface costs	95000
Interface stoppages	12
Blockages	12

Table 12.10: Parameter values.

Parameter	Value
Maximum number of stages	200
Total runtime (min)	300
Global Tabu List length	50
Max. iterations per stage	10
Max. stage runtime (min)	8
Local Tabu List length	5

The parameters were conveniently tuned by trying different sets of values. For example, when the neighborhood was explored intensively, too much time was spent in assessing a large number of schedules without letting the technique drift to different regions within the solution space. Alternatively, if only a reduced number of neighbors were assessed, potentially interesting schedules were not studied and the searching process quickly wandered towards different regions. Similarly, the maximum runtime for each stage and the maximum number of iterations without improving the main criterion for a particular stage were adjusted so that the overall process was efficient enough. If the runtime was to short, there was not enough time to improve the main criterion whereas if the runtime was to long, not many stages were run and therefore it was impossible to achieve a satisfactory schedule within a reasonable amount of time. Eventually, with the parameters given in Table 12.10 the technique offered a suitable performance as described later.

The Tabu Search was run 20 times, being 300 minutes long each run. As an example, in Figs. 12.6, 12.7, 12.8 and 12.9, there can be observed the evolution of two criteria for a particular run. Figs. 12.6 and 12.7 show how interface costs evolved over a particular run, and, for that same run, the evolution of the number of forbidden interfaces are given in Figs. 12.8 and 12.9.

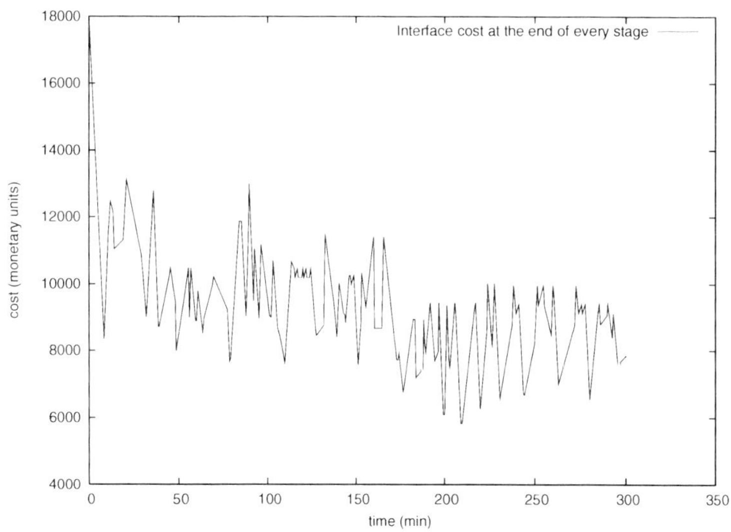

Fig. 12.6: Evolution of the interface costs over stages.

For these 20 runs a confidence interval was calculated for the time to obtain a first satisfactory schedule and another one for the number of satisfactory schedules obtained (0.05 significance level). It is noteworthy that every time a satisfactory schedule is obtained, the target values are altogether reduced by multiplying them by a factor (0.95 in this case). That is the reason why it is interesting obtaining more than a single schedule. For all runs, al least

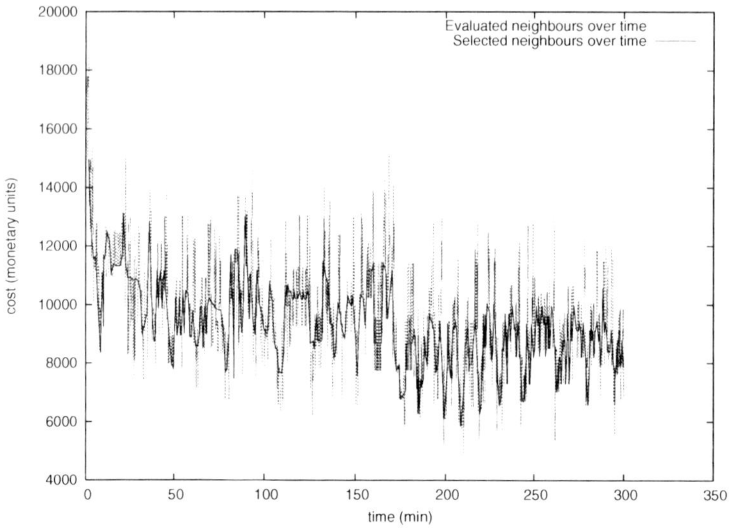

Fig. 12.7: Evolution of the interface costs over time.

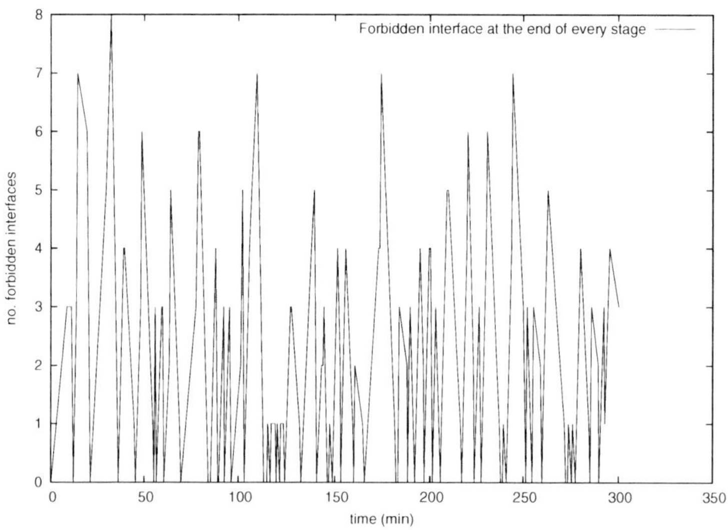

Fig. 12.8: Evolution of the number of forbidden over stages.

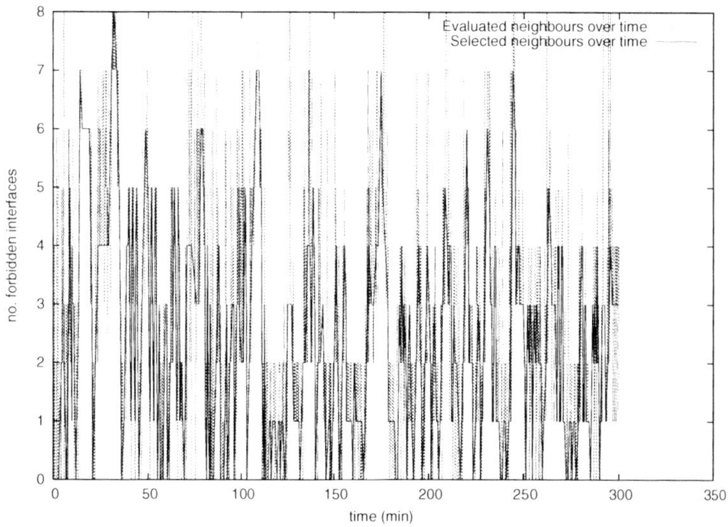

Fig. 12.9: Evolution of the number of forbidden interfaces.

Table 12.11: Performance measure.

Index	Confidence interval
Number of satisfactory schedules	(2.82, 3.48)
Time to find the first schedule	(18.94, 33.16)

two satisfactory schedules where found, which means that the technique is effective.

With regard to the efficiency, the *time to find a fist schedule* was never longer than an hour. From the time schedulers receive their data until the beginning of the horizon, they have 5 days to prepare a schedule. With this tool, it is possible to obtain a schedule with plenty of time to assess whether some features excluded from the model have not been taken into account and the schedule needs some alteration.

12.7 Conclusions and Future Work

Pipeline scheduling is a complex task which has not received much attention. For this problem a simulator has been devised. This simulator represents the pipeline with enough accuracy for assessing the quality of a schedule in terms of the scheduling problem, being an effective tool for this purpose. Besides, a Tabu Search has been implemented in order to generate and select different schedules with the objective of achieving a satisfactory schedule, where satisfactory means that the schedule offers values equal or smaller than certain target values.

As further work, it could be of interest to broaden the scope of application to other pipelines, with more complex topologies, including reversible pipes and pipe crossings. It may also be interesting trying to improve the efficiency of the technique by means of devising new stages.

In all, in this chapter, after describing some general remarks about the importance and the particularities of pipeline scheduling, a set of modules to build simulation models has been presented. Besides, a Tabu Search has been described, which in combination with the simulation models, has resulted effective and efficient in terms of obtaining satisfactory schedules.

References

1. C. Zhao-ying. Decisión support system for management of oil pipelines. *Intelligence in Economics and Management*, pages 103–106, 1986.
2. Christopher A. Hane and H. Donald Ratliff. Sequencing inputs to multi-commodity pipelines. *Annals of Operations Research*, 57(1):73–101, December 1995.
3. R. L. Milidu, A. A. Pessoa, and E. S. Laber. Pipeline transportation of petroleum with no due dates. In *Proceedings of the LATIN 2002,Theoretical Informatics: 5th Latin American Symposium*, April 2002.
4. R. Rejowski and J. M. Pinto. Scheduling of a multiproduct pipeline system. *Computers & Chemical Engineering*, 27(8-9):1229–1246, September 2003.
5. R. Rejowski and J. M. Pinto. Efficient MILP formulations and valid cuts for multiproduct pipeline scheduling. *Computers & Chemical Engineering*, 28(8): 1511–1528, July 2004.
6. D. C. Cafaro and J. A. Cerda. Continuous-time approach to multiproduct pipeline scheduling. In *Proceedings of European Symposium on Computer Aided Process Engineering*, pages 65–73, 2003.
7. Diego C. Cafaro and Jaime Cerda. Optimal scheduling of multiproduct pipeline systems using a non-discrete milp formulation. *Computers & Chemical Engineering*, 28(10):2053–2068, September 2004.
8. E. Camponogara and P. S. De Souza. A-teams for oil transportation problem through pipelines. In *Proceedings of the International Conference of Information Systems Analysis and Synthesis*, pages 718–725, 1986.
9. M. Sasikumar, P. Ravi Prakash, Shailaja M. Patil, and S. Ramani. Pipes: A heuristic search model for pipeline schedule generation. *Knowledge-Based Systems*, 10(3):169–175, October 1997.
10. S. I. Ali, K. Nakatani, and S. D Liman. A conceptual model and its implementation for petroleum products pipeline. In *Decision Sciences Institute 1998 Proceedings*, volume 3, pages 1274–1276, 1998.
11. J. M. De la Cruz, B. De Andrés-Toro, A. Herrán, E. B. Porta, and P. F. Blanco. Multiobjective optimization of the transport in oil pipelines networks. In *Proceedings of the 2003 Emerging Technologies and Factory Automation*, pages 566–573, 2003.

12. J. M. De la Cruz, J. L. R. Martín, A. H. González, and P. Fernández. Hybrid heuristic mathematical programming in oil pipelines networks. *Evolutionary Computation*, 2:1479–1486, 2004.
13. E. F. Camacho, Ridao M. A., and Ternero J. A. Power optimization of multi-fluid transportation systems. In *Proceedings of the Symposium on Information Control Problems in Manufacturing Technology*, pages 811–816, 1989.
14. Inc. Lanner Group. *Learning Witness*. 1998.

13

Particle Swarm Scheduling for Work-Flow Applications in Distributed Computing Environments

Ajith Abraham[1,2], Hongbo Liu[2,3] and Mingyan Zhao[3]

[1] Centre for Quantifiable Quality of Service in Communication Systems, Faculty of Information Technology, Mathematics and Electrical Engineering, Norwegian University of Science and Technology, NO-7491 Trondheim, Norway. ajith.abraham@ieee.org, http://www.softcomputing.net

[2] School of Computer Science and Engineering, Dalian Maritime University, 116026 Dalian, China. lhb@dlut.edu.cn

[3] School of Software, Dalian University of Technology, 116620 Dalian, China. mingy_zhao@163.com

Summary. Recently, the scheduling problem in distributed data-intensive computing environments has been an active research topic. This Chapter models the scheduling problem for work-flow applications in distributed data-intensive computing environments (FDSP) and makes an attempt to formulate the problem. Several meta-heuristics inspired from particle swarm optimization algorithm are proposed to formulate efficient schedules. The proposed variable neighborhood particle particle swarm optimization algorithm is compared with a multi-start particle swarm optimization and multi-start genetic algorithm. Experiment results illustrate the algorithm performance and its feasibility and effectiveness for scheduling work-flow applications.

Key words: Particle Swarm Optimization, Scheduling, Work-Flow Applications, Variable Neighborhood Search, Multi-Start Genetic Algorithm, Distributed Computing Environments

13.1 Introduction

With the development of the high performance computing (HPC), computational grid, etc., some complex applications are designed by communities of researchers in domains such as chemistry, meteorology, high-energy physics, astronomy, biology and human brain planning (HBP) [1], [2]. For implementing and utilizing successfully those applications, one of the most important task is to find appropriate schedules before the application is executed. The goal is to find an optimal assignment of tasks in the applications with respect

A. Abraham et al.: *Particle Swarm Scheduling for Work-Flow Applications in Distributed Computing Environments*, Studies in Computational Intelligence (SCI) **128**, 327–342 (2008)
www.springerlink.com

to the costs of the available resources. However, the scheduling problem in distributed data-intensive computing environments seems quite different from the conventional situation. Scheduling jobs and resources in data-intensive applications need to meet the specific requirements, including process flow, data access/transfer, completion cost, flexibility and availability. All kinds of components in the application can interact with each other directly or indirectly. Scheduling algorithm in traditional computing paradigms barely consider the data transfer problem during mapping computational tasks, and this negligence would be costly in the case of distributed data-intensive applications [3].

Priority scheduling plays a crucial role in the differentiated services architecture for the provisioning of Quality-of-Service (QoS) of network-based applications. Jin and Min [17] proposed a novel analytical model for priority queuing systems subject to heterogeneous Long Range Dependent (LRD) self-similar or Short Range Dependent (SRD) Poisson traffic. Authors [17] applied the generalized Schilder's theorem to deal with heterogeneous traffic and further develop the analytical upper and lower bounds of the queue length distributions for individual traffic flows.

Sabrina et al. [18], discuss issues in designing resource schedulers for processing engines in programmable networks. Authors developed two CPU scheduling algorithms that could schedule CPU resource adaptively among all the competing flows. One of the packet scheduling algorithm is called start time weighted fair queueing that does not require packet processing times and the other one is called prediction based fair queueing, which uses a prediction algorithm to estimate CPU requirements of packet.

Rodrigues et al. [19] proposed a branch and bound approach based on constraint-based search (CBS) for scheduling of continuous processes. Tasks time-windows are submitted to a constraint propagation procedure that identifies existing orderings among tasks and linear programming is used to determine the optimal flow rate for each bucket whenever all buckets are ordered in the branch and bound.

In this chapter, we introduce the scheduling problem for work-flow applications in distributed data-intensive computing environments. Rest of the Chapter is organized as follows. We model and formulate the problem in Section 13.2. We present an approach based on particle swarm algorithm based heuristics in Section 13.3. In Section 13.4, experiment results and discussions are provided. Finally, we conclude our work in the chapter.

13.2 Problem formulation

The scheduling problem in distributed data-intensive computing environments has been an active research topic, and therefore many terminologies have been suggested. Unfortunately, some of these technical terms are neither clearly stated nor consistently used by different researchers, which frequently makes readers confused. For clarity purposes, some key terminologies are re-defined for formulating the problem.

- Machine (computing unit)
  Machine (computing unit) is a set of computational resources with limited capacities. It may be a simple personal machine, a workstation, a supercomputer, or a cluster of workstations. The computational capacity of the machine is depend on its number of CPUs, amount of memory, basic storage space and other specializations. In other words, each machine has its calculating speed, which can be expressed in number of Cycles Per Unit Time (CPUT).
- Data Resource
  Data resources are the datasets, which effect the scheduling. They are commonly located on various storage repositories or data hosts. Data resources are connected to the computational resources (machines) by links of different bandwidths.
- Job and Operation
  A job is considered as a single set of multiple atomic operations/tasks. Each operation will be typically allocated to execute on one single machine without preemption. It has input and output data, and processing requirements in order to complete its task. One of the most important processing requirements is the work-flow, which is the ordering of a set of operations for a specific application. These operations can be started only after the completion of the previous operations from this sequence, which is the so-called workflow constraints. The operation has the processing length in number of cycles.
- Work-flow Application
  A work-flow application consists of a collection of interacting components that need to be executed in a certain partial order for solving successful a certain problem. The components involve a number of dependent or independent jobs, machines, the bandwidth of the network, etc. They have specific control and data dependencies between them.
- Schedule and Scheduling Problem
  A schedule is the mapping of the tasks to specific time intervals of machines. A scheduling problem is specified by a set of machines, a set of jobs/operations, optimality criteria, environmental specifications, and by other constraints. The Scheduling Problem for work-flow applications in distributed Data-intensive computing environments is abbreviated as "FDSP".

To formulate the scheduling problem, suppose a work-flow application comprises of q Jobs $\{J_1, J_2, \cdots, J_q\}$, m Machines $\{M_1, M_2, \cdots, M_m\}$ and k Data hosts $\{D_1, D_2, \cdots, D_k\}$. In the application considered, the processing speeds of the machine are $\{P_1, P_2, \cdots, P_m\}$. Each job consists of a set of operations $J_j = \{O_{j,1}, O_{j,2}, \cdots, O_{j,p}\}$. For convenience, we will decompose all the jobs to atomic operations and re-sort the operations as $\{O_1, O_2, \cdots, O_n\}$.

The processing lengths of the operation are $\{L_1, L_2, \cdots, L_n\}$. All the operations are in the specific work-flow, and they will be carried orderly on the machines with data retrieval, data input and data output.

The operations in the work-flow can be represented as or transformed to a Directed Acyclic Graph (DAG), where each node in the DAG represents an operation and the edges denote control/data dependencies. The relation between the operations can be represented by a flow matrix $F = [f_{i,j}]$, in which the element $f_{i,j}$ stores the weight value if the edge $< O_i, O_j >$ is in the graph, otherwise it is set to "-1". Fig. 13.1 depicts a work-flow of 9 operations. The recursive loop between O_1 and O_9 can be neglected when the scheduling focus on the stage within the loop. Its flow matrix F is represented as follows:

$$\begin{bmatrix} -1 & 8 & 3 & 9 & -1 & -1 & -1 & -1 & -1 \\ -1 & -1 & -1 & -1 & 5 & 6 & -1 & -1 & -1 \\ -1 & -1 & -1 & -1 & -1 & 2 & 12 & 11 & -1 \\ -1 & -1 & -1 & -1 & -1 & -1 & -1 & 7 & -1 \\ -1 & -1 & -1 & -1 & -1 & -1 & -1 & -1 & 13 \\ -1 & -1 & -1 & -1 & -1 & -1 & -1 & -1 & 4 \\ -1 & -1 & -1 & -1 & -1 & -1 & -1 & -1 & 1 \\ -1 & -1 & -1 & -1 & -1 & -1 & -1 & -1 & 8 \\ -1 & -1 & -1 & -1 & -1 & -1 & -1 & -1 & -1 \end{bmatrix}$$

The data host dependencies of the operations are determined by the retrieval matrix $R = [r_{i,j}]$. The element $r_{i,j}$ is the retrieval time, which O_i executes retrieval processing on the data host D_j. There are the other matrices $A = [a_{i,j}]$ and $B = [b_{i,j}]$, where the element $a_{i,j}$ in the former is the distance between between the machine M_i and M_j, and the element $b_{i,j}$ in the latter is the distance between the machine M_i and the data host D_j. For each operation, its completion time is the sum of three components: the input data time, the retrieval data time, and the execution time on the assigned machine. It is to be noted that the input data time can be started to accumulate only after the completion of the previous operations in the work-flow.

Given a feasible solution $S = \{S_1, S_2, \cdots, S_n\}$, S_i is the serial number of the machine, which the operation O_i is assigned on. Define C_{O_i} ($i \in \{1, 2, \cdots, n\}$) as the completion time that the machine M_{S_i} finishes the operation O_i. For the operation O_i, its completion time C_{O_i} can be calculated by Eq. (13.1).

$$C_{O_i} = \sum_{\substack{l=1 \\ f_{l,i} \neq -1}}^{n} f_{l,i} a_{S_l,S_i} + \sum_{h=1}^{k} r_{i,h} b_{S_i,h} + L_i/P_{S_i} \qquad (13.1)$$

To formulate the objective, $\sum C_{M_i}$ represents the time that the machine M_i completes the processing of all the operations assigned on it. Define $C_{max} = max\{\sum C_{M_i}\}$ as the makespan, and $C_{sum} = \sum_{i=1}^{m}(\sum C_{M_i})$ as the flowtime. The scheduling problem is thus to both determine an assignment

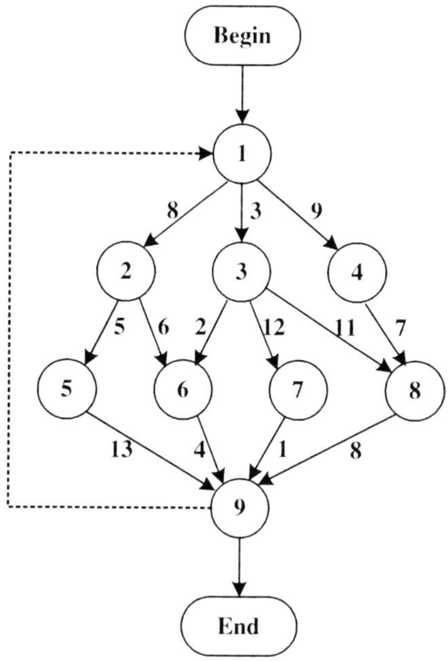

Fig. 13.1: A works-flow application with 9 operations.

and a sequence of the operations on all machines that minimize some criteria. Most important optimality criteria are to be minimized:

1. the maximum completion time (makespan): C_{max};
2. the sum of the completion times (flowtime): C_{sum}.

Minimizing C_{sum} asks the average operation is finished quickly, at the expense of the largest operation taking a long time, whereas minimizing C_{max}, asks that no operation takes too long, at the expense of most operations taking a long time. Minimization of C_{max} would result in maximization of C_{sum}. The weighted aggregation is the most common approach to the problems. According to this approach, the objectives, $F_1 = min\{C_{max}\}$ and $F_2 = min\{C_{sum}\}$, are aggregated as a weighted combination:

$$F = w_1 min\{F_1\} + w_2 min\{F_2\} \tag{13.2}$$

where w_1 and w_2 are non-negative weights, and $w_1 + w_2 = 1$. These weights can be either fixed or adapt dynamically during the optimization. The fixed weights, $w_1 = w_2 = 0.5$, are used in this article. In fact, the dynamic weighted aggregation mainly takes C_{max} into account [4] because C_{sum} is commonly much larger than C_{max} and the solution has a large weight on C_{sum} during minimizing of the objective. Alternatively, the weights can be changed gradually according to the Eqs. (13.3) and (13.4). The changes in the dynamic weights ($R = 200$) are illustrated in Fig. 13.2.

$$w_1(t) = |sin(2\pi t/R)| \tag{13.3}$$

$$w_2(t) = 1 - w_1(t) \tag{13.4}$$

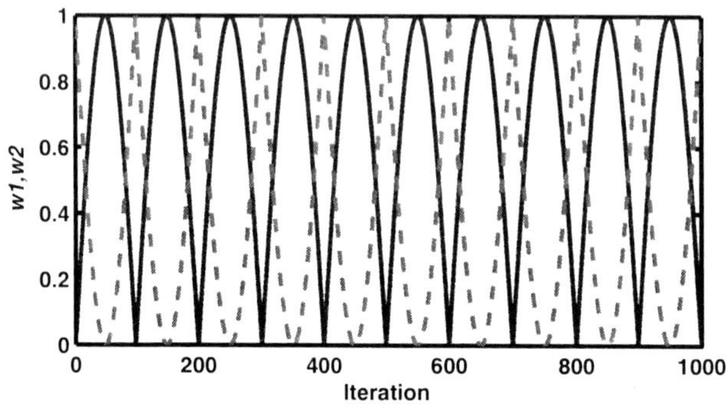

Fig. 13.2: Changes in dynamic weights.

13.3 Particle Swarm Heuristics for FDSP

13.3.1 Canonical Model

Particle swarm algorithm is inspired by social behavior patterns of organisms that live and interact within large groups. In particular, it incorporates swarming behaviors observed in flocks of birds, schools of fish, or swarms of bees, and even human social behavior, from which the Swarm Intelligence(SI) paradigm has emerged [5, 6]. It could be implemented and applied easily to solve various function optimization problems, or the problems that can be transformed to function optimization problems.

As an algorithm, its main strength is its fast convergence, which compares favorably with many global optimization algorithms [7–9]. The canonical PSO model consists of a swarm of particles, which are initialized with a population of random candidate solutions. They move iteratively through the d-dimension problem space to search the new solutions, where the fitness, f, can be calculated as the certain qualities measure.

Each particle has a position represented by a position-vector $\mathbf{x}_i$ (i is the index of the particle), and a velocity represented by a velocity-vector $\mathbf{v}_i$. Each particle remembers its own best position so far in a vector $\mathbf{x}_i^{\#}$, and its j-th dimensional value is $x_{ij}^{\#}$. The best position-vector among the swarm so far is then stored in a vector $\mathbf{x}^*$, and its j-th dimensional value is x_j^*. During the iteration time t, the update of the velocity from the previous velocity to the new velocity is determined by Eq.(13.5). The new position is then determined by the sum of the previous position and the new velocity by Eq.(13.6).

$$v_{ij}(t+1) = wv_{ij}(t) + c_1r_1(x_{ij}^{\#}(t) - x_{ij}(t)) + c_2r_2(x_j^*(t) - x_{ij}(t)). \quad (13.5)$$

$$x_{ij}(t+1) = x_{ij}(t) + v_{ij}(t+1). \quad (13.6)$$

where w is called as the inertia factor, r_1 and r_2 are the random numbers, which are used to maintain the diversity of the population, and are uniformly distributed in the interval [0,1] for the j-th dimension of the i-th particle. c_1 is a positive constant, called as coefficient of the self-recognition component, c_2 is a positive constant, called as coefficient of the social component.

From Eq.(13.5), a particle decides where to move next, considering its own experience, which is the memory of its best past position, and the experience of its most successful particle in the swarm. In the particle swarm model, the particle searches the solutions in the problem space with a range $[-s, s]$ (If the range is not symmetrical, it can be translated to the corresponding symmetrical range.) In order to guide the particles effectively in the search space, the maximum moving distance during one iteration must be clamped in between the maximum velocity $[-v_{max}, v_{max}]$ given in Eq.(13.7):

$$v_{ij} = sign(v_{ij})min(|v_{ij}|, v_{max}). \quad (13.7)$$

$$x_{i,j} = sign(x_{i,j})min(|x_{i,j}|, x_{max}) \quad (13.8)$$

The value of v_{max} is $p \times s$, with $0.1 \leq p \leq 1.0$ and is usually chosen to be s, i.e. $p = 1$. The pseudo-code for particle swarm optimization algorithm is illustrated in Algorithm 1.

The termination criteria are usually one of the following:

- Maximum number of iterations: the optimization process is terminated after a fixed number of iterations, for example, 1000 iterations.
- Number of iterations without improvement: the optimization process is terminated after some fixed number of iterations without any improvement.
- Minimum objective function error: the error between the obtained objective function value and the best fitness value is less than a pre-fixed anticipated threshold.

The role of inertia weight w, in Eq.(13.5), is considered critical for the convergence behavior of PSO. The inertia weight is employed to control the impact of the previous history of velocities on the current one. Accordingly, the parameter w regulates the trade-off between the global (wide-ranging) and local (nearby) exploration abilities of the swarm. A large inertia weight facilitates global exploration (searching new areas), while a small one tends

Algorithm 13.1: Particle Swarm Optimization Algorithm

Initialize the size of the particle swarm n, and other parameters.
Initialize the positions and the velocities for all the particles randomly.
While (the end criterion is not met) do
    $t = t + 1$;
    Calculate the fitness value of each particle;
    $\mathbf{x}^* = argmin_{i=1}^{n}(f(\mathbf{x}^*(t-1)), f(\mathbf{x}_1(t)), f(\mathbf{x}_2(t)), \cdots, f(\mathbf{x}_i(t)), \cdots, f(\mathbf{x}_n(t)))$;
    For $i = 1$ to n
        $\mathbf{x}_i^{\#}(t) = argmin_{i=1}^{n}(f(\mathbf{x}_i^{\#}(t-1)), f(\mathbf{x}_i(t))$;
        For $j = 1$ to $Dimension$
            Update the j-th dimension value of $\mathbf{x}_i$ and $\mathbf{v}_i$
                according to Eqs.(13.5), (13.6), (13.7), (13.8);
        Next j
    Next i
End While.

to facilitate local exploration, i.e. fine-tuning the current search area. A suitable value for the inertia weight w usually provides balance between global and local exploration abilities and consequently results in a reduction of the number of iterations required to locate the optimum solution. Initially, the inertia weight is set as a constant. However, some experiment results indicates that it is better to initially set the inertia to a large value, in order to promote global exploration of the search space, and gradually decrease it to get more refined solutions [10, 11]. Thus, an initial value around 1.2 and gradually reducing towards 0 can be considered as a good choice for w. A better method is to use some adaptive approaches (example: fuzzy controller), in which the parameters can be adaptively fine tuned according to the problem under consideration [12, 13].

The parameters c_1 and c_2, in Eq.(13.5), are not critical for the convergence of PSO. However, proper fine-tuning may result in faster convergence and alleviation of local minima. As default values, usually, $c_1 = c_2 = 2$ are used, but some experiment results indicate that $c_1 = c_2 = 1.49$ might provide even better results. Recent work reports that it might be even better to choose a larger cognitive parameter, c_1, than a social parameter, c_2, but with $c_1 + c_2 \leq 4$ [16].

The particle swarm algorithm can be described generally as a population of vectors whose trajectories oscillate around a region which is defined by each individual's previous best success and the success of some other particle. Various methods have been used to identify some other particle to influence the individual. Eberhart and Kennedy called the two basic methods as "gbest model" and "lbest model" [5]. In the lbest model, particles have information only of their own and their nearest array neighbors' best (lbest), rather than that of the entire group.

In the gbest model, the trajectory for each particle's search is influenced by the best point found by any member of the entire population. The best particle acts as an attractor, pulling all the particles towards it. Eventually all particles will converge to this position. The lbest model allows each individual to be influenced by some smaller number of adjacent members of the population array. The particles selected to be in one subset of the swarm have no direct relationship to the other particles in the other neighborhood.

Typically lbest neighborhoods comprise exactly two neighbors. When the number of neighbors increases to all but itself in the lbest model, the case is equivalent to the gbest model. Some experiment results testified that gbest model converges quickly on problem solutions but has a weakness for becoming trapped in local optima, while lbest model converges slowly on problem solutions but is able to "flow around" local optima, as the individuals explore different regions. The gbest model has faster convergence. But very often for multi-modal problems involving high dimensions it tends to suffer from premature convergence.

13.3.2 Variable Neighborhood Particle Swarm Optimization Algorithm (VNPSO)

Variable Neighborhood Search (VNS) is a relatively recent metaheuristic which relies on iteratively exploring neighborhoods of growing size to identify better local optima with shaking strategies [14, 15]. More precisely, VNS escapes from the current local minimum x^* by initiating other local searches from starting points sampled from a neighborhood of x^*, which increases its size iteratively until a local minimum is better than the current one is found. These steps are repeated until a given termination condition is met. The metaheuristic method, Variable Neighborhood Particle Swarm Optimization (VNPSO) algorithm, was originally inspired by VNS [20]. In PSO, if a particle's velocity decreases to a threshold v_c, a new velocity is assigned using Eq.(13.9):

$$v_{ij}(t) = w\hat{v} + c_1 r_1 (x_{ij}^{\#}(t-1) - x_{ij}(t-1)) + c_2 r_2 (x_j^*(t-1) - x_{ij}(t-1)) \quad (13.9)$$

$$\hat{v} = \begin{cases} v_{ij} & \text{if } |v_{ij}| \geq v_c \\ u(-1,1) v_{max}/\eta & \text{if } |v_{ij}| < v_c \end{cases} \quad (13.10)$$

The VNPSO algorithm scheme is summarized as Algorithm 2. The performance of the algorithm is directly correlated to two parameter values, v_c and η. A large v_c shortens the oscillation period, and it provides a great probability for the particles to leap over local minima using the same number of iterations. But a large v_c compels the particles in the quick "flying" state, which leads them not to search the solution and forcing them not to refine the search. The value of η changes directly the variable search neighborhoods for the particles. It is to be noted that the algorithm is different from the multi-start technique.

We also implemented the Multi-Start Particle Swarm Optimization (MSPSO) (illustrated in Algorithm 3) and the Multi-Start Genetic Algorithm (MSGA) to compare the empirical performances.

Algorithm 13.2: Variable Neighborhood Particle Swarm Optimization

---

Initialize the size of the particle swarm n, and other parameters.
Initialize the positions and the velocities for all the particles randomly.
Set the flag of iterations without improvement $Nohope = 0$.
While (the end criterion is not met) do
  $t = t + 1$;
  Calculate the fitness value of each particle;
  $\mathbf{x}^* =$
$argmin_{i=1}^{n}(f(\mathbf{x}^*(t-1)), f(\mathbf{x}_1(t)), f(\mathbf{x}_2(t)), \cdots, f(\mathbf{x}_i(t)), \cdots, f(\mathbf{x}_n(t)))$;
  If $\mathbf{x}^*$ is improved then $Nohope = 0$, else $Nohope = Nohope + 1$.
  For $i = 1$ to n
    $\mathbf{x}_i^{\#}(t) = argmin_{i=1}^{n}(f(\mathbf{x}_i^{\#}(t-1)), f(\mathbf{x}_i(t))$;
    For $j = 1$ to d
      If $Nohope < 10$ then
        Update the j-th dimension value of $\mathbf{x}_i$ and $\mathbf{v}_i$
          according to Eqs.(13.5),(13.7),(13.6),(13.8);
      else
        Update the j-th dimension value of $\mathbf{x}_i$ and $\mathbf{v}_i$
          according to Eqs.(13.10),(13.9),(13.6),(13.8).
    Next j
  Next i
End While.

For applying PSO successfully for the FDSP problem, one of the key issues is the mapping of the problem solution to the PSO particle space, which directly affects its feasibility and performance. We setup a search space of n dimension for an $(n - Operations, m - Machines)$ FDSP problem. Each dimension was limited to $[1, m+1)$. For example, consider a little scale $(7 - Operations, 3 - Machines)$ FDSP, Fig. 13.3 shows a mapping between a one possible assignment instance to a particle position coordinates in the PSO domain. Each dimension of the particle's position maps one operation, and the value of the position indicates the machine number to which this task/-operation is assigned to during the course of PSO. So the value of a particle's position should be an integer but after updating the velocity and position of the particles, the particle's position may appear real values such as 1.4, etc. It is meaningless for the assignment. Therefore, in the algorithm we usually round off the real optimum value to its nearest integer number. By this way, we convert a continuous optimization algorithm to a scheduling problem. The particle's position is a series of priority levels of assigned machines according to the order of operations. The sequence of the operations will be not changed during the iteration.

Algorithm 13.3: Multi-start Particle Swarm Optimization

```
Initialize the size of the particle swarm n, and other parameters.
Initialize the positions and the velocities for all the particles randomly.
Set the flag of iterations without improvement Nohope = 0.
While (the end criterion is not met) do
    t = t + 1;
    Calculate the fitness value of each particle;
    x* = argmin_{i=1}^{n}(f(x*(t-1)), f(x_1(t)), f(x_2(t)), ..., f(x_i(t)), ..., f(x_n(t)));
    If x* is improved then Nohope = 0, else Nohope = Nohope + 1.
    For i = 1 to n
        x_i^#(t) = argmin_{i=1}^{n}(f(x_i^#(t-1)), f(x_i(t)));
        For j = 1 to d
            If Nohope < 10 then
                Update the j-th dimension value of x_i and v_i
                  according to Eqs.(13.5),(13.7),(13.6),(13.8);
            else
                Re-initialize the positions and the velocities
                  for all the particles randomly.
        Next j
    Next i
End While.
```

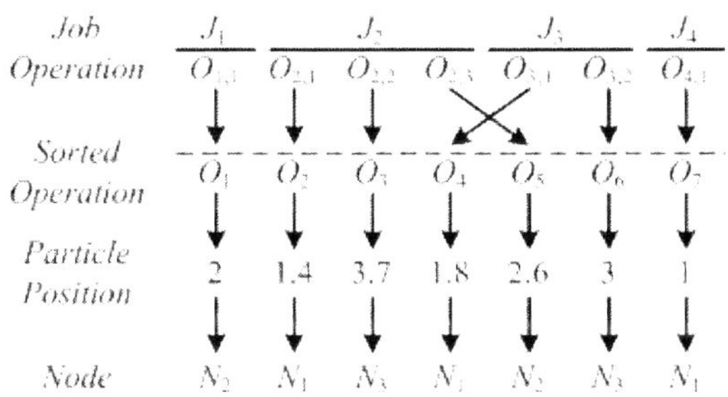

Fig. 13.3: The Mapping between particle and FJSP.

Since the particle's position indicates the potential schedule, the position can be "decoded" to the feasible solution. It is to be noted that the position matrix may violate the work-flow constraints. The starting point of operations must be started only after the completion of the previous latest operation in the work-flow. The best situation is the starting point of the operation in alignment with the ending point of its previous latest operation. After all the operations have been processed, we get the feasible scheduling solution and then calculate the cost of the solution.

13.4 Experiment Results and Algorithm Performance Demonstration

To illustrate the effectiveness and performance of the particle swarm searching algorithm, three representative instances based on practical data have been selected. In our experiments, the algorithms used for comparison were VNPSO, MSPSO (Multi-start PSO) and MSGA (Multi-start GA). In VNPSO, η and v_c were set to 2 and 1e-7 before 15,000 iterations, while they were set to 5 and 1e-10 after 15,000 iterations. Other specific parameter settings of the different algorithms are described in Table 13.1. The algorithms were run 20 times with different random seeds. Each trial had a fixed number of 2,000 iterations. The average fitness values of the best solutions throughout the optimization run were recorded. Usually another emphasis will be to generate the schedules at a minimal amount of time. So the completion time for 20 trials were used as one of the criteria to improve their performance.

Table 13.1: Parameter settings for the algorithms.

Algorithm	Parameter name	Parameter value
	Size of the population	20
GA	Probability of crossover	0.9
	Probability of mutation	0.09
	Swarm size	20
	Self-recognition coefficient c_1	1.49
PSOs	Social coefficient c_2	1.49
	Inertia weight w	$0.9 \rightarrow 0.1$
	Clamping Coefficient ρ	0.5

We illustrate a small scale FDSP problem involving an application with 9 operations, 3 machines and 3 data hosts represented as $(O9, M3, D3)$ problem. The speeds of the 3 machines are 4, 3, 2 CPUT, respectively, i.e., $P = \{4, 3, 2\}$. The length of the 9 operations are 6,12,16,20,28,36,42,52,60 cycles, respectively, i.e., $L = \{6, 12, 16, 20, 28, 36, 42, 52, 60\}$. The flow matrix is F as depicted in Section 13.2, and all other information are as follows:

$$R = \begin{bmatrix} 6 & 18 & 76 \\ 50 & 4 & 51 \\ 1 & 85 & 15 \\ 19 & 11 & 1 \\ 39 & 12 & 0 \\ 73 & 0 & 1 \\ 57 & 29 & 77 \\ 36 & 0 & 74 \\ 61 & 82 & 30 \end{bmatrix}$$

$$A = \begin{bmatrix} 0 & 21 & 95 \\ 21 & 0 & 41 \\ 95 & 41 & 0 \end{bmatrix}$$

$$B = \begin{bmatrix} 0 & 45 & 91 \\ 45 & 0 & 59 \\ 91 & 59 & 0 \end{bmatrix}$$

Fig. 13.4 illustrates the performance of the three algorithms during the search processes for $(O9, M3, D3)$ problem. The best scheduling solution in the 20 MSGA runs is $\{3, 1, 2, 3, 1, 1, 1, 3, 1\}$, in which the makespan is 23654 and the flowtime is 34075.

The best scheduling solution obtained in the 20 MSPSO and VNPSO runs is $\{3, 1, 2, 3, 1, 1, 2, 3, 2\}$, in which the makespan is 15011 and the flowtime is 30647. While MSPSO provides the best results 8 times in 20 runs, VNPSO provides the best result 10 times in the same runs respectively.

Fig. 13.5 provides an optimal schedule for $((O9, M3, D3)$ problem, in which "W" means the waiting time. As depicted in Fig. 13.5, the operations O_2 and O_3 both have to wait for 1611 time units before they are processed in the scheduling solution.

Further, we tested the three algorithms for two more FDSP problems, i.e. $(O10, M3, D3)$ and $(O12, M4, D3)$. Empirical results are illustrated in Table 13.2. In general, VNPSO performs better than the other two approaches, although its computational time is worse than MSPSO. VNPSO could be an ideal approach for solving the large scale problems when other algorithms failed to give a better solution.

Table 13.2: Comparison of performance for different FDSPs.

Instance	*Items*	MSGA	MSPSO	VNPSO
$(O9, M3, D3)$	average	28864	24152	28.8000
	time	200.2780	133.6920	181.2970
$(O10, M3, D3)$	average	21148	19594	16389
	time	210.6230	138.5890	140.3920
$(O12, M4, D3)$	average	16146	14692	14412
	time	235.1080	152.5520	154.4420

13.5 Conclusions

In this chapter, we modeled and formulated the scheduling problem for work-flow applications in distributed data-intensive computing environments (FDSP). A particle swarm optimization based variable neighborhood search

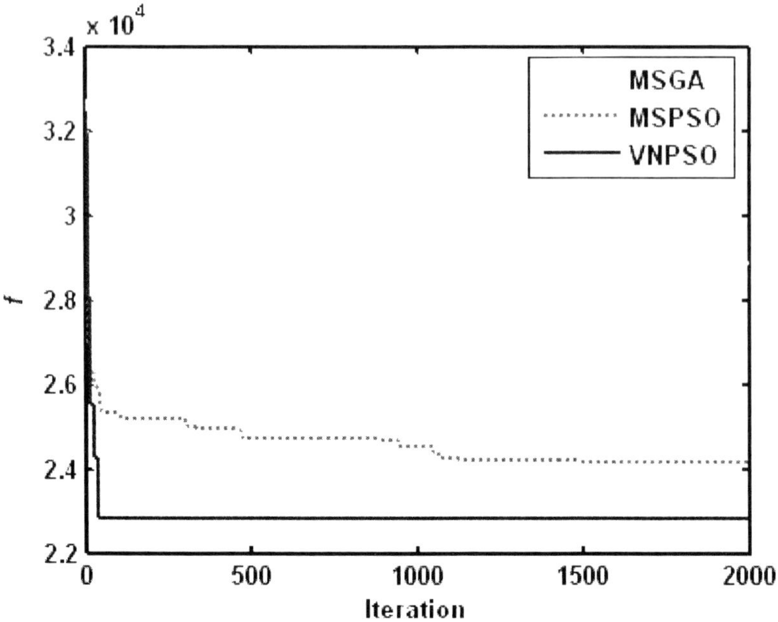

Fig. 13.4: Performance for the FDSP $(O9, M3, D3)$

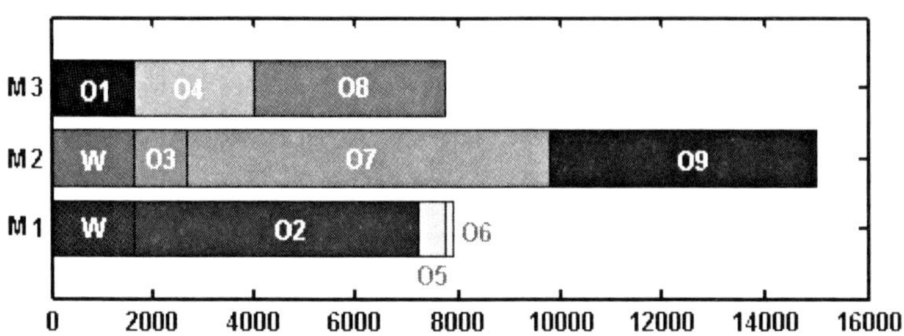

Fig. 13.5: A scheduling solution for the FDSP $(O9, M3, D3)$

is proposed to solve the problem. Empirical results demonstrate that the proposed VNPSO algorithm is feasible and effective. VNPSO can be applied in distributed data-intensive applications to meet the specified requirements, including work-flow constraints, data retrieval/transfer, job interaction, minimum completion cost, flexibility and availability.

Our future research is targeted to generate more FDSP instances and investigate more optimization/meta-heuristic approaches.

13.6 Acknowledgements

This work was partly supported by NSFC (60373095), DLMU (DLMU-ZL-200709) and the Research Council, NTNU and UNINETT.

References

1. I. Foster and C. Kesselman (Eds.) " The Grid: Blueprint for a New Computing Infrastructure". *Morgan-Kaufmann*, 1998.
   S. Venugopal, and R. Buyya. "A Set Coverage-based Mapping Heuristic for Scheduling Distributed Data-Intensive Applications on Global Grids". *Technical Report*, GRIDS-TR-2006-3, Grid Computing and Distributed Systems Laboratory, The University of Melbourne, Australia, March 8, 2006.
2. N. Zhong, J. Hu, S. Motomura, J. Wu, and C. Liu. "Building A Data-mining Grid for Multiple Human Brain Data Analysis". *Computational Intelligence*, 2005, 21(2), pp. 177.
3. F. Dong, and S.G. Akl. "Scheduling Algorithms for Grid Computing: State of the Art and Open Problems". *Technical Report*, 2006-504, School of Computing, Queen's University, Canada, January 2006.
4. K.E. Parsopoulos, and M.N. Vrahatis. "Recent Approaches to Global Optimization Problems through Particle Swarm Optimization". *Natural Computing*, 2002, 1, pp. 235–306.
5. J. Kennedy, and R. Eberhart. *Swarm Intelligence*. Morgan Kaufmann, CA, 2001.
6. M. Clerc. *Particle Swarm Optimization*. ISTE Publishing Company, London, 2006.
7. R.C. Eberhart, and Y. Shi. "Comparison Between Genetic Algorithms And Particle Swarm Optimization". *Proceedings of IEEE International Conference on Evolutionary Computation*, 1998, pp. 611–616.
8. D.W. Boeringer, and D.H. Werner "Particle Swarm Optimization versus Genetic Algorithms for Phased Array Synthesis". *IEEE Transactions on Antennas and Propagation*, 2004, 52(3), pp. 771–779.
9. A. Abraham, H. Guo, and H. Liu. "Swarm intelligence: Foundations, Perspectives And Applications". *Swarm Intelligent Systems*, Nedjah N, Mourelle L (eds.), Nova Publishers, USA, 2006.
10. J. Kennedy J and R. Mendes. "Population structure and particle swarm performance". *Proceeding of IEEE conference on Evolutionary Computation*, 2002, pp. 1671–1676.
11. H. Liu, B. Li, Y. Ji and T. Sun. "Particle Swarm Optimisation from lbest to gbest". *Applied Soft Computing Technologies: The Challenge of Complexit*, Springer Verlag, 2006, pp. 537–545.
12. Y. H. Shi and R. C. Eberhart. "Fuzzy adaptive particle swarm optimization". *Proceedings of IEEE International Conference on Evolutionary Computation*, 2001, pp. 101–106.
13. H. Liu and A. Abraham. "Fuzzy Adaptive Turbulent Particle Swarm Optimization". *Proceedings of the Fifth International conference on Hybrid Intelligent Systems*, 2005, pp. 445–450.

14. P. Hansen and N. Mladenović. "Variable neighbourhood search:Principles and applications". *European Journal of Operations Research*, 2001, 130, pp. 449–467.
15. P. Hansen and N. Mladenović. "Variable neighbourhood search". *Handbook of Metaheuristics*, Dordrecht, Kluwer Academic Publishers, 2003.
16. M. Clerc, and J. Kennedy. "The Particle Swarm-explosion, Stability, and Convergence in A Multidimensional Complex Space". *IEEE Transactions on Evolutionary Computation*, 2002, 6, pp. 58–73.
17. X. Jin and G. Min, Performance analysis of priority scheduling mechanisms under heterogeneous network traffic Journal of Computer and System Sciences, Volume 73, Issue 8, pp. 1207-1220, 2007.
18. F. Sabrina, C.D. Nguyen, S. Jha, D. Platt and F. Safaei, Processing resource scheduling in programmable networks Computer Communications, Volume 28, Issue 6, pp. 676-687, 2005.
19. L.C.A. Rodrigues, R. Carnieri and F. Neves Jr., Scheduling of continuous processes using constraint-based search: An application to branch and bound Computer Aided Chemical Engineering, Volume 10, pp. 751-756, 2002.
20. A. Abraham, H. Liu, and T.G. Chang, Variable Neighborhood Particle Swarm Optimization Algorithm, Genetic and Evolutionary Computation Conference (GECCO-2006), Seattle, USA, 2006.

Index